Methods in Life Sciences

VOLUME 1

Microchemical Analysis of Nervous Tissue

NEVILLE N. OSBORNE

Microchemical Analysis of Nervous Tissue

PERGAMON PRESS

OXFORD · NEW YORK · TORONTO · SYDNEY

Pergamon Press Ltd., Headington Hill Hall, Oxford
Pergamon Press Inc., Maxwell House, Fairview Park, Elmsford, New York 10523
Pergamon of Canada Ltd., 207 Queen's Quay West, Toronto 1
Pergamon Press (Aust.) Pty. Ltd., 19a Boundary Street, Rushcutters Bay, N.S.W. 2011, Australia

---

First edition 1974

Library of Congress Cataloging in Publication Data

Osborne, Neville N.
Microchemical analysis of nervous tissue.

(Methods in life sciences, v. 1)
1. Brain chemistry. 2. Microchemistry. I Title.
[DNLM: 1. Microchemistry. 2. Neurochemistry. W1 ME9615S / WL104 081]
QP356.3.082 1974 611′.0188 74–17043
ISBN 0–08–018100–7

TO JANE

Contents

Preface

In the pioneer years of neurochemistry, homogenisation of whole brain and the chemical analysis of its content was the technique generally employed. Many of the data obtained in this way are to be found listed in different textbooks. Although this approach still offers more possibilities, today the trend is towards a more detailed analysis. The availability of more and more sensitive methods together with the rapid development in molecular biology in the past years have also changed the neurochemical point of view. One of the most fascinating aspects in modern science is to use brain-power for the objective analysis of the molecular aspects of just this very brain-power.

It is generally accepted that a good method is the backbone of any experimental result. It is often also thought that results obtained with micromethods are inaccurate and that working with microprocedures is reserved for only a few really artistic experimenters. This monograph, however, demonstrates clearly that these two prejudices are wrong. On the contrary, micromethods need only be handled as carefully and skilfully as with normal macromethods and the results will have the same distribution error. The time necessary for learning a micromethod is also more or less the same as a normal experienced experimenter needs to learn the correct handling of any new method. Furthermore, micromethods often have the added advantage in that the performance of a technique in microscale is much less time-consuming and often less expensive than the corresponding macro procedure.

It will probably be only a question of time till physiologists have mapped out to a great extent all types of spikes from all the different brain areas and cell types, and yet there will still be but little information on the biochemistry behind these spikes. Only the teamwork between physiologists and neurochemists familiar with as many neurochemical micromethods as possible can assist in finding the necessary correlation between the physiological observations and the neurochemical background. The choice of the optimal experimental approach is therefore extremely important. Starting with kilograms of brain tissue and preparing their different subcellular particles is only meaningful when finding out the common basic principles, but is meaningless for analysing a physiologically defined function of a brain area or an identified neuron. This will be the domain of the neurochemical micromethods alone.

With regard to the molecular aspects of modern neurochemistry, the use of micromethods, including the isolation of defined brain areas and defined single cells or parts of defined cells, and their subsequent biochemical analysis, is the most promising approach. This is demonstrated in this monograph, which also includes a review of the literature of neurochemical analysis performed with different micromethods. It is my firm conviction that micromethods of the type described here or even more sophisticated and more sensitive methods which may be developed on requirement can help to bridge the gap between the different hypotheses for explaining the complex function of the nervous system on a molecular basis and the neurochemical facts which have been obtained so far.

V. Neuhoff

Acknowledgements

FIRST and foremost I should like to thank Professor G. A. Kerkut for suggesting I write this monograph and for reading the manuscript. My sincere thanks are also due to Professor V. Neuhoff, not only for writing the Preface and reading the manuscript, but also for his support and encouragement throughout. Drs. H. H. Althaus, W. Dames, H. Haljamäe, R. Rüchel and D. Wolfrum must be gratefully mentioned because of their many helpful comments, suggestions and advice on various sections of the book, and thanks also to Dr. T. W. Waehneldt for photographing a number of specimens, and to Herr H. Ropte for enthusiastically and expertly producing the photographs and diagrams. Further I wish to acknowledge the valuable assistance of other members of the neurochemical group, in particular Fräulein E. Priggemeier, and the secretarial aid of Frau I. von Bischoffshausen and Frau B. Rosetz. Last, but by no means least, I thank my wife Jane, not only for inspiring me to write the monograph, but also for spending many hours typing and proof reading in the different stages of development.

CHAPTER 1

General Introduction

WHY a monograph on microprocedures in neurochemistry? This is not difficult to justify when one considers that the human brain has approximately 10^{10} nerve cells, while the tiny brain of the ant (*Formica lugubris*) has about 100,000. These vast populations of neurons present a formidable challenge to the biologist trying to understand how the nervous system works. From the mass of electrophysiological and electron microscopical data which has accumulated it can now generally be concluded that nerve cells are independent units (see e.g. Bullock, 1967; Bullock and Horridge, 1965; Eccles, 1964; Segundo, 1970; Horridge, 1968). Furthermore, each neuron has several parts: (1) receptive loci specialized in transducing the dozens of imputs which impinge on them in several ways; (2) pacemaker loci which inject spontaneous rhythms; (3) mixing and integrating loci; (4) threshold loci for initiating all-or-none nerve impulses in bursts and trains from 1 to 1000 per second; and (5) transmitter loci at each of the far ends of the nerve cell, where they influence up to several dozen others. Clearly, biochemical information to be gained from classical studies using relatively large amounts of nervous tissue (and therefore large numbers of cells which may have very different properties) is of limited value. This problem is complicated by the existence of vast numbers of glial cells which form a close and integrated association with the neurons. Obviously the physiology, morphology, functional role and biochemistry of individual neurons have to be studied and the neurons in the nervous system related to one another before a real insight is gained into the intricate mechanism of the nervous system. However, progress has been slow, basically for two main reasons. Firstly, the majority of neurons are difficult to characterise and study as entities because of their small size, and, secondly, there is a lack of suitable microprocedures which would permit the study of different biochemical parameters in individual neurons.

One way of circumventing these difficulties is either to separate disaggregated nervous tissue, thus obtaining populations of neurons and glial (Rose, 1968), or to fractionate homogenates of nervous tissue and secure relatively pure fractions of a constituent part of the different neurons, e.g. the nerve endings (Whittaker, 1973). Studies of this kind have many advantages, but they, too, suffer from certain drawbacks, such as the possibility that changes could occur in the constituents, caused by the elaborate separation or fractionation procedures employed; moreover, any differences there may be in the properties of similar structures obtained from the brain cannot be observed. Another approach is to analyse small defined areas of the nervous system or individual neurons where possible. This presupposes the presence of suitable microchemical procedures. Perhaps a distinction should be made here between macro- (normal), micro- and ultra-procedures, though one might think that such a distinction is meaningless, since they represent a scale continuum.

In theory they do; however, in practice there is a change from macroscale (i.e. brain homogenates) to micro (i.e. one very large neuron, or microquantities of nervous tissue), and over this range many macroprocedures can be modified and scaled down. The next step, the ultra-microprocedure (parts of a single (20 μ) minute nerve cell), is a 'quantum jump' and often requires elaborate apparatus and new approaches.

The purpose of this monograph is to describe some microprocedures recently developed in this laboratory. Special attention will be paid to the choice of biological material and the various procedures used for the isolation by dissection of defined components of the nervous system. The microbiochemical methods described will be those related especially to the study of amines, amino acids, phospholipids and proteins. Many other extremely sensitive microprocedures (plus-ultra-microprocedures) have been developed within the last thirty years (see Chap. 5) and though their description is beyond the scope of the monograph, a brief review of some of these methods and their applications is presented. Perhaps it should be pointed out that emphasis is often laid only on the applicability of microprocedures for studying small amounts of tissue, e.g. isolated cells, discrete areas of brain, biopsy material, etc., whereas they also have other important virtues. Some micromethods, for example, are less time-consuming than normal procedures, and are for this reason employed even when the material available is unlimited. Moreover, the cost of analysing material by micromethods can often be very much less than that of similar normal scale studies.

References

BULLOCK, T. H. (1967) Signals and neuronal coding. In: *The Neurosciences: a Study Program* (Eds. G. C. Quarton, T. Melnechuk, and F. O. Schmitt), pp. 347–452. The Rockefeller University Press, New York.

BULLOCK, T. H. and HORRIDGE, G. A. (1965) *Structure and Function in the Nervous Systems of Invertebrates*. W. H. Freeman, San Francisco.

ECCLES, J. C. (1964) *The Physiology of Synapses*. Academic Press, New York.

HORRIDGE, G. A. (1968) *Interneurons*. W. H. Freeman, San Francisco.

ROSE, S. P. R. (1968) The biochemistry of neurones and glia. In: *Applied Neurochemistry* (Eds. A. N. Davison and J. Dobbing), pp. 332–355.

SEGUNDO, J. P. (1970) Functional possibilities of nerve cells for communication and for coding. *Acta Neurol. Latinoamer*. **14**, 340–344.

WHITTAKER, V. P. (1973) The biochemistry of synaptic transmission. *Naturwissenschaften*, **60**, 281–289.

CHAPTER 2

Choice of Biological Material for Microanalysis

As previously mentioned, the fact that the mammalian brain contains a great number of neurons presents a problem. The difficulty lies in the choice of appropriate experimental objects. Ideally one needs a nervous system that produces a reasonably complex repertoire of behaviour and has only a few cells, each of which can be recognised so that suitable experiments can be carried out on them. In this respect certain invertebrate nervous systems offer a number of advantages, in that they are organised in an orderly manner, have fewer nerve cells than the vertebrates, have specialised giant neurons and can be individually characterised. There are certain vertebrate preparations which do contain populations of giant neurons, though they are difficult to characterise individually. Another important advantage of the invertebrate neurons is that they can retain their functional activity after dissection and survive for several hours or even days (see e.g. Strumwasser, 1967). This therefore makes it possible to perform *in vitro* experiments on invertebrate nervous systems, monitoring the activity of individual neurons by means of intra- or extracellular recording while the environment of the cell can be controlled or changed by adding or substituting ions, inhibitors, or drugs, etc. These and the many other advantages in using invertebrate neurons in the analysis of individual cells are summarised in Table 1.

There is not only an enormous variety of invertebrate cell preparations (see Table 2), but also of invertebrate preparations of giant synapses and giant axons (see Table 3) which

TABLE 1. ADVANTAGES OF INVERTEBRATE OVER VERTEBRATE CELL PREPARATIONS FOR MICROCHEMICAL ANALYSIS

1. Invertebrate neurons can be easily dissected from the surrounding nervous tissue.
2. The neurons can overcome slight environmental changes which often prove disastrous for vertebrate cells.
3. The cell bodies are generally larger (diameter of up to 1 mm in *Aplysia*), which means that more material is available for biochemical analysis.
4. The same neuron is easy to identify in different preparations, reducing the variability of the system.
5. The neurons retain their functional activity after dissection for several hours or even days.
6. The functional activity of the neuron can be tested and altered by changing the environment of the cell. The alterations in functional activity can then be related to the cell's biochemistry.
7. The study of individual identifiable neurons can be followed through from very young animals weighing, for example, a few grams in the case of *Aplysia*, to adult animals of 700–800 g.
8. Glial material is easy to isolate, making it possible to study individual neuron–glial relationships.
9. Some invertebrates have also large axons and synapses, both of which can be isolated and analysed for their chemical contents.
10. There are smaller numbers of neurons and simpler organisations of the different pathways in the invertebrates to simplify the study.

are suitable for biochemical analysis. In addition, glial cells can often be dissected from invertebrates (see Kuffler and Nicholls, 1966) and analysed by biochemical procedures, thus allowing the study of the relationship between specific glial and nerve cells. Despite all these advantages, some scientists hesitate to compare the properties of neurons of the vertebrate central nervous systems with those of the invertebrate nervous system. Physical, biochemical and pharmacological differences do exist between vertebrate and invertebrate neurons, though they seem to consist mainly of differences in pharmacological sensitivity and some chemical characteristics rather than fundamental functional mechanisms and the mode of response of the cells to transmitters and drugs. Although electrophysiological properties of all neurons appear to be identical and in general comparable with those of electrogenetic cells, there are detailed electrophysiological differences between invertebrate

TABLE 2. INVERTEBRATE AND VERTEBRATE CELL BODIES SUITABLE FOR MICROANAYLSIS

Species	Preparation	Cell body diameter
Crustacea (invertebrate)		
Astacus fluviatilis (crayfish)	Stretch receptor (slow-adapting)	50– 80 μ
Astacus astacus (crayfish)	Stretch receptor (slow-adapting)	50– 85 μ
Homarus americanus (lobster)	Stretch receptor (slow-adapting)	75– 120 μ
Insecta (invertebrate)		
Periplaneta americana (cockroach)	Thoracic ganglion	50– 120 μ
Mollusca (invertebrate)		
Aplysia californica (sea slug)	Visceral ganglion	400– 800 μ
Helix aspersa (snail)	Visceral ganglion	40– 320 μ
Helix pomatia (snail)	Visceral ganglion	60– 360 μ
	Cerebral ganglion	40– 180 μ
	Parietal ganglion	260– 400 μ
Tritonia gilberti (sea slug)	Pleural ganglia	500– 800 μ
	Parietal ganglia	500–1000 μ
Loligo (squid)	Pedal ganglion	150– 170 μ
Octopus (octopus)	Cerebral ganglion	20– 80 μ
Annelida (invertebrate)		
Lumbricus (earthworm)	Ventral nerve cord	30– 60 μ
Hirudo medicinalis (leech)	Ventral nerve cord	30– 65 μ
Fishes (vertebrates)		
Goldfish	Mauthner's cells	30– 40 μ
Puffer fish	Supramedullary cells	200– 400 μ
Amphibians (vertebrate)		
Frog	Spinal ganglion	15– 20 μ
	Sympathetic ganglion	10– 35 μ
Mammals (vertebrate)		
Rabbit	Deiters' cells	50– 100 μ
	Cortical cells	20– 40 μ
	Cells of nucleus supraopticus	30– 45 μ
	Spinal ganglion cells	60– 150 μ
	Anterior horn cells	20– 50 μ
	Granular cells of cerebellum	10– 20 μ
	Hippocampus cells	10– 30 μ
Cat	α-Motoneurons of lumbar spinal cord region	40– 80 μ

and vertebrate nerve cells. Furthermore, the anatomy of invertebrate neurons differs in many ways from that of the vertebrates (Cohen, 1970). It is for these reasons that the biochemical analysis of mammalian cell preparations is of particular importance.

TABLE 3. INVERTEBRATE GIANT AXONS AND GIANT SYNAPSES SUITABLE FOR MICROANALYSIS

Species	Preparations	Giant axon of synapse diameter
Crustacea		
Homarus americanus (lobster)	Inhibitory and motor axons in leg	30– 50 μ
Carcinus meanus (crab)	Inhibitory axon in leg	20– 30 μ
Procambarus clarkii (crayfish)	Giant axons in ventral nerve cord	100– 220 μ
Cambarus (crayfish)	Giant axons in ventral cord	150– 200 μ
	Synapses in septal segment	150– 240 μ
Insecta		
Periplaneta americana (cockroach)	Giant axons in ventral nerve cord	20– 45 μ
Mollusca		
Aplysia californica (sea slug)	Axons in peripheral nerves	25– 50 μ
Loligo peatii (squid)	Giant nerve fibre axon	500– 700 μ
	Stellate ganglion, axon	50– 900 μ
Loligo forbesi (squid)	Giant nerve fibre axon	700– 900 μ
Annelida		
Lumbricus (earthworm)	Giant axon in ventral nerve cord	75– 100 μ
	Synapses in septal segmental axon	150– 200 μ
Myxicoly (marine worm)	Giant axon in ventral nerve cord	100–1700 μ

The autonomic ganglia, which comprise a few thousand nerve cells organised in physiological units, have proved suitable material for identifying neurons with different pharmacological and biochemical properties. Single neurons isolated by microdissection and subcellular fractions of the ganglia have been analysed by microprocedures so as to follow changes involved in the process of synaptic plasticity (e.g. Giacobini, 1970). Populations of large nerve cell bodies also exist in the following mammalian tissues: cerebellum, hippocampus, anterior horn, spinal ganglion, nucleus supraopticus, cortex and Deiters' nucleus (Table 2), and can be hand-dissected in excellent morphological condition (Fig. 1) from pieces of nervous tissue. However, care must be taken in order not to damage the chemical integrity of the neurons, since dyes (e.g. methylene blue) often have to be used to help in the identification process and also because vertebrate neurons are exceedingly susceptible to minute environmental changes. It may be worthwhile noting that the elegant experiments of Hydén showing changes in brain protein during learning were carried out on isolated hippocampus cells (Hydén, 1967; Hydén and Lange, 1970). In addition to Hydén (1972), Lowry and Passoneau (1972) and Giacobini (1970) have all been pioneers in the chemical analysis of isolated vertebrate neurons, and their work has contributed a great deal to our understanding of a number of important aspects in neurobiology. The

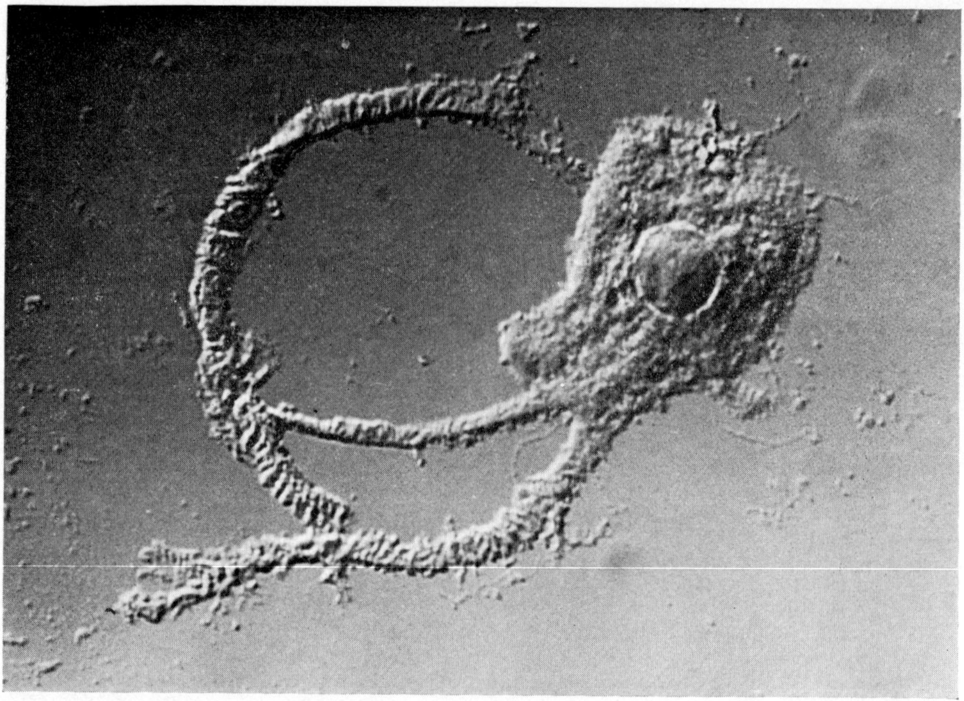

FIG. 1.

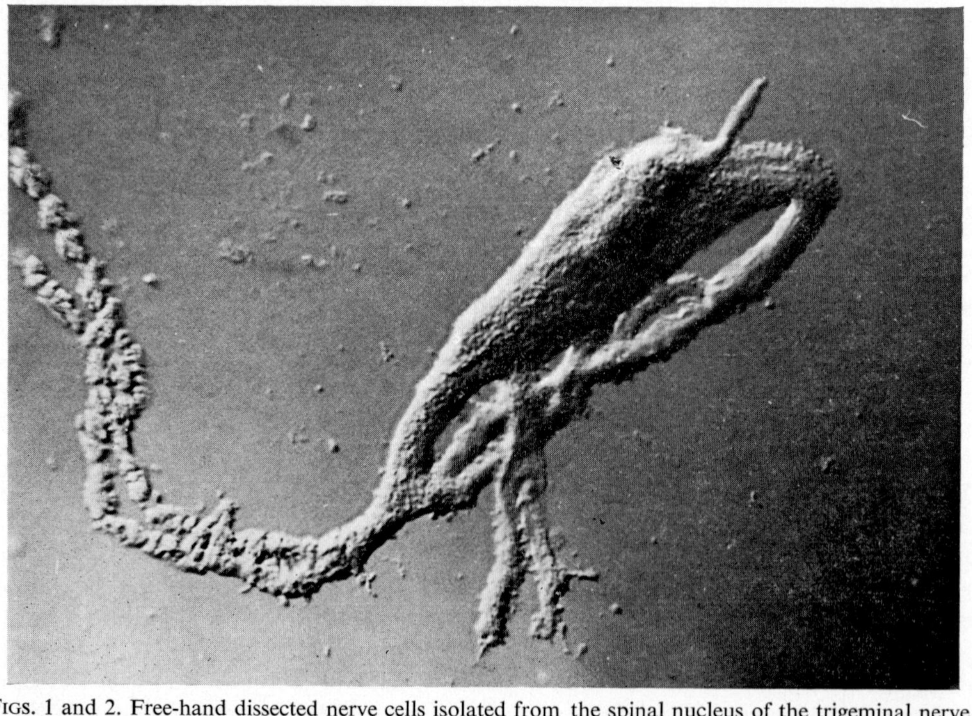

FIGS. 1 and 2. Free-hand dissected nerve cells isolated from the spinal nucleus of the trigeminal nerve, slightly stained with methylene blue solution. Photographed under Normarski optics. (Photographs by courtesy of V. Neuhoff.) Fig. 1, × 1800. Fig. 2, × 1000.

point should be made, nevertheless, that studies on vertebrate nervous tissues, though exceedingly elegant, suffer from the drawback that the individuality of the cells cannot be exploited for technical reasons.

References

COHEN, M. J. (1970) A comparison of invertebrate and vertebrate central neurons. In: *The Neurosciences* (Ed. F. O. Schmitt), pp. 798–812. The Rockefeller University Press, New York.

GIACOBINI, E. (1970) Biochemistry of single neuronal models. In: *Biochemical Psychopharmacology* (Eds. E. Costa and E. Giacobini), Vol. 2, pp. 9–64. Raven Press.

HYDÉN, H. (1967) RNA in brain cells. In: *The Neurosciences* (Eds. G. C. Quarton, T. Melnechik and F. O. Schmitt), pp. 248–266. The Rockefeller University Press, New York.

HYDÉN, H. (1972) Macromolecules and behaviour. In: *Arthur Thomson Lectures* (Eds. G. B. Ansell and P. B. Bradley), pp. 3–75. Macmillan, London.

HYDÉN, H. and LANGE, P. W. (1970) Protein changes in nerve cells related to learning and conditioning. In: *The Neurosciences* (Ed. F. O. Schmitt), pp. 278–289. The Rockefeller University Press, New York.

KUFFLER, S. W. and NICHOLLS, J. G. (1966) The physiology of neuroglial cells. *Ergeb. Physiol.* **57**, 1–90.

LOWRY, S. W. and PASSONNEAU, J. V. (1972) *A Flexible System in Enzymatic Analysis.* Academic Press, New York.

STRUMWASSER, F. (1967) Types of information stored in single neurons. In: *Invertebrate Nervous Systems* (Ed. C. A. G. Wiersma), pp. 291–319. University of Chicago Press, Chicago.

CHAPTER 3

Isolation of Nervous Tissue for Analysis

THE isolation of characterised neurons, nervous tissue parts and populations of cells is often the most critical step in the biochemical analysis of the nervous component. When working with vertebrate tissue, the most hazardous period is usually between the moment the blood supply is cut off and the dissection. It cannot be overstressed that either the functional integrity of the nerve cell should be maintained after dissection, or the metabolism of the neuron should be stopped abruptly by means of rapid freezing. The first alternative is essential for direct neurophysiological and neuropharmacological correlations and it can be achieved comparatively easily when using invertebrate neurons for the reasons stated earlier. The second alternative is often necessary when analysing vertebrate nervous tissue, especially labile metabolites.

1. Isolation of Characterised Invertebrate Neurons

The problems encountered in dissecting individual neurons from the various isolated ganglia are generally caused by the illumination of the ganglion, the identification of the neuron involved and its removal, intact, from the ganglion. An essential first step is to pin the ganglion down in a relatively stretched out position. It should not be stretched too much, otherwise the cells will tend to 'pop out' when the connective tissue layers are removed. In the instance of the snail (*Helix pomatia*) circumoesophageal ganglia, a small bath (volume 0.7 ml) containing a nylon sheet at the bottom and filled with snail saline is most suitable. This stiff plastic sheet is necessary to retain the insect pins used for holding the ganglia down. Either transillumination, reflected light, or dark-field illumination can then be used for the identification and dissection of individual neurons, which is carried out under microscopic vision. The most useful method by far is dark-field illumination. One simple approach is to focus a pencil of light on to the cell concerned. This can be done by attaching a tapered glass rod to a microscope lamp, and bringing the glass tip (about 3 mm diameter) very close to the preparation, thus illuminating the cell alone. An alternative to this is the Dark-field Dissecting Stand obtainable from Brinkmann Instruments Company (Westbury, New York). Proper illumination is essential for identifying individual neurons within the ganglia.

The localisation of characterised neurons will depend upon the familiarity of the experimenter with the ganglia, though a number of published maps of ganglia are available, e.g. *Aplysia californica* (Coggeshall *et al.*, 1966; Frazier *et al.*, 1957; Kandel *et al.*, 1957), *Helix pomatia* (Kerkut, 1969; Osborne, 1973), the cockroach (Cohen and Jacklet, 1965), *Hirudo medicinalis* (Nicholls and Baylor, 1969) and *Tritonia* (Willows, 1967, 1968). It should be kept in mind that these maps are based on a variety of physiological characteristics such as electrical activity, neurohormone response, sensitivity to drugs, etc., as well as on morphology. Morphology, which depends on size, colour and position, can be used to identify cells, though often variations do occur which can be due to age differences among different animals. For example, in *Aplysia* more small cells appear as the animal gets larger, often making the giant neurons difficult to identify. When identification by morphology alone is found difficult, it is advisable to use neurophysiological methods to provide confirmation. Generally it is preferable to make the identification of characterised neurons before cutting through the final connective tissue sheath, and so prevent disturbance of the morphology.

A couple of pairs of fine forceps together with a microscalpel are all that is required to free a characterised cell from the surrounding nervous tissue. A constant check should be made that the tips of the forceps fit exactly together and are sharply pointed, best achieved by filing them with emery stone. Using two pairs of forceps, one in each hand, connective tissue surrounding the cell can be gently teased away until the cell 'pops out'. The microscalpel together with a single pair of forceps are then used to cut the cell's axon, so freeing it entirely (see Fig. 3). Needless to say, considerable caution has to be taken throughout the

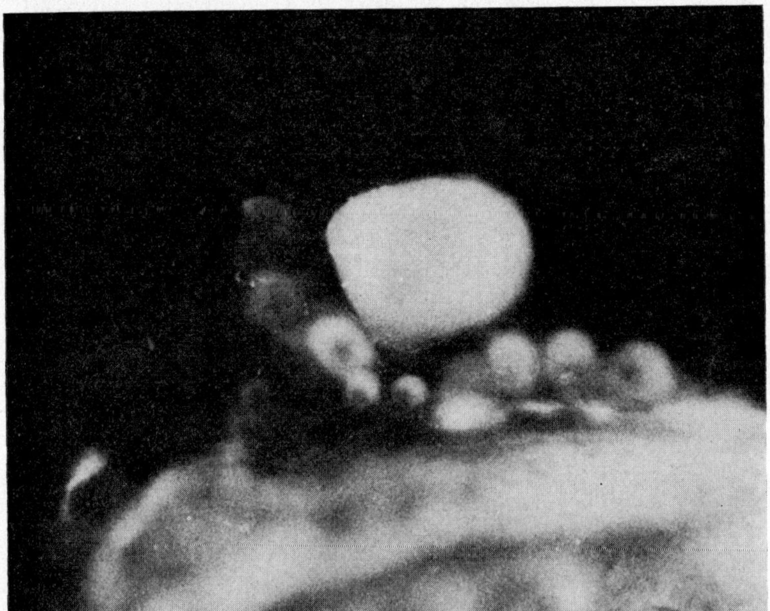

FIG. 3. A free-hand dissected giant neuron (GSC) isolated from the metacerebral ganglion of the snail *Helix pomatia*. The neuron measures about 150 μ along its major axis and can be compared with some smaller cells. The photograph was taken using dark-field illumination.

dissection to avoid any damage to the cell. The cell can then be transferred to a microtube by means of a constricted pipette (see page 18) attached to the mouth by rubber tubing, which allows complete control of the transfer process. Alternative methods have been used to transfer cells which include merely using the scalpel tip, not a method to be advocated. Another procedure is to expose the cell without cutting the axon. A loop of fine wire (enamelled michrome 'Trophet C', size 001, Wilbur Driver Co., Newark, New Jersey,

USA) is then dropped over the cell and drawn tightly around the axon at the base of the cell. The axon is severed below the wire knot and the wire and cell transferred to a microtube (McCaman and Dewhurst, 1970).

2. Isolation of Cell Components from Fresh Tissue

Giacobini (1956), working with spinal and sympathetic ganglion cells and anterior horn cells of spinal cord, was one of the first to dissect nerve cell bodies from fresh tissue. Subsequently Hydén (1959) used the same procedure on mammalian brain. Such free-hand dissections are normally carried out on slides with a layer of about 1 mm paraffin wax (histological quality) in a drop of a suitable Ringer medium (see Edström and Neuhoff, 1973). The surface tension of a water drop on a paraffin layer is high enough to prevent the drop spreading, and the evaporation is therefore low. The microdissection is carried out under a stereomicroscope, and where this lasts for longer periods the stereomicroscope should be equipped with a cooling stage to give temperatures of between 0°C and 10°C.

Small pieces of tissue approximately 1 mm square containing the cells to be studied are placed in the drop of suitable Ringer solution on top of the paraffin wax-covered slide. The uppermost surface of the piece of tissue should be observed under a stereomicroscope and differences in colour and texture noted. Then a drop of solution of methylene blue in isotonic Ringer solution (1 part in 10,000) is applied to the surface. As soon as the dye is taken up, preferentially by the neurons (see Fig. 4), the methylene blue solution is washed

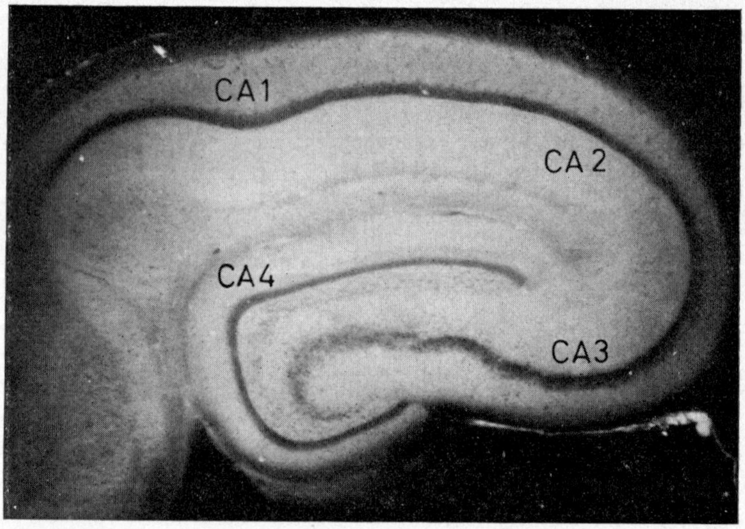

FIG. 4. This figure shows the area of nerve cells (dark region) in the hippocampus of rabbit, after staining with a solution of methylene blue. A thick section of nervous tissue, kept cool, was stained for 3 min with 1% methylene blue solution in order to observe the neurons. The defined areas of certain regions in the hippocampus are also indicated (see page 194).

away and substituted by Ringer alone. Single cells can then be dissected free from the tissue with fine steel wires fixed to a glass rod (see page 18). The free nerve cells which float in the drop of Ringer can be seen clearly under the stereomicroscope and can be transferred by use of hair loops (see page 18). Once the technique has been mastered, nerve cells with many dendrites can be isolated individually in less than a minute. Some neurons isolated from human brain tissue are shown in Figs. 1 and 2.

Clumps of glial and neuropile can also be prepared by this procedure. A piece of fresh nervous tissue is placed on a drop of medium on a slide and single nerve cell bodies adhering to the neuropile/glial may then be lifted out of the tissue with a stainless steel microspatula and forceps. Giacobini (1959) isolated glial cells of spinal and sympathetic ganglia through such an approach.

Lasek and Dowler (1971) have developed a useful technique for obtaining nuclei from freshly dissected invertebrate (*Aplysia*) neurons. Their method is as follows. Single characterised neurons are isolated in a way similar to that described earlier, and placed in a depression slide which contains a suitable Ringer. A small hole, 50–150 μ, is made in the cell membrane with watchmaker's forceps and the neuron gently squeezed. The cell is then left alone and the nucleus left to emerge. This occurs within 15 min. Any excess cytoplasm surrounding the nucleus is then removed by 'sweeping' the nuclear membrane with a fine needle or forceps. Intact nuclei can now be transferred from depression slides to a test-tube by means of a large-bore capillary pipette and analysed.

3. Isolation of Cell Components from 'Fixed' Tissue

This technique was first developed by Edström (1964). A coverslip with a section of material is inverted and placed on a thick glass slide which has a wide groove running across it. The groove forms the floor and ends of the chamber while the coverslip with the section on its underside forms the roof. The sides are left open. The chamber is then filled

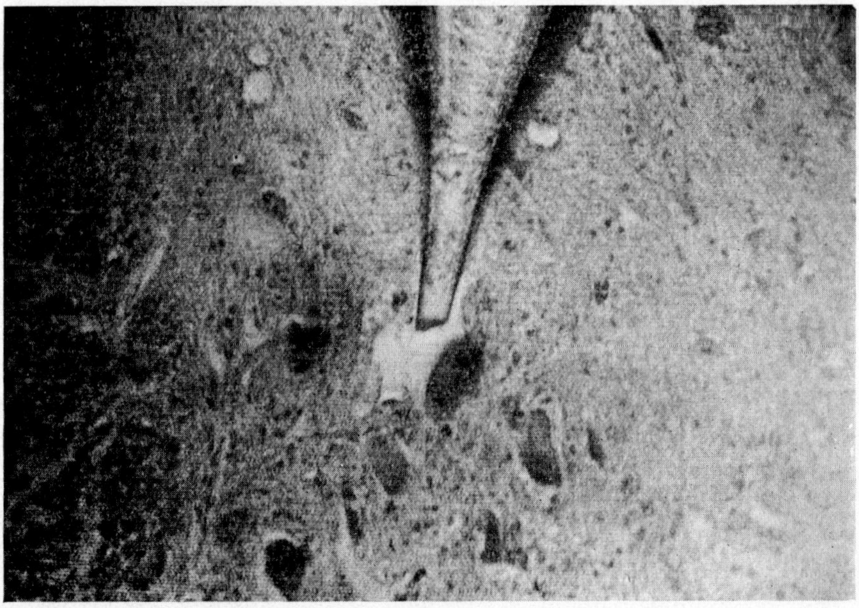

FIG. 5. Histological section of nervous tissue from the medulla of rat, 20 μ thick and stained with methylene blue. The microscope slide containing the section is lying on oil. Cells from the section are then dissected free from the surrounding glia with a dissecting needle which is manipulated by a De Fonbrune attachment (see Fig. 8). This figure clearly illustrates the dissecting needle used to free the neurons, which are darkly stained. × 200.

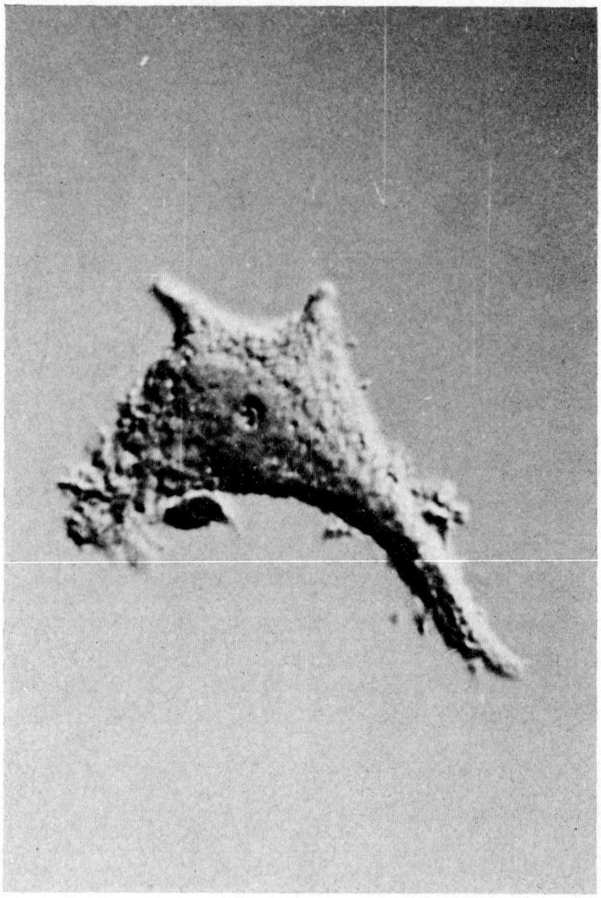

FIG. 6. This figure shows an isolated nerve cell from a deparaffinised section of nerve tissue stained with methylene blue and photographed under Normarski optics. The cell was isolated in the following way. The glass slide bearing the section of deparaffinised brain tissue (from the pons of man) was inverted over another glass slide which formed a glass chamber open at both ends. The chamber was then filled with liquid paraffin and, under observation in a phase contrast microscope (magnification about × 1200), the nerve cells were dissected by means of a micromanipulator using a curved glass needle. Cells thus freed remained attached to the upper glass slide (see Fig. 5). The individual cells could then be transferred with a fine capillary pipette to a new slide for examination. (Photograph by courtesy of V. Neuhoff.)

with paraffin oil which is retained in the open-sided chamber by surface tension. This is placed on a phase contrast microscope which has its focusing device associated with the tube of the microscope rather than the stage. Two fine glass needles, bent upwards, are mounted on a micromanipulator and inserted into the chamber through the open sides. Small neurons and cell parts can then be isolated and pushed away while still sticking to the coverslip (see Fig. 5). A micropipette which has a slightly bigger diameter than the isolated components can be attached to the micromanipulators and used to suck up the cell components together with a drop of paraffin oil (see Edström and Neuhoff, 1973; Neuhoff *et al.*, 1968).

Cells can be isolated either from paraffin embedded sections (Fig. 7) or from deparaffinised sections (Fig. 6). The procedure used for fixing the tissue and dewaxing the

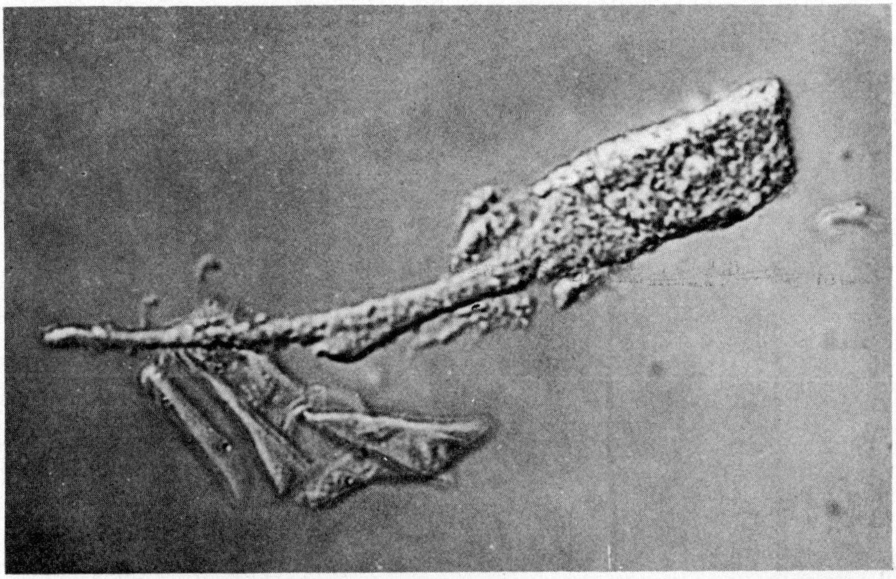

FIG. 7. This figure shows an isolated nerve cell from a section of nervous tissue (medulla oblongata of man) embedded in paraffin and dissected in the manner described in Fig. 6. Some pieces of paraffin wax can be seen clearly at the bottom left-hand corner near the axon. × 1200. (Photograph by courtesy of V. Neuhoff.)

sections will depend upon the parameter investigated. For example, in studying the RNA content of cells (Edström, 1953; Edström and Neuhoff, 1973), Carnoy's solution is a suitable fixative.

The de Fonbrune micromanipulator (Etablissements Beaudouine, Paris) is particularly useful in conjunction with an oil chamber (see Edström and Neuhoff, 1973). This micromanipulator (see Fig. 8) consists of two parts: the lever control and the receptor. The lever can be moved in such a way that the dissecting needles can easily be manipulated in three dimensions. The lever controls the receptor pneumatically; the two parts are connected to each other with rubber tubing so that they can be arranged independently of each other.

4. Isolation of Cell Components from Freeze-dried Tissue

Lowry (see Lowry and Passonneau, 1972) was the first to develop the technique for dissecting cell bodies from freeze-dried sections. Small pieces of nervous tissue (or whole 'brain' in the case of some invertebrates) are placed on a metal microtome chuck and frozen in liquid isopentane or a propane/propylene (Calor gas) mixture. The tissue is then allowed to warm up from −20°C to −10°C in a refrigerated Cryostat and sectioned. The cutting temperature is critical. If it is too high the section will be compressed, and if it is too low the section will fragment. The optimal temperature for cutting is about −15°C. In order to obtain a series of sections it is advisable to prevent ice forming on the cutting knife by brushing it frequently with a small camel-hair brush. The thickness of the sections produced will depend upon the size of the cell bodies in the tissue. For rabbit root ganglion and anterior horn 20 μ is suitable. As they are cut, individual sections can be transferred to microscope slides or other types of holders with a fine pair of forceps, and placed on the platform of a Pearse–Edwards Tissue Dryer (Edwards High Vacuum Ltd.) kept at −30°C.

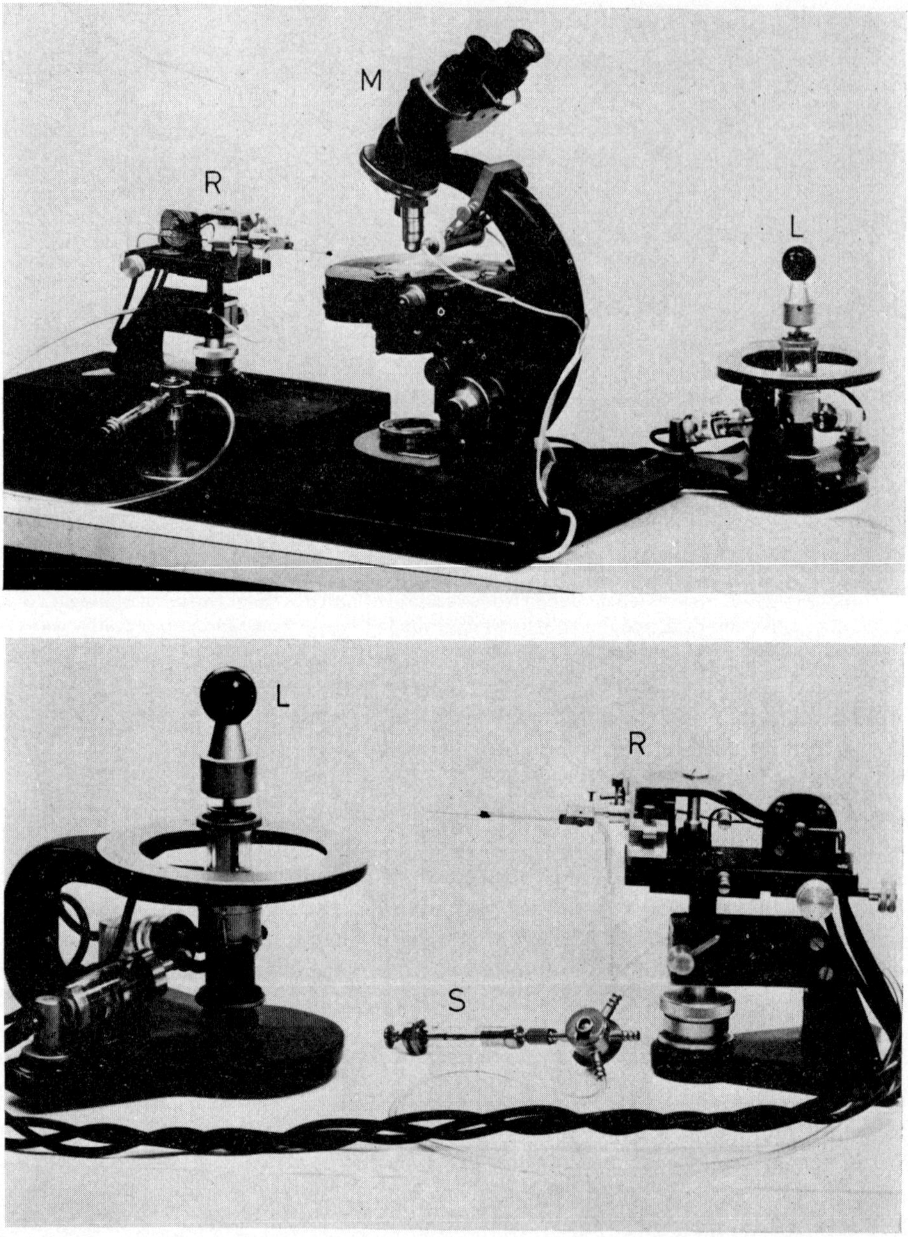

FIG. 8. The set-up required for the dissection of neurons from freeze-dried or fixed sections from nervous tissues. On the stage of a phase-contrast microscope (M) is a section of tissue inverted over an oil chamber. A De Fonbrune micromanipulator, which consists of a receptor part (R) containing a dissection needle capillary (see Figs. 11 and 5), can be moved in all directions by means of the lever part (L). A syringe system (S) for introducing solution into the needle capillaries is also attached.

The sections are then freeze-dried at this temperature and 10^{-3} torr for a period of 2 days. Some phosphorus pentoxide is also placed in the freeze-dryer to absorb the water molecules. From the time the tissue is frozen until the sections are completely dry the samples have to be kept well below freezing point.

Individual cell bodies may then be dissected from the freeze-dried sections by teasing out free-hand with the use of fine glass needles under microscopic vision. The cells can be transferred with 'hair points', which are short pieces of hair mounted in a glass holder (see page 18) and held down by hair loops. It is advisable to carry out the dissection procedure in a room free from draughts, because the freeze-dried sections can easily blow away.

5. Isolation of Cell Components from Frozen Impregnated Tissue

This method described by Giller and Schwartz (1971) has the virtue of stabilising the nervous tissue and so facilitating the dissection while allowing the enzyme activities to be retained. The nervous tissue is covered with 70% ethylene glycol and then frozen with powdered solid carbon dioxide. The ethylene glycol solution does not solidify under these conditions, although the nervous tissue is hardened to the consistency of solid cheese. Throughout the dissection the nervous tissue is kept hard by constantly renewing the supply of solid carbon dioxide. The cells can easily be seen with either transillumination or dark-field light under optical magnification, and the individual cell components dissected in the manner described earlier. Though Giller and Schwartz (1971) used this procedure for isolating cells from *Aplysia* ganglia, it may also be applied to pieces of vertebrate nerve tissue. In the latter case the dissection is simpler if phase contrast microscopy is employed.

6. Isolation of Discrete Areas of Nervous Tissue

In this procedure tissue containing the area to be investigated is first frozen with solid carbon dioxide and placed in a refrigerator so that the temperature of the tissue reaches −20°C. This temperature is maintained throughout the process of isolation. A slice (about 1 mm in thickness) is then placed on a cooled microscope slide and the precise area punched. A suitable tool for this punching out process is shown in Fig. 9. It consists of a stainless

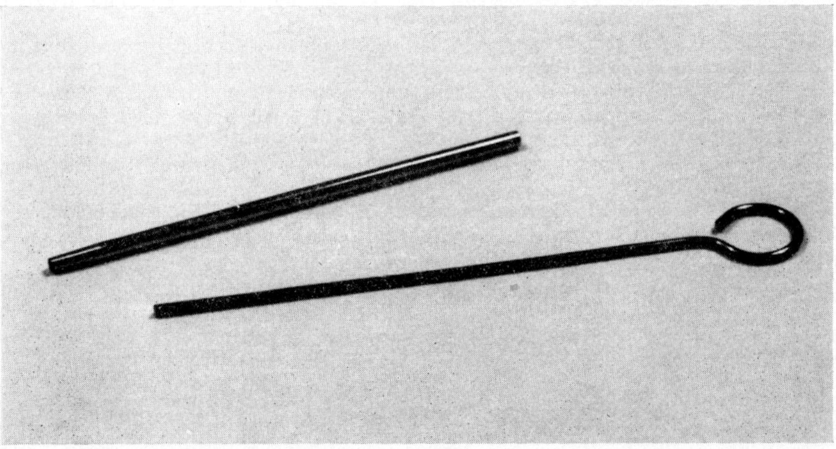

FIG. 9. Tool used for punching out cylinders of nervous tissue, in pieces of *ca.* 1 mg. It consists of two parts: a hollow pipe for punching the nervous tissue, and a plunger to push the nervous tissue out of the pipe.

steel piece of tubing (internal diameter 1 mm), one end of which is sharpened to punch out a cylinder of frozen tissue (the tissue cylinder weighs about 1 mg). The punched out piece of tissue remains in the tubing and is pushed out rapidly with a metal plunger.

References

COGGESHALL, R. E., KANDEL, E. R., KUPPERMAN, I. and WAZIRI, R. (1966) A morphological and functional study on a cluster of identifiable neurosecretory cells in the abdominal ganglia of *Aplysia californica*. *J. Cell. Biol.* **31**, 363–368.

COHEN, M. J. and JACKLET, J. W. (1965) Neurons of insects: RNA changes during injury and regeneration. *Science*, **148**, 1237–1239.

EDSTRÖM, J. E. (1953) Ribonucleic acid mass and concentration in individual nerve cells. A new method for quantitative determinations. *Biochem. Biophys. Acta*, **12**, 361–386.

EDSTRÖM, J. E. (1964) Microextraction and microelectrophoresis for determination and analysis of nucleic acids in isolated cellular units. In: *Methods in Cell Physiology* (Ed. D. M. Prescott), Vol. I, pp. 417–447. Academic Press, New York.

EDSTRÖM, J. E. and NEUHOFF, V. (1973) Micro-electrophoresis for RNA and DNA base analysis. In: *Micromethods in Molecular Biology* (Ed. V. Neuhoff), pp. 215–256. Springer Verlag, Berlin.

FRAZIER, W. T., KANDEL, E. R., KUPFERMAN, I., WAZIRI, R. and COGGESHALL, R. E. (1957) Morphological and functional properties of identified neurons in the abdominal ganglion of *Aplysia californica*. *J. Neurophysiol.* **30**, 1287–1351.

GIACOBINI, E. (1956) Histochemical demonstration of AChE activity in isolated nerve cells. *Acta Physiol. Scand.* **36**, 276–290.

GIACOBINI, E. (1959) The distribution and localisation of cholinesterase in nerve cells. *Acta Physiol. Scand.* Suppl. 156, pp. 1–54.

GILLER, E. and SCHWARTZ, J. H. (1971) Choline acetyltransferase in identified neurons of the abdominal ganglion of *Aplysia californica*.

HYDÉN, H. (1959) Quantitative assay of compounds in isolated, fresh nerve cells and glial cells from control and stimulated animals. *Nature*, **184**, 433–435.

KANDEL, E. R., FRAZIER, W. T., WAZIRI, R. and COGGESHALL, R. E. (1957) Direct and common connections among identified neurons in *Aplysia*. *J. Neurophysiol.* **30**, 1352–1376.

KERKUT, G. A. (1969) Neurochemistry of invertebrates. In *Handbook of Neurochemistry* (Ed. A. Lajtha), Vol. II, pp. 539–562. Plenum Press, New York.

LASEK, R. and DOWLER, W. J. (1971) Analysis of nuclear DNA in individual nuclei of giant neurons. *Science*, **172**, 278–280.

LOWRY, O. H. and PASSONNEAU, J. V. (1972) *A Flexible System of Enzymatic Analysis*. Academic Press, New York.

McCAMAN, R. E. and DEWHURST, S. A. (1970) Choline acetyltransferase in individual neurons of *Aplysia californica*. *J. Neurochem.* **17**, 1421–1426.

NEUHOFF, V., MÜLLER, D. and terMEULEN, V. (1968) Präparation von Ganglienzellen für cytophotometrische Untersuchungen. *Z. Mik. Tech.* **69**, 65–72.

NICHOLLS, J. G. and BAYLOR, D. A. (1969) The specificity and functional role of individual cells in a simple central nervous system. *Endeavor*, **29**, 3–7.

OSBORNE, N. N. (1973) The analysis of amines and amino acids in micro-quantities of tissue. In: *Progress in Neurobiology* (Eds. G. A. Kerkut and J. W. Phillis), Vol. 1, part 4, pp. 301–329. Pergamon Press, Oxford.

WILLOWS, A. O. D. (1967) Behavioural acts elicited by the stimulation of single identifiable brain cells. *Science*, **157**, 570–574.

WILLOWS, A. O. D. (1968). Behavioural acts elicited by stimulation of single identifiable nerve cells. In: *Physiological and Biochemical Aspects of Nervous Integration* (Ed. F. D. Carlson), pp. 217–244. Prentice-Hall, New York.

CHAPTER 4

Some Instruments and Glassware used in Microprocedures

1. Dissecting Tools

(a) SCALPEL

Halved razor blades on untempered steel are broken transversely into many pieces by crushing them between the teeth of a pair of pliers. Each fragment measures about 1 mm × 0.5 mm, with the cutting edge across the 0.55 mm part. The individual fragment can then be mounted on to suitable handles using Dekhotinski cement (Cenco, Breda, Netherlands) to produce a microdissecting knife as shown in Fig. 10.

FIG. 10. Photograph of various sizes of microknives required for dissection purposes.

(b) GLASS NEEDLES

These can be prepared simply from 3.0 mm diameter pyrex glass rods, by heating a small area intensely in a microburner and drawing one end out so as to give a needle-fine point.

When preparing dissection glass needles for use, especially in conjunction with an oil chamber, a more complex procedure has to be adopted, as described by Edström and Neuhoff (1973). Needles are prepared from a soft glass rod with a diameter of 3 mm, which is drawn out to give a piece 5 cm long and 1 mm in diameter. The end of the thin part is then placed in contact with the heated platinum wire of a de Fonbrune microforge (Etablissements Beaudouine, Paris) and the molten glass pulled at an angle of 50–60° from the axis to form a cone-like shape. The tip is then further sharpened in the microforge, during which time the air blower of the microforge is brought into use to obtain a localised heating (see Fig. 11).

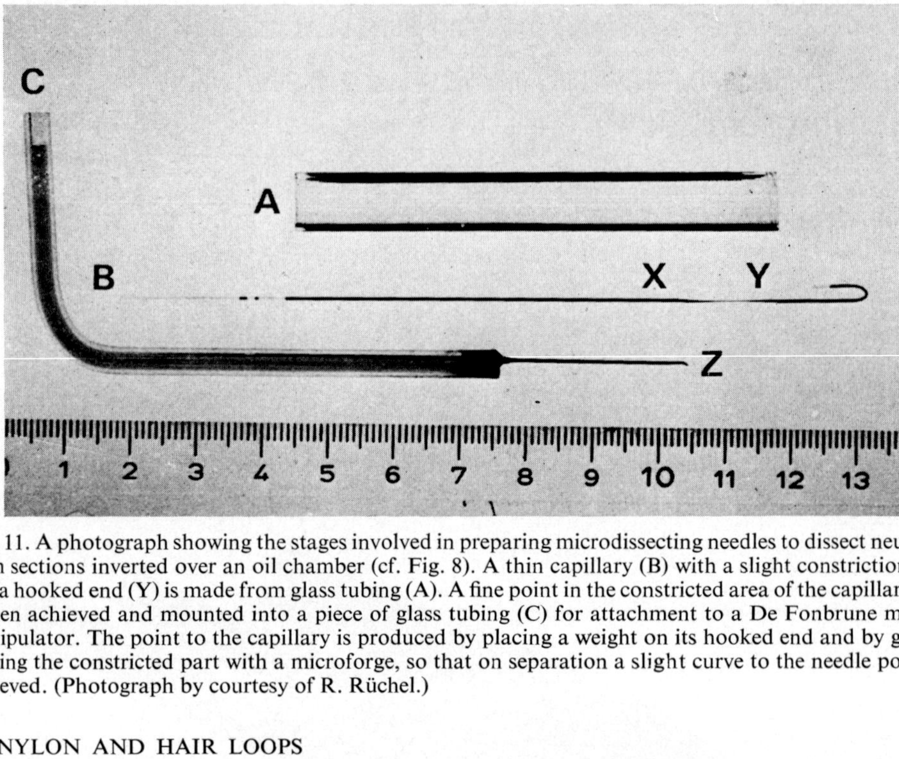

FIG. 11. A photograph showing the stages involved in preparing microdissecting needles to dissect neurons from sections inverted over an oil chamber (cf. Fig. 8). A thin capillary (B) with a slight constriction (X) and a hooked end (Y) is made from glass tubing (A). A fine point in the constricted area of the capillary (Z) is then achieved and mounted into a piece of glass tubing (C) for attachment to a De Fonbrune micromanipulator. The point to the capillary is produced by placing a weight on its hooked end and by gently heating the constricted part with a microforge, so that on separation a slight curve to the needle point is achieved. (Photograph by courtesy of R. Rüchel.)

(c) NYLON AND HAIR LOOPS

Soft glass tubing with a diameter of 4 mm is heated in a microburner and pulled out to form a Pasteur-like pipette. The end of the pipette should be very fine, yet with an internal diameter large enough to allow a thread of nylon or hair to be inserted. Different thicknesses of nylon thread (9–15 denier) or hair are inserted into the end to form a loop and sealed in place by applying a drop of cement (e.g. Plexiglass or post-office sealing wax) with a glass rod on to the end of the pipette so that the hollow tip becomes filled and blocked (Fig. 12).

(d) HAIR POINTS

These are made by sealing short lengths of hair (especially good is nostril hair) or nylon thread into glass holders in the way described (see Fig. 12).

2. Pipettes

(a) GENERAL CONSIDERATIONS

Pipettes can be produced in a variety of ways and in a large range of sizes particularly suitable for microbiochemical procedures (see Edström, 1964; El-Badry, 1963; Lowry

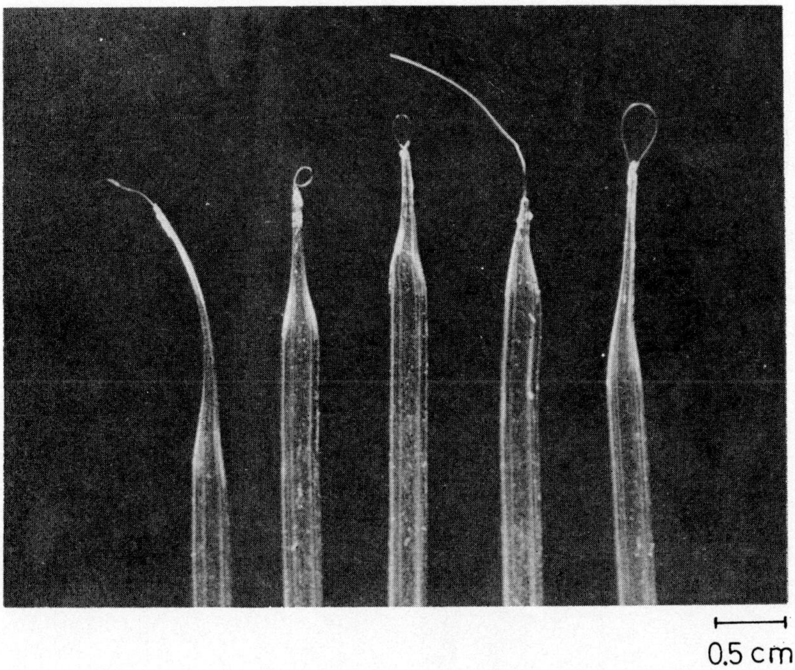

FIG. 12. Photograph of various nylon hair loops and hair points necessary for dissecting individual neurons.

and Passonneau, 1972; Mattenheimer, 1966; Neuhoff, 1973). A well-constructed pipette should have a sharp constriction (see Fig. 13) narrow enough to stop a meniscus of solution at a reproducible point. It should also be wide enough to prevent obstruction but without excess force being necessary to drive the meniscus past a point on the shaft or a constriction. Moreover, it should never be wide enough to allow the solution to run out without any pressure being applied. The tip of the pipette must also be in relation to the size of the pipette. It should taper quite abruptly to a fine point so that any liquid adhering to it will be of a minute volume in comparison to the delivered volume. It is also important to have the shaft diameter or upper constriction of the pipette and the size of the tip orifice balanced to provide a suitable rate of delivery.

(b) THEIR CONSTRUCTION

There are a number of commercial sources for standard types and larger sizes of constriction pipettes (e.g. Microchemical Specialities Co., California; H. E. Pedersen, Copenhagen, Denmark; or Calbiochem, California). However, for routine purposes, where smaller sizes of pipettes often of modified design are required, they have to be hand made. As in all glass-making procedures, the degree of success depends on the quality of the flame which has to be of correct size and temperature. Pipettes are best made from Pyrex tubing, though soft glass can also be used. Tubing with an outer diameter of 4 mm and wall thickness of 1 mm is cut into 10–15-cm lengths, each of which will make two pipettes. The centre part of the length of tubing is heated and drawn out to a diameter of 1–3 mm in such a way that the tapering is gradual. The tubing should always be removed from the flame to pull into shape. The centre portion is once again heated and narrowed to a diameter of between

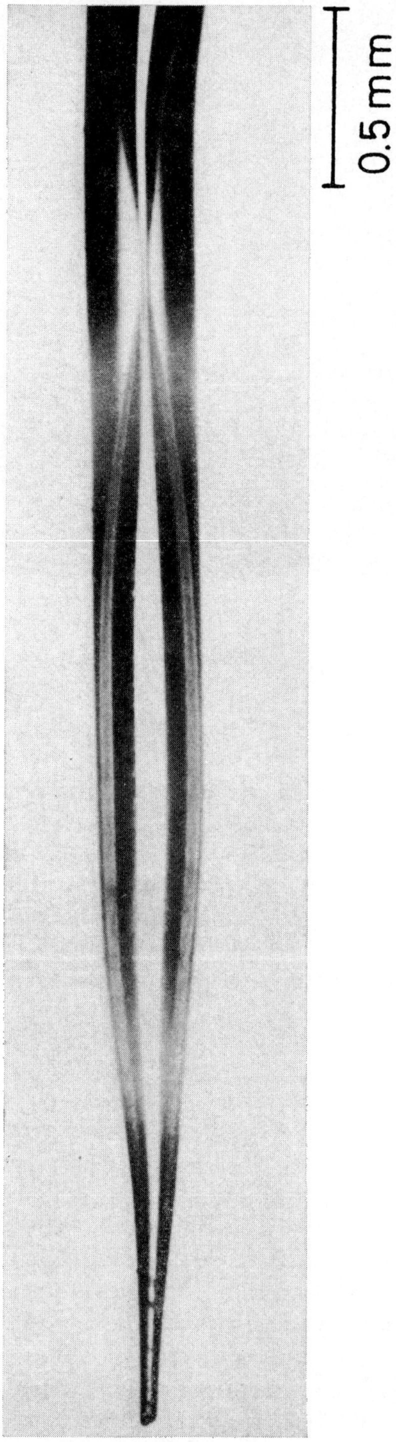

FIG. 13. A photograph of a microconstriction pipette for minute volumes of solution. It is made from quartz glass.

0.2 and 0.8 mm and then separated in the middle (see Fig. 14). The degree of drawing out of the pipettes after each heating finally determines their dimensions, i.e. diameter and length of shaft. This type of pipette, which is suitable for transferring small amounts of solution, is routinely used in many of the described microtechniques, e.g. for loading Drummond microcaps, or for the application of samples for microchromatography. They

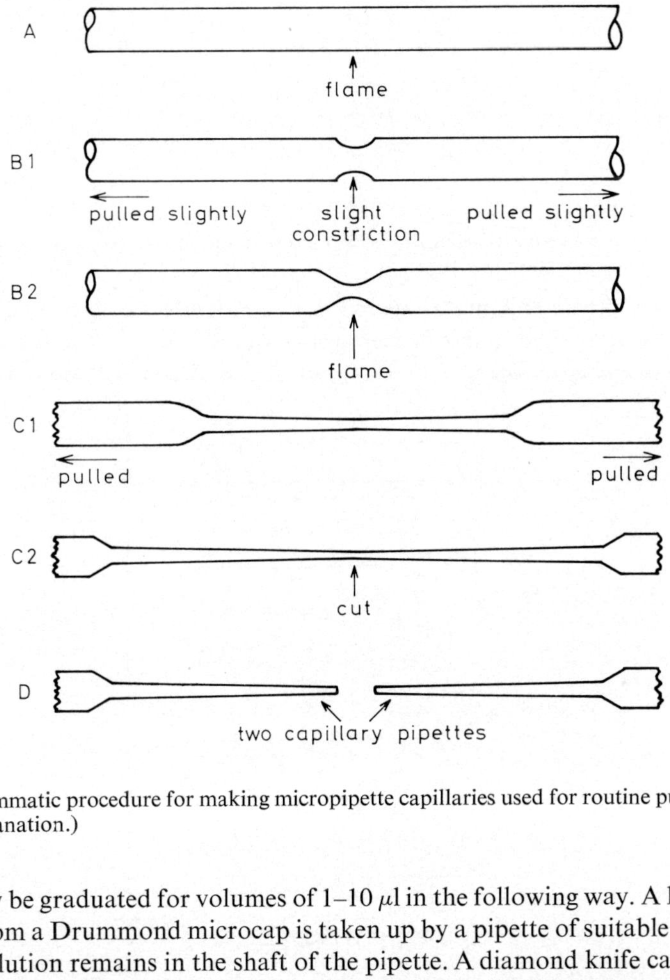

FIG. 14. Diagrammatic procedure for making micropipette capillaries used for routine purposes. (See text for further explanation.)

can also easily be graduated for volumes of 1–10 μl in the following way. A known volume of solution from a Drummond microcap is taken up by a pipette of suitable dimensions so that all the solution remains in the shaft of the pipette. A diamond knife can then be used to mark the position of the solution on the shaft so completing the graduation. The volume of the Drummond microcap can be checked simply by use of either radioactive material of known specific activity or fluorometrically by quinine hydrobromide in 0.1 N sulphuric acid.

Should graduated pipettes of more than 10 μl be required, it is best to alter the structure of the pipette slightly so as to produce a reservoir on the side nearest to the thicker part of the pipette (see Fig. 15). When pipettes of less than 1 μl are required, it is advisable to use quartz instead of glass or Pyrex, not only because of its strength, but also because it lends itself better to the making of these pipettes. A slightly different procedure for their manufacture has to be followed. Quartz tubing approximately 10 cm long and 4 mm in

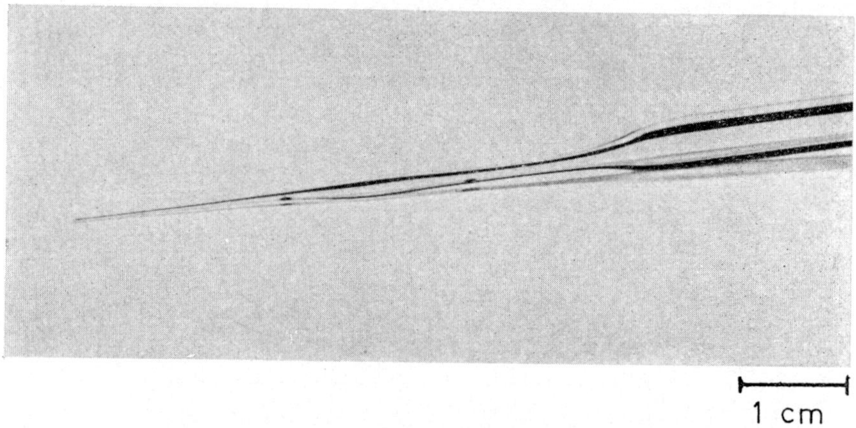

FIG. 15. A photograph of a constriction pipette with a reservoir in the shaft for larger volumes of solution.

diameter (internal diameter 2 mm) is heated in a flame so that a 1-cm section in the centre is brought to red heat. The tubing is then drawn quickly out to a suitable diameter (see Fig. 16) to give a shaft measuring between 6 and 10 mm. A constriction is now made in the

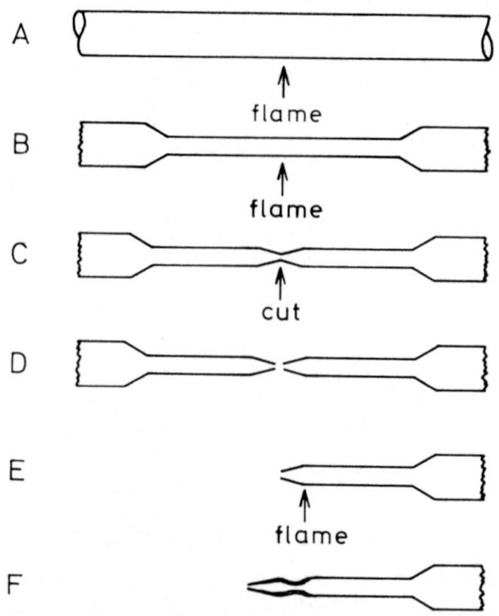

FIG. 16. Diagrammatic procedure for making nanolitre-volume micropipettes. (See text for further explanation.)

shaft using a smaller flame and holding the free end as a handle to produce a pipette as shown in Fig. 13. The smaller flame must have an exceedingly narrow hot zone which can be achieved by using a torch-tip with an orifice of less than 0.5 mm diameter and increasing the oxygen/hydrogen: gas ratio to its maximum. The volume of solution contained between the tip and constriction in the shaft of the pipette can then be determined either with radioactive material of known specific activity or fluorometrically by quinine hydrobromide in 0.1 N sulphuric acid.

(c) USING THE PIPETTES

The pipettes are filled and emptied by pressure applied from the mouth. This is done by attaching flexible rubber tubing (Silicon Schlauch) to the thick part of the pipette and a mouthpiece to the other end of the tubing (see Fig. 17). It is important not to have too stiff a piece of rubber tubing which would require undue force to manipulate, thus in-

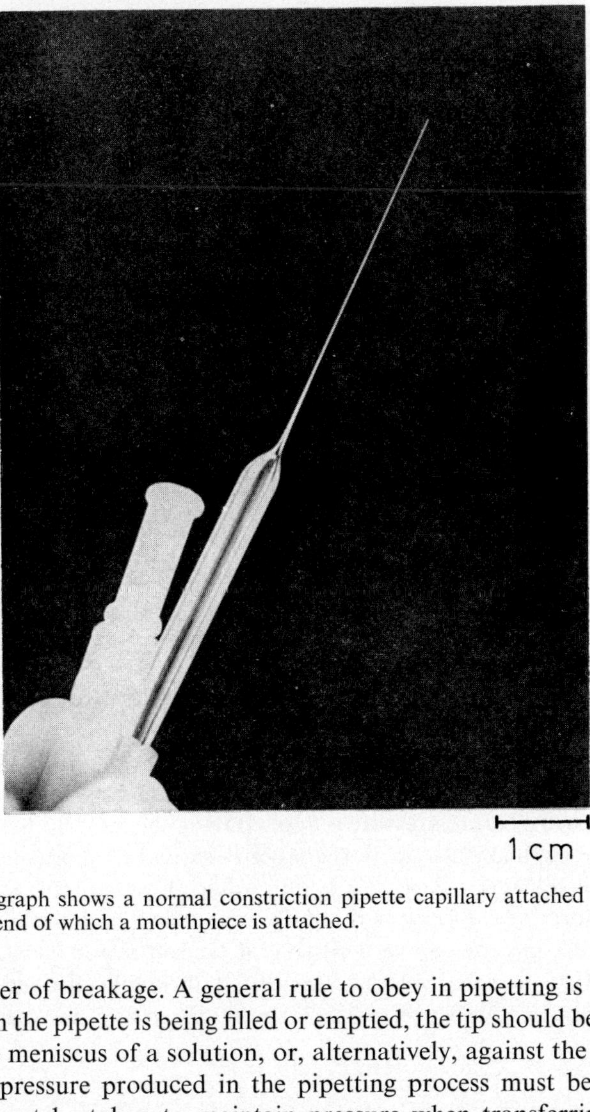

FIG. 17. This photograph shows a normal constriction pipette capillary attached to a piece of rubber tubing, to the other end of which a mouthpiece is attached.

creasing the danger of breakage. A general rule to obey in pipetting is to use only the tip of the shaft. When the pipette is being filled or emptied, the tip should be placed just under the surface of the meniscus of a solution, or, alternatively, against the wall of the vessel. In addition, the pressure produced in the pipetting process must be even and gentle. Particular care must be taken to maintain pressure when transferring until the tip is removed from the meniscus, otherwise fluid will be drawn back into the pipette shaft by capillarity.

It is best to aim to prepare a pipette with a shaft length of about 2–3 cm and a thick part measuring about 3 cm (see Fig. 14). These dimensions are especially convenient when transferring a solution into a microcapillary (Drummond microcap), since too long a pipette requires the experimenter's hands to be wide apart in an unrelaxed position, so allowing room for a pipetting error. When transferring substances to capillaries (Drummond microcaps), it is best to sit in a relaxed position with the heels of the hands resting against each other and the elbows supported by a bench top. The capillary can then be held in one hand and the pipette inserted into it with the other hand as shown in Fig. 18. The pipette

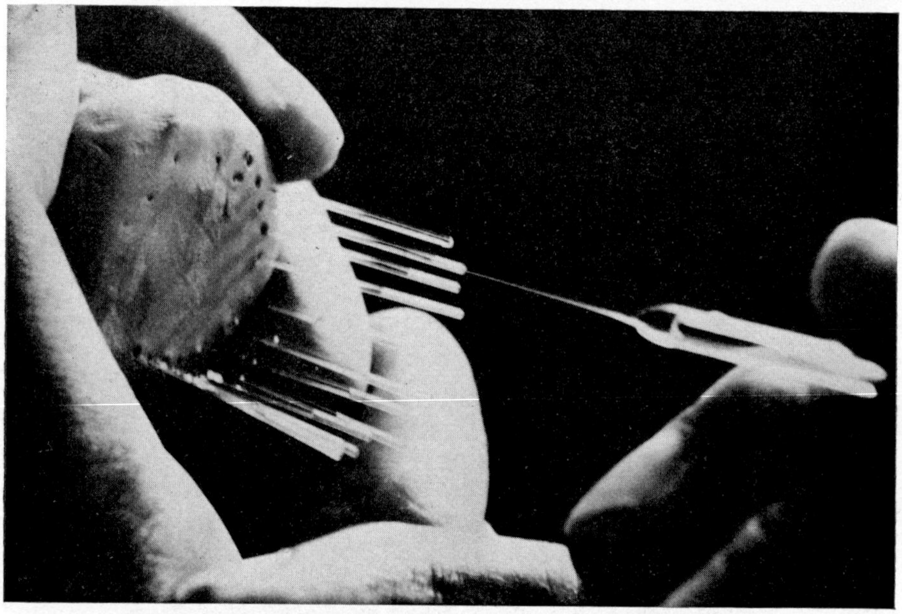

FIG. 18. The procedure for filling or removing a sample from capillaries using a microconstriction capillary pipette. Please note that for convenience the capillaries are placed in a plasticine pad.

shaft outer diameter should be about half of the internal diameter of the capillary, so that the shaft can be easily inserted and air bubbles cannot be trapped between the pipette and the capillary wall. When transferring a solution into a capillary, the tip of the pipette is inserted into the bottom of the empty capillary, or, if it already contains some solution, barely into the meniscus, before pressure is applied from the mouthpiece for the release of the solution. As the capillary fills up so the pipettes should be gradually drawn out. The reverse procedure is applied to pipette a solution out of a capillary; the pipette shaft is gradually inserted into the capillary as the solution is drawn up into the pipette.

The same graduated pipette can be used several times if rinsed correctly. A good procedure is simply to rinse repeatedly with concentrated nitric acid, distilled water, and finally with absolute alcohol. A number of pipettes can be stored in a wooden block, as shown in Fig. 19.

3. Microtubes

Microtubes required for homogenisation, centrifugation and reacting of substances in nervous tissues weighing 0.1–10 mg are best made just before use to avoid the possibility of contamination. Glass tubing (Pyrex is recommended, though soft glass can also be used) of 4 mm diameter (2 mm internal diameter) or 3 mm diameter (1.3 mm internal diameter) is cut into pieces measuring 6 and 3 cm long respectively. The end of the tubing is

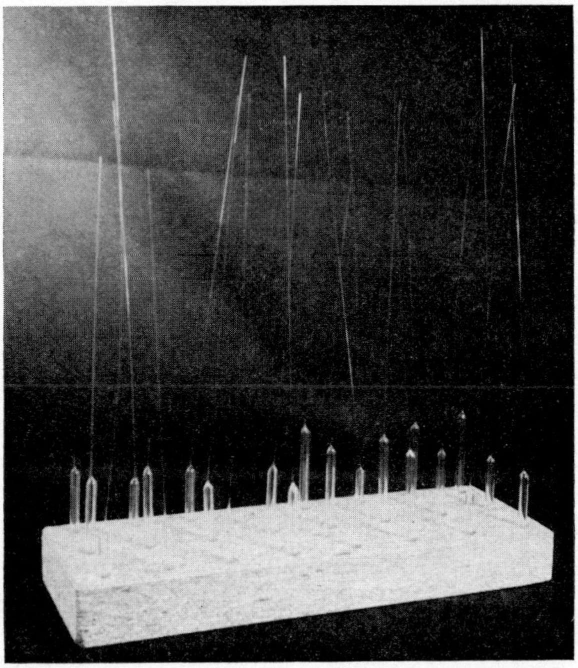

FIG. 19. Wooden block stand containing a set of capillary pipettes.

then carefully heat-sealed so that the bottom of the inside of the tube and the walls are at 90° to each other. The flat base of the microtube is particularly important for homogenisation and is obtained by sealing the tube at the edge of a very hot flame and rotating it constantly.

When soft glass is used the sealed tube should be tempered in a cooled flame. This increases the strength of the seal considerably.

References

EDSTRÖM, J. E. (1964) Microextraction and microelectrophoresis for determination and analysis of nucleic acids in isolated cellular units. In: *Methods in Cell Physiology* (Ed. D. M. Prescott), Vol. I, pp. 417–447. Academic Press, New York.

EDSTRÖM, J. E. and NEUHOFF, V. (1973) Microelectrophoresis for RNA and DNA base analysis. In: *Micromethods in Molecular Biology* (Ed. V. Neuhoff), pp. 215–256. Springer Verlag, Berlin.

EL-BADRY, H. M. (1963) Micromanipulators and micromanipulation. In: *Monographien aus dem Gebiet der qualitativen Mikroanalyse* (Ed. A. A. Beneditti-Pichler), Vol. 3. Springer, Wien.

LOWRY, O. H. and PASSONNEAU, J. V. (1972) *A Flexible System of Enzymatic Analysis.* Academic Press, New York.

MATTENHEIMER, H. (1966) *Mikromethoden für das Klinisch-chemische und Biochemische Laboratorium*, 2 Aufl. Walter de Gruyter & Co., Berlin.

NEUHOFF, V. (1973) Production of capillary pipettes. In: *Micromethods in Molecular Biology* (Ed. V. Neuhoff), Vol. 14, pp. 407–409. Springer Verlag, Berlin.

CHAPTER 5

Available Microbiochemical Procedures and Their Application

A number of sensitive procedures are available for the study of various biochemical parameters at the cellular level. Table 4 summarises some of these methods and detailed information on them and other micromethods can be found in the articles by Hydén (1960, 1972), Giacobini (1968, 1969), Lowry (1953, 1963), Lowry and Passonneau (1972), Zeuthen (1961), Edström (1964) and Neuhoff (1973). A brief review of some of these biochemical procedures and a few examples to show their application will be discussed to demonstrate their use. It should perhaps be made clear that most of the studies reported have been concerned with the correlation of intracellular biochemical reactions of the cell with its actual morphology, biophysical nature, physiological nature, etc. Such problems are clearly best studied on individual characterised neurons where all these parameters can be analysed separately. For the reasons discussed on page 3, it has become evident in recent years that certain invertebrate preparations offer a number of advantages for such studies. It is, however, only lately that invertebrate neurons have been fully exploited through microbiochemical methods.

1. Nucleic Acids

Recent reports suggest that in addition to the established role of nucleic acids in the biosynthesis of proteins, RNA (ribonucleic acid) and protein synthesis may be involved in the accumulation of sensory information in the brain. This might indicate a possible approach to the elucidation of the brain function on a molecular basis (Hydén, 1964). At the cellular level three main approaches have been used to study the functional role of the nucleic acids. They have been concerned with the base composition, the content, and the distribution of the various species in the different cell components.

(a) METHODS

Edström (1953, 1964) described a procedure for the base analysis of RNA in individual neurons and axons. The original method utilises an alkaline-treated fibre (made from cellulose, for example), impregnated with a strongly acidic, viscous buffer to which hydrolysates are applied containing 500–1000 pg of RNA. After high-voltage electrophoresis, the absorbing zones on the fibre are photographed in an ultraviolet microscope at 257 mμ, and the photographic plate is scanned in a double-beam automatic recording densitometer (see example in Fig. 20). The molar base proportions are determined by measuring each

TABLE 4. SUMMARY OF SOME MICROBIOCHEMICAL METHODS AVAILABLE

Enzyme analysis at the cellular level	Cartesian diver procedure	Linderström-Lang (1937), Giacobini (1957, 1969)
	Cycling procedure	Lowry *et al.* (1961), Lowry and Passonneau (1972)
	Radiochemical procedure	Buckley *et al.* (1967), McCaman (1968)
	Microelectrophoresis	Cremer *et al.* (1972), Neuhoff and Lezius (1967), Neuhoff *et al.* (1970)
Nucleic acid analysis at the cellular level	Total RNA and DNA content	Hydén (1943), Edström and Kawiak (1961)
	Ribonucleic acid analysis	Edström (1953, 1964)
	Species analysis	Egyhazi *et al.* (1968, 1969)
Protein analysis at the cellular level	Disc electrophoresis of water-soluble proteins	Hydén *et al.* (1966), Neuhoff (1968)
	Electrophoresis of total proteins (SDS system)	Wilson (1971), Gainer (1971), Rüchel *et al.* (1973)
Dry weight determination at the cellular level	Electronmicroscopy	Bahr *et al.* (1961), Bahr and Zeitler (1965)
	X-ray absorption	Brattgård *et al.* (1957)
	Interferometry	Bahrer (1952), Carlson (1965)
	Quartz fibre	Lowry (1963), Lowry and Passonneau (1972)
Ion measurements at the cellular level	Sodium, potassium	Ramsay *et al.* (1951), Müller (1958)
	Sodium, potassium, calcium, magnesium, etc.	Vurek and Bowman (1965)
		Haljamäe and Wood (1971)
Phospholipid analysis at the cellular level	Microchromatography and fluorometry	Schiefer and Neuhoff (1971)
Transmitter substances (amines and amino acids) at the cellular level	Microchromatography of amines	Osborne (1971)
	Radiochemical procedures for amines	Hildebrand *et al.* (1971), McCaman *et al.* (1973)
	Microchromatography of dansyl derivatives for amines and amino acids	Neuhoff *et al.* (1969)
		Neuhoff and Weise (1970)
		Osborne *et al.* (1971), Briel *et al.* (1972)

area under the density curve, and are expressed as percentages of the sum of the base areas. This technique, with a simple modification (see Edström, 1964), also allows DNA base content and composition to be determined. The fibres normally employed for electrophoresis or microphoresis, as termed by Edström (1956), are made from cellulose or rayon silk. However, for radioactive procedures it is best to use strips made from cellophane (Koenig and Brattgård, 1963), although the sensitivity of the procedure is slightly decreased since more than 3000 pg starting material is required. The procedure described by Koenig and Brattgård for preparing the cellophane strips is as follows.

A piece of cellophane (Bemberg AG, Wuppertal-Barmen, Germany), approximately 1.8 × 0.8 cm, is immersed first in distilled water for 2 min, in 1.5 M NaOH for a further 2 min, then in 2.25 M NaOH for 30 sec at 12°C, and is finally given three rinses, each of 2 min duration, in distilled water. The cellophane is then immersed in buffer (18 ml conc. H_2SO_4, 10 ml H_2O, 66 g glycerol, 72 g D-glucose heated to 100°C for 20 min in a water

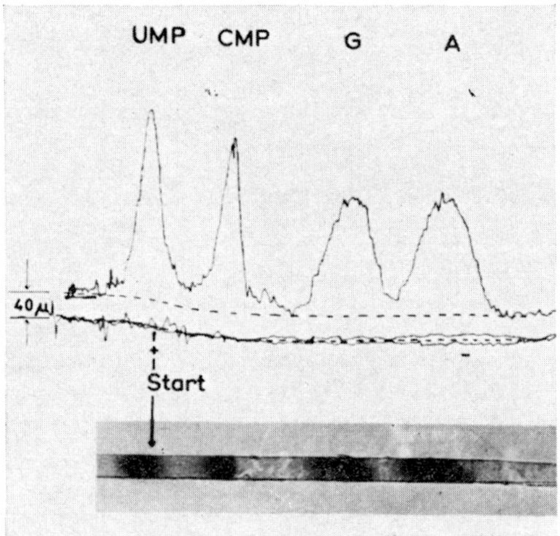

FIG. 20. A microphoretic separation scanned at 257 mµ of a hydrolysate of RNA products from isolated hippocampus cells. UMP, CMP, G and A stand for uridinemonophosphate, cytidinemonophosphate, guanine and adenine. (Figure by courtesy of V. Neuhoff.)

bath) for 15 min at 38°C. Subsequently the cellophane is removed from the buffer, embedded in a block of paraffin, and strips 40–80 µ in thickness are cut in a microtome. The strips are then carefully rinsed in distilled water, placed once again in the buffer and stored in a refrigerator.

For detailed information on apparatus required for electrophoresis (microphoresis), the buffer system employed, the procedure of applying the hydrolysate to the fibre, the electrophoresis, the photographic and photometry measurements of the fractionated sample, and the analysis of the base composition of the nucleic acids in the sample, refer to the recent excellent accounts by Edström (1964) and Edström and Neuhoff (1973).

Though a number of microprocedures can be scaled down to allow measurement of nucleic acid content in the various components of the nervous system, the methods of Edström and his collaborators (Edström, 1958; Pigon and Edström, 1959; Edström and Kawiak, 1961) are best known. Essentially the procedure is as follows. For the analysis of RNA content, ribonuclease is used to extract the nucleic acids. The extract is then collected on a specially prepared quartz glass, placed in an 'oil chamber', treated with a glycerol-containing buffer, and eventually photographed in ultraviolet light at 257 mµ with a reference system of the different optical densities. The photographs, magnified if necessary, are then investigated by photometry and the RNA content is ultimately determined by means of a device for integration of the absorption. This technique also works for DNA extracted with deoxyribonuclease (Edström and Kawiak, 1961), though the extraction is slightly different from that used for RNA determination.

Few methods exist for studying the various nucleic acid species at the cellular level, and these normally involve the use of electrophoresis. One method, recently developed by Wolfrum *et al.* (1973), overcomes a number of the difficulties usually encountered when fractionating nucleic acids by electrophoresis, and allows 10^{-9} g of RNA to be analysed (see page 184).

(b) STUDIES

Hydén and collaborators (see Hydén, 1972) have shown that RNA (Table 5) and base content (Table 6) vary in different cell types. Furthermore, the base content and the rate of synthesis of RNA in nerve cells and glial were found to be different (Table 7).

TABLE 5. RNA CONTENT IN DIFFERENT TYPES OF NEURONS (from Hydén, 1972)

Type of neuron	RNA (pg/cell) mean values
Ganglion cells of the supraoptic nucleus, rabbit	70
Spinal ganglion cells, rabbit	1070
Anterior horn cells, rabbit	530
Hypoglossal cells, rabbit	200
Anterior horn cells, man 60–70 years	540
Anterior horn cells, man 40–50 years	670
Deiters' cells, rabbit	
Large type	1550
Small type	700

TABLE 6. THE MOLAR PERCENTAGE OF BASES IN NERVE CELL AND GLIA RNA (from Hydén, 1972)

	N. Deiters' rabbit	N. supraopt. rat	N. Deiters' rat	F. reticul. rat	Glob. pall. man	Glia
Adenine	19.7	18.9	20.7	19.5	17.1	25.3
Uracil	18.0	18.2	18.5	16.5	16.1	19.2
Guanine	33.5	36.0	33.3	30.0	29.8	29.0
Cytosine	28.8	26.9	27.5	34.0	37.0	26.5

TABLE 7. *In vivo* INCORPORATION OF ^{3}H-ADENINE AND ^{3}H-CYTIDINE INTO RNA OF HYPOGLOSSAL NEURONS AND GLIA FROM RABBITS OVER 4 HOURS

The incorporation is expressed as μmol labelled base/g RNA (mean value ± S.E.M.) (from Daneholt and Brattgård, 1966)

Base	Nerve cells	Glia
Adenine	2.6 ± 0.12	5.8 ± 0.66
Uracil	0.90 ± 0.11	2.0 ± 0.30
Guanine	0.43 ± 0.05	1.0 ± 0.11
Cytosine	3.2 ± 0.16	2.8 ± 0.36
A:G/C:U	2.9:0.48/3.6:1	2.9:0.50/1.4:1

The experiments of Edström *et al.* (1962) answered some of the questions concerning the metabolism of axonal proteins by showing RNA to be present in the axon as well as in myelin sheaths of Mauthner neurons of the goldfish. The RNA in the axon and sheath was subsequently characterised as a ribosomal type, and a proximal–distal decrease of RNA content over the length of the axon was also demonstrated (Edström, 1964). After spinal cord transaction there was a marked increase in the adenine: guanine ratio of RNA, this increase being maximal after 2–3 days (Edström, 1964). Such results are of interest in the study of the origin and function of axoplasm, and agree with the findings in a series of papers by Koenig (1965a, b, 1967a, b) on the local axonal synthesis of enzymes.

An investigation into the types of RNA present in single identified neurons of *Aplysia* was made by Peterson (1970). The long *in vitro* functioning of the giant neurons permits adequate labelled precursor to be incorporated, so allowing gel electrophoresis or sedimentation analysis of the RNA, not only in a single neuron, but also in a single nucleus or cytoplasm. Analysis of the RNA showed a number of distinct peaks, most of which were related to ribosomal RNA. In an effort to discern possible precursor–product relations and the localisation of these RNA species, RNA from the nucleus and the cytoplasm of a single specific neuron (R_2) was analysed after varying periods of labelling with tritiated precursors. The results support the view that a large transcribed precursor is subsequently processed to form finished ribosomal RNA.

Berry (1969) studied the effects of electrical stimulation on the metabolism of RNA in the R_2 cell of *Aplysia*. He found that when the cell is at rest, tritiated uridine is incorporated into RNA at a constant rate. This rate increases under long periods of synaptic stimulation, the increase being proportional to the number of action potentials produced by the neuron. Similar results have been observed by Peterson and Kernell (1970), who showed that after prolonged stimulation of the R_2 neuron at a level strong enough to elicit post-synaptic spikes, there is a marked increase in the amount of labelled RNA in the nucleus as well as in the cytoplasm. Electrophoresis studies suggested that the increase in labelled RNA corresponds to the molecular weights of ribosomal as well as of non-ribosomal RNA (Kernell and Peterson, 1970).

The ultrastructure and RNA synthesis in identified neurons of *Lymnea stagnalis* after electrical stimulation for short periods of time have been studied by Bocharova *et al.* (1972). At the end of a stimulation period of 40 min, which generated about 1500–2500 action potentials, incorporation of tritiated uridine into nuclear RNA was only 40–50% of that in the control. At this time no alteration in the ultrastructure of the stimulated neuron could be detected. However, there was a dramatic increase (150–200%) in incorporation of labelled precursor into RNA in neurons examined 40–70 min after the end of the stimulation period; changes were also observed in the nucleus, demonstrating the activation of RNA synthesis in the nucleoli and RNA transport from nucleus to cytoplasm.

It is of interest to compare the increase in the rate of RNA synthesis observed in molluscan neurons following electrical stimulation to what occurs in the stretch receptor neuron of the lobster (Edström and Grampp, 1965). The stretch receptor, studied by Edström and Grampp (1965), exhibited no changes in RNA content, composition or synthesis during extended periods of spike generation and firing, though significant increases in the adenine/uracil and purine/pyridine were found. Moreover, the complete inhibition of protein and RNA synthesis by adequate concentrations of puromycin and actinomycin D respectively had no effect upon the stretch receptor's pulse activity (Toschi and Giacobini, 1965) or respiration (Giacobini, 1969).

Jarlstedt (1963) investigated the effect of increased sensory activity of Purkinje cells in the cerebellum of the rabbit on the RNA content. This was done by irrigating the outer ear with cold and warm water. When the rabbit's left ear was irrigated with warm water for 30 min, there was a unilateral 30–60% increase in the RNA of the Purkinje cells in the modulus as well as in those of the lobulus centralis. When he used cold water, Jarlstedt found that the RNA content of the Purkinje cells in the lobulus centralis on the contralateral side increased by 25–35%. A number of other stimuli (Table 8) and drugs (Table 9) have been shown to influence the RNA content and base composition in individual vertebrate cell types (Hydén, 1972). In an effort to study behaviour effects upon the RNA content in the various cell components, Hydén and Egyhazi (1962, 1963 and 1964) trained rats to balance on a wire 1 m long and then later did experiments on 'reversal of handedness'. Results of such experiments showed (see Tables 10 and 11) that the RNA base composition and content changed in individual cell types after the rats had 'learned' their tasks, suggesting a modification in the different types of RNA.

TABLE 8. EFFECTS OF SOME STIMULI ON NERVE CELL RNA (from Hydén, 1972)

Stimulus	Cell type	RNA increase in pg per cell	%	*P* value
Intermittent horizontal rotation 25 min/day, 7 days	Deiters' neurons, rat	680–750	10	0.02
Intermittent horizontal rotation 25 min/day, 7 days	Deiters' neurons, rabbit	1550–1750	10	0.01
Intermittent vertical rotation 30 min	Deiters' neurons, rat	680–850	25	0.001
Sodium chloride 1.5%, 30 days	Neurons of N. supra-opticus, rat	68–121	80	0.01
Thirst for 7 days	Neurons of N. supra-opticus, rat	52–129		0.001

TABLE 9 EFFECT OF 0.3 mg/kg OF TRANCYPROMINE ON NERVE CELL AND GLIA RNA. Values given in pg per cell. N = number of animals, P = probability level, n = number of analyses (from Hydén, 1972)

	Nerve cells				Glial cells			
	Mean	*P* value	*N*	*n*	Mean	*P* value	*N*	*n*
One hour	2012 ± 97	0.01	3	33	58 ± 6.4	0.01	3	17
Six days	1914 ± 87	0.01	3	30				
Controls	1550 ± 62		24	30	123 ± 17.5		17	41

2. Enzyme Activities

When measuring enzyme activities, the handling of the nervous tissue component prior to the analysis is a very critical stage in the proceedings, since the accuracy of the results obtained depends on the localisation of the enzyme in the cell. The nature of the enzyme and the binding of the enzyme molecule to specific cellular units are of prime importance.

The studies of Giacobini (1959, 1969) on AChE (acetylcholineserase), ChAc (cholinesterase) and MAO (monoamine oxidase) have demonstrated the necessity of careful attention to these points.

TABLE 10. CHANGES IN THE RNA CONTENT AND BASE COMPOSITION OF CORTICAL NEURONS FROM THE CONTROL (LEFT) SIDE AND FROM THE LEARNING (RIGHT) SIDE (from Hydén, 1972)

	Controls Mean	Learning Mean	% change	*P* value
Adenine	18.4 ± 0.48	20.1 ± 0.11	+ 9.2	0.02
Uracil	18.3 ± 0.48	19.6 ± 0.56	+ 7.1	0.05
Guanine	26.5 ± 0.64	28.7 ± 0.90	+ 8.3	0.01
Cytosine	36.8 ± 0.97	31.5 ± 0.75	−14.4	0.01
$\frac{G + C}{A + U}$	1.72 ± 0.054	1.51 ± 0.026	−12.2	0.02
$\frac{A + G}{C + U}$	0.81 ± 0.27	0.95 ± 0.035	+17.3	0.01
RNA pg	22 ± 2.3	31 ± 2.5		0.002

TABLE 11. MICROELECTROPHORETIC ANALYSIS OF THE NUCLEAR RNA AND THE GLIAL RNA OF THE DEITERS' NUCLEUS IN LEARNING EXPERIMENTS IN RATS
The purine and pyridine bases are expressed in molar proportions in percentage of the sum (from Hydén, 1967)

	Nucleus			Glia		
	Control	Learning	RNA fraction	Control	Learning	RNA fraction
Adenine	21.4 ± 0.44	24.1 ± 0.39	38.1 ± 3.25	25.3 ± 0.16	28.3 ± 0.45	43.0 ± 2.83
Cytosine	31.9 ± 0.77	31.0 ± 0.96	26.7 ± 6.95	26.5 ± 0.43	24.3 ± 0.36	14.0 ± 3.12
Guanine	26.2 ± 0.45	26.7 ± 0.87	28.8 ± 5.75	29.0 ± 0.31	28.8 ± 0.31	27.0 ± 2.24
Uracil	20.5 ± 1.01	18.2 ± 1.11	6.4 ± 8.08	19.2 ± 0.27	18.6 ± 0.21	16.0 ± 1.88
A/U	1.06 ± 0.056	1.32 ± 0.084	5.9 ± 7.5	1.32 ± 0.020	1.52 ± 0.030	2.70 ± 0.364
$\frac{G + C}{A + U}$	1.38 ± 0.042	1.37 ± 0.049	1.25 ± 0.218	1.25 ± 0.030	1.13 ± 0.028	0.69 ± 0.076
No. of animals	5	8		5	6	
No. of cells	500	900				
No. glia samples				33	42	

(a) METHODS

A number of methods exist for analysing enzyme activities at the cellular level (see Table 4). The elegant technique of the Cartesian diver respirometer of Linderström-Lang (1937) and Holter (1961) has been brought to the single-cell level by the ampulla diver of Zeuthen (1961), resulting in a number of studies at the cellular level (e.g. Giacobini, 1968; Hydén and Pigon, 1960). The procedure has disadvantages. It is rather an involved and time-consuming technique, requiring repeated manual adjustments by the experimenter. Furthermore, only a limited number of samples can be examined in one single experiment, and the more sensitive type of instrument, which may have a self-recording magnetic diver, so automatically registering both temperature and volume variations, can be rather expensive.

The two methods now particularly used for enzyme analysis are fluorometry, where the technique of cycling and recycling increases the sensitivity (for details see Lowry and Passonneau, 1972), and isotopic procedures. The latter method has certain unique advantages in quantitative microstudies: (a) radioactive methods are specific and their high sensitivity can be increased by using substrates of higher specific activities; (b) any drugs or other substances which may be present in the incubation medium often have no effect on the assay; (c) rapid and simultaneous analysis of a relatively large number (50–200) of samples can be made; (d) special instruments are mostly not required, except for the conventional scintillation counter; (e) isotopically labelled substrates may be used to develop new and specific methods for analysing almost all enzyme activities. Microprocedures where isotopes are used usually amount to measurements of the labelled product (Fig. 21), labelled gaseous release (Fig. 22) or labelled products of a secondary chemical reaction (Fig. 23). For each of these procedures special substrates have to be employed (see McCaman, 1968, for details and references). The practicality of these isotope methods depends generally upon their sensitivity and efficiency. Both of these criteria are related to adequate separation techniques of the labelled reaction product from the labelled substrate. Furthermore, the procedure involved should allow quantitative recovery of the labelled product.

Substrate		*Product*
1] ^{14}C-Serotonin	→	^{14}C-5-Hydroxyindole acetaldehyde
2] ^{14}C-Carnitine (Oleyl CoA)	→	^{14}C-Oleylcarnitine
3] ^{3}H-Adrenaline (S-adenosylmethionine)	→	^{3}H-Metanephrine
4] ^{14}C-Adenosylmethionine (Dihydroxybenzoate)	→	^{14}C-3-Methoxy-4-hydroxybenzoate
5] Cytidine diphospho-^{14}C-choline (Diglyceride)	→	^{14}C-Phosphatidyl choline
6] Cytidine diphospho-^{14}C-ethanolamine (Diglyceride)	→	^{14}C-Phosphatidyl ethanolamine
7] ^{14}C-L-α-Glycerolphosphate (Oleyl CoA)	→	^{14}C-Phosphatidic acid

FIG. 21. Some examples of enzyme reactions which are measured isotopically after solvent extraction. (From McCaman, 1968.)

Substrate		*Product*
1] ^{14}C-Glutamic acid	→	$^{14}C\text{-}O_2$
2] ^{14}C-Acetate (Choline, CoA, ATP)	→	^{14}C-Acetylcholine
3] ^{14}C-Hydroxytryptophan	→	$^{14}C\text{-}O_2$
4] ^{14}C-6-Phosphogluconate (TPN)	→	$^{14}C\text{-}O_2$

FIG. 22. Some examples of enzyme reactions which are measured isotopically by gaseous diffusion methods. (From McCaman, 1968.)

	Substrate		*Product* (secondary reaction)
1]	^{14}C-Acetyl CoA (choline)	→	^{14}C-Acetylcholine
2]	^{14}C-Acetyl CoA (carnitine)	→	^{14}C-Acetylcarnitine
3]	^{14}C-Adenosylmethionine (noradrenaline)	→	^{14}C-Adrenaline
4]	^{14}C-Choline (ATP)	→	^{14}C-Choline phosphate
5]	^{14}C-Ethanolamine (ATP)	→	^{14}C-Ethanolamine phosphate

FIG. 23. Some examples of enzyme reactions which are measured isotopically by procedures involving a secondary chemical reaction. In the first two examples, the labelled products are converted to insoluble derivatives to facilitate their selective isolation from the respective labelled substrates. Conversely, substrates not metabolised are selectively removed in the procedures used for the last three examples. (After McCaman, 1968.)

Fluorometry, combined with cycling and recycling as described by Lowry (see Lowry and Passonneau, 1972) represents one of the most sensitive ways of analysing enzyme activities at the cellular level. Special use is made of the cell's oxidised DPNH or TPNH and reduced DPN^+ or TPN^+, since almost every substance in the living cell can oxidise or reduce them (pyridine nucleotide system) either directly or indirectly. A two-enzyme system quoted as an example by Lowry (1964) is given below:

$$\text{glucose} + \text{ATP} \xrightarrow{\text{hexokinase}} \text{glucose-6-P} + \text{ADP}$$

$$\text{glucose-6-P} + \text{TPN}^+ \xrightarrow[\text{dehydrogenase}]{\text{glucose-6-P}} \text{6-P-gluconate} + \text{TPNH}$$

These two reactions can be combined into a single analytical step, and the TPNH formed, which is highly fluorescent, becomes a measure of either glucose, ATP or hexokinase. The sensitivity of the process can be increased by a factor of 10 by destroying excess TPN^+ with a weak alkali and then adding peroxide in strong alkali. The peroxide converts the TPNH to TPN^+ and the strong alkali converts TPN^+ to a highly fluorescent product which is easy to measure (see Lowry and Passonneau, 1972). Measurement of the TPNH by cycling and recycling can further increase the sensitivity of the process. In a cycling procedure TPN^+ serves as a catalyst for a two-enzyme process. The system contains ammonia, α-ketoglutarate and the enzyme glutamic dehydrogenase, which quickly oxidises the TPNH to TPN^+ and forms glutamate:

$$\text{TPNH} + \alpha\text{-ketoglutarate} + \text{NH}_4{}^+ \xrightarrow[\text{dehydrogenase}]{\text{glutamic}} \text{glutamate} + \text{TPN}^+$$

The system also contains glucose-6-P and its dehydrogenase, which quickly reduce the TPN^+ back to TPNH and form 6-P-gluconate:

$$\text{TPN}^+ + \text{glucose-6-P} \xrightarrow[\text{dehydrogenase}]{\text{glucose-6-P}} \text{6-P-gluconate} + \text{TPNH}$$

Adequate amounts of each of the two enzymes are added to the system for each molecule of TPNH to go round the cycle as many as 10,000 times in 30 min. At a stated time the enzymes are killed by heat and one of the products 6-P-gluconate is measured by adding 6-P-gluconate dehydrogenase and excess new TPN^+ so that the following reaction occurs:

$$\text{6-P-gluconate} + \text{TPN}^+ \xrightarrow[\text{dehydrogenase}]{\text{6-P-gluconate}} \text{ribulose-5-P} + \text{TPNH}$$

The net result is that for the original molecule of TPNH, there are now more than 10,000 molecules. If this is still not sufficient, excess TPN^+ is destroyed with weak alkali and the TPNH recycled with a new reagent. With this procedure of cycling and recycling a great number of substances can be measured, often allowing detection of as little as 10^{-18} moles. Reference should be made to Lowry and Passonneau (1972) and the article by Maischinsky *et al.* (1971) for detailed procedures of the analysis of different enzymes.

(b) STUDIES

The AChE activity of individual sympathetic ganglion cells was determined by Giacobini *et al.* (1967a) using the Cartesian diver, and by Koslow and Giacobini (1969) using an isotopic technique. The results of both investigations were similar and showed that in the L7 ganglion of the cat the AChE activity of normal cells varies widely (up to fifty-fold). The cells of low AChE activity represent the majority of the population in normal ganglia, while neurons with very high activity were rare and some neurons even lacked the enzyme (see Fig. 24). Three weeks after denervation, the mean activity per cell decreased; cells of

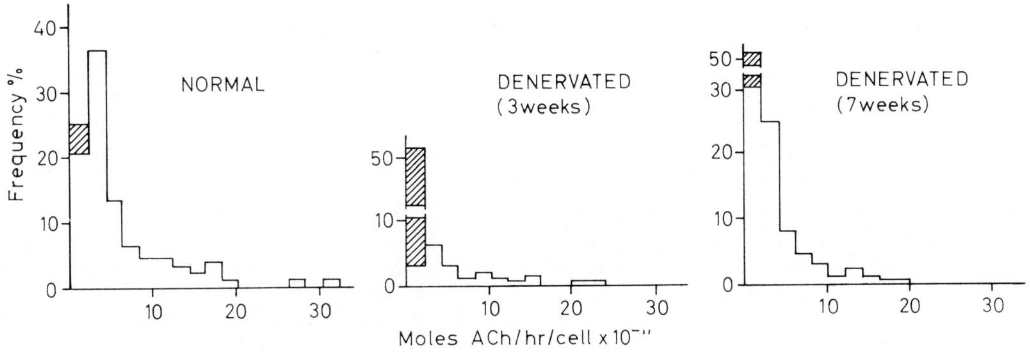

FIG. 24. This figure shows the frequency distribution of acetylcholinesterase (AChE) activity in individual sympathetic cell bodies from the L7 ganglia of the cat before and after denervation. The shaded area represents the inactive cells. (From Koslow and Giacobini, 1970.)

AChE activity were numerous and the percentage of active cells decreased from 85% to 40–60% (see Fig. 24). Seven weeks following the denervation, the AChE in single cells increased (Fig. 24) because the number of low activity cells was greater than normal and the number of inactive cells decreased.

Buckley *et al.* (1967) measured the ChAc activity in several hundred of cell bodies from the L7 ganglion using a radioactive procedure. The ChAc activity in normal cells of L7 varies from 14×10^{-13} to 1×10^{-13} moles ACh/hr, and 12.5% of the cells investigated showed enzyme activity. After denervation, the proportion of cells containing ChAc activity remained the same.

The MAO activity of cells from normal and decentralised L7 ganglia has also been investigated (Consolo *et al.*, 1968; Giacobini *et al.*, 1970). The enzyme was present in 13% of normal cells investigated and in 60% of cells from decentralised ganglia. However, the cells from decentralised ganglia exhibited a significant increase in the average enzyme activity of their cell bodies (Fig. 25).

Single medullary neuron cell bodies and fragments of nuclei and cytoplasm from giant cells in the puffer fish (see Table 12) have been examined for enzyme activity by Lehrer *et al.* (1968). Lactic dehydrogenase and malic acid dehydrogenase were found at approximately equal levels in both the nucleus and cytoplasm, whereas the distributions of hexokinase and glucose-6-phosphate dehydrogenase were different. This variation in the concentrations of enzymes in different parts of a nerve cell substantiates the results of Giacobini (1959). In this instance the Cartesian method was used to study the concentration of AChE in cytoplasm, nucleoplasm, nucleus, axon, dendrite and cell body of anterior horn cells in the spinal cord of the cat. The results showed the enzyme to be most highly

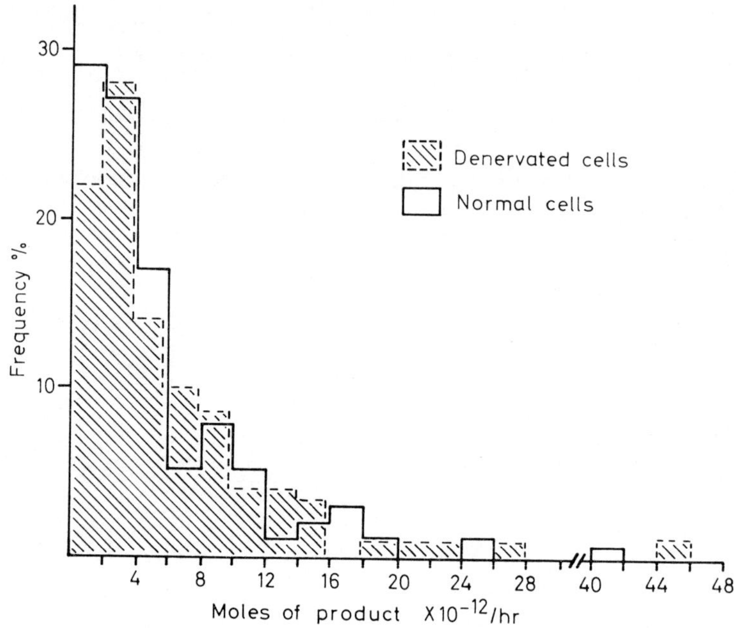

FIG. 25. This figure shows the frequency distribution of monoamine oxidase (MAO) activity in individual sympathetic cell bodies from L7 ganglia of the cat before and after denervation. (From Giacobini *et al.*, 1970.)

concentrated in the cell body and dendrites, followed by the axon and finally the nucleus. The nucleolus showed no activity.

TABLE 12. ENZYMES IN SINGLE NEURON CELL BODIES (from Lehrer *et al.*, 1968)
Figures in parenthesis are standard errors of mean.

	Hexokinase	β-D-glucuronidase (β-D-GLR)	Lactic dehydrogenase (LDH)	Malic dehydrogenase (MDH)	Glucose-6-phosphate dehydrogenase (G-6-PDH)
Didon (puffer fish)					
Nuclei	1.77 (0.10)	0	5.20 (0.47)	64.5 (2.4)	3.87 (0.20)
Cytoplasm	3.51 (0.25)	68.2 (9.7)	5.20 (0.4)	71.1 (2.0)	7.20 (0.28)
Whole cell bodies	2.58 (0.09)	48.3 (5.7)	7.0 (0.71)	80.8 (3.1)	5.05 (0.54)
Cerebellum granularis	3.14 (0.18)	36.5 (4.5)	54.5 (1.2)	197.0 (7.6)	2.05 (0.09)
Spheroides (puffer fish)					
Whole cell bodies	1.87 (0.20)	79.0 (5.8)	26.9 (2.0)	102.0 (7.0)	1.68 (0.81)
Rat					
Cerebellum granularis	4.36 (0.34)	26.5 (0.5)	62.6 (0.8)	318.0 (7.4)	1.69 (0.06)

Enzyme activities in nerve cells, neuronal and capillary glia of Deiters' nucleus have been investigated by Hydén and Pigon (1960), using a micromanometric method which makes use of the Cartesian microdiver. Table 13 shows how the different nerve components oxidise some substrates. The cytochrome oxidase activity in both types of glial is almost three times as high as in the nerve cells and the rate of succinate oxidisation is highest in the capillary glia and lowest in the nerve cells. Oxidisation of α-ketoglutarate is also highest in the capillary glial, while the other two components have the same activity. Glutamate oxidisation was very much higher in the nerve cells in comparison with the glia tissue. These studies confirmed the earlier work of Lowry *et al.* (1957) who showed clearly contrast of enzyme activity in cell bodies and glial cells. His work demonstrated cell bodies and the oligodendrocytes (glia) to be low in hexokinase, phosphoglucoisomerase, lactic and malic dehydrogenase, glutamic dehydrogenase, and glutamic aspartic transaminase and relatively rich in glucose-6-phosphate, 6 phosphogluconate dehydrogenase, isocitric dehydrogenase and succinic dehydrogenase.

TABLE 13. SOME ENZYME ACTIVITIES IN NERVE CELLS, NEURONAL AND CAPILLARY GLIA OF DEITERS' NUCLEUS

Enzyme activity expressed as 10^{-4} μl O_2/hr. Mean values ± S.E.M. (*from Hydén and Pigon, 1960; †from Hamberger, 1961; ‡from Hydén, 1972)

	Nerve cell	*n*	Neuronaglia	*n*	Capillary glia	*n*
Glutamate oxidation	2.2 ± 0.3†	21	1.1 ± 0.2‡	9	1.6 ± 0.4‡	10
α-Ketoglutarate oxidation	2.2 ± 0.4‡	9	2.1 ± 0.2‡	12	4.0 ± 0.7‡	13
Cytochrome oxidase	4.2 ± 0.6*	16	11.5 ± 0.8*	17	11.6 ± 1.1‡	10
Succinate oxidation	2.2 ± 0.3*	17	4.2 ± 0.5‡	9	6.9 ± 0.7‡	9

The microfluorometric procedure of Lowry *et al.* (1957) has also been applied to the study of pyridine nucleotide concentrations in both vertebrate and invertebrate nerve preparations (see Table 14). The data showed clearly that the concentration of NADPH is similar to that of the oxidised one (NAPD), whereas the NAD content is generally higher than NADH by a factor of 2.5–11. Giacobini and Grasso (1966) and Giacobini (1968), using the Lowry *et al.* (1957) procedure, extended their studies on the stretch receptor in an effort to relate physiological stimulation with variations in pyridine nucleotides. The results of their studies are summarised in Table 15, and show that after physiological stimulation both the NADP:NADPH ratio and the NAD:NADH ratio fall by at least threefold, though the sum of the NADs (oxidised and reduced) remains almost constant.

A number of isotopical substrates have recently been used to measure various enzyme activities in individual invertebrate neurons. Making use of ^{14}C–acetylcoenzyme-A as a

TABLE 14. NUCLEOTIDE CONCENTRATIONS IN VERTEBRATE AND INVERTEBRATE NERVE PREPARATIONS

Preparation	NADP	NADPH	NAD	NADH	Ratio NADP/NADPH	Ratio NAD/NADH
	(μmoles/kg wet weight)					
Rabbit brain (Lowry *et al.*, 1957)	10	12	241	107	0.8	2.5
Rabbit brain (Lowry *et al.*, 1961)	7	2	321	94	3.5	3.4
Mouse brain (anaesthetised) (Lowry *et al.*, 1964)	—	—	332	29	0.6	11.0
Crayfish stretch receptor (Giacobini and Grasso, 1966)	10.7	15.4	255	58.1	0.69	4.4

TABLE 15. VARIATIONS OF PYRIDINE NUCLEOTIDES IN CRAYFISH STRETCH RECEPTOR NEURON AT REST AND AFTER PROLONGED STIMULATION (from Giacobini and Grasso, 1966; and Giacobini, 1968)

Pyridine nucleotide (μmoles/kg wet wt.)	At rest	After 12×10^4 impulses	After 50×10^4 impulses
NADP	10.7 ± 2.3	4.6 ± 3.2	4.2 ± 1.5
NADPH	15.4 ± 1.4	14.6 ± 0.6	20.1 ± 0.9
NADP/NADPH	0.69	0.31	0.21
NAD	255 ± 44	180 ± 36	160 ± 19
NADH	58.1 ± 6	90 ± 14.8	105 ± 11
NAD/NADH	4.4	2	1.5
NADP + NADPH	26.1	19.2	22.3
NAD + NADH	313.1	270	265
$\frac{\text{NAD + NADH}}{\text{NADP + NADPH}}$	12	14	12
NADP + NAD	265.7	184.6 (− 30%)	164.2 (− 38%)
NADPH + NADH	73.5	104.6 (+ 42%)	125.1 (+ 69%)
$\frac{\text{NADP + NAD}}{\text{NADPH + NADH}}$	3.6	1.8	1.3
Total pyridine nucleotide	339.2	289.2	289.3

substrate, McCaman and Dewhurst (1970) measured the choline acetyltransferase in nine characterised neurons in the pleural–parietal–visceral ganglia of *Aplysia californica*. They found that four of the neurons had no enzyme activity, while those which contained detectable enzyme activity displayed a 25-fold variation in the activity per cell (see Table 16). The existence of very high levels of enzyme in the three neurons (R_2, L_{10} and L_{11}) situated in the parietal–visceral ganglion agrees quantitatively with the observations of Giller and Schwartz (1969, 1971a).

In another study, McCaman and Dewhurst (1971) measured the enzyme activities of AChE and COMT in individual neurons of *Aplysia*. To estimate AChE activity they used ^{14}C–acetylcholine as a substrate, whereas ^{14}C–S-adenosylmethionine was used as a substrate to measure the activity of COMT. All the neurons studied had measurable levels of activity for both enzymes (see Table 16). They concluded that since the various neurons exhibited approximately the same level of activity, none of these enzymes could be used to classify neurons as 'cholinergic' or 'aminergic'. Similar results concerning the enzyme activities of AChE in *Aplysia* neurons have also been reported by Giller and Schwartz (1971b). Recently Weinreich *et al.* (1972) measured the activities of aromatic amino acid decarboxylase in individual identified neurons of *Aplysia*, using ^{14}C–5-HTP as a substrate and measuring the amount of ^{14}C–dopamine or ^{14}C–serotonin produced. The results showed the decarboxylase enzyme to have a ubiquitous distribution throughout the nervous system, being present in all neurons (see Table 17) and nerves analysed. It could thus be concluded that aromatic decarboxylase activity cannot be used to classify neurons as 'aminergic'.

Other enzymes measured in individual *Aplysia* neurons include choline phosphokinase (Dewhurst, 1972) and enzymes converting adenine to c-AMP (Cedar *et al.*, 1972). Choline phosphokinase activity was analysed by using ^{14}C–choline as a substrate and measuring

TABLE 16. ENZYME ACTIVITY IN SINGLE MOLLUSCAN NEURONS

Cell	Enzyme	Activity in *in vitro* (pmole/cell/hr)	Reference
Aplysia			
R_1	Choline acetyltransferase	0	McCaman and Dewhurst, 1970
L_1	Choline acetyltransferase	0	
R_2	Choline acetyltransferase	588 ± 44	
LPGC	Choline acetyltransferase	507 ± 57	
R_{14}	Choline acetyltransferase	0	
R_{15}	Choline acetyltransferase	0	
L_{10}	Choline acetyltransferase	322	
L_{11}	Choline acetyltransferase	199	
R_1	Acetylcholinesterase	2.6	McCaman and Dewhurst, 1971
L_1	Acetylcholinesterase	3.7	
L_7	Acetylcholinesterase	6.5	
L_{11}	Acetylcholinesterase	6.5	
R_{14}	Acetylcholinesterase	7.2	
R_{15}	Acetylcholinesterase	7.7 ± 1.4	
R_2	Acetylcholinesterase	12.8 ± 1.2	
LPGC	Acetylcholinesterase	14.1	
L_7	Catechol-O-methyltransferase	6.62	Dewhurst, 1972
L_{10}	Catechol-O-methyltransferase	2.66	
L_{11}	Catechol-O-methyltransferase	5.7 ± 1.1	
R_{14}	Catechol-O-methyltransferase	4.01	
R_{15}	Catechol-O-methyltransferase	4.90	
R_2	Catechol-O-methyltransferase	34 ± 4.1	
LPGC	Catechol-O-methyltransferase	34.0 ± 3.3	
L_1	Choline phosphokinase	120	
L_2	Choline phosphokinase	439 ± 188	
L_3	Choline phosphokinase	1123	
L_4	Choline phosphokinase	670	
L_6	Choline phosphokinase	468	
L_7	Choline phosphokinase	583 ± 127	
LPGC	Choline phosphokinase	1863	
R_1	Choline phosphokinase	238 ± 80	
R_2	Choline phosphokinase	1344 ± 200	
R_{14}	Choline phosphokinase	568	
R_{15}	Choline phosphokinase	491	
	Conversion of 5-hydroxytryptophan to serotonin		
	in vitro (Cottrell and Powell, 1971)	*in vivo* (Osborne, 1972)	
Helix			
GSC	60 pmole/nl cell/hr	2.3 pmole/nl cell/hr	
posterior buccal cell	0	0	

the content of ^{14}C–phosphorylcholine produced. The results showed that all the neurons tested contained choline phosphokinase (see Table 16) activity, which is not surprising since the enzyme product is utilised in the biosynthesis of choline-containing phospholipids. The formation of c-AMP in one neuron type (R_2) was analysed and found to be about fourteen times less active than the abdominal ganglion (Cedar *et al.*, 1972). The rate of function of c-AMP was stimulated by serotonin and dopamine (Cedar and Schwartz, 1972), both likely transmitter substances in *Aplysia*.

TABLE 17. DISTRIBUTION OF DOPA DECARBOXYLAS ACTIVITY IN INDIVIDUAL NEURONS AND OTHER STRUCTURES OF APLYSIA (from Weinreich *et al.*, 1972)

Structure	Enzyme activity (pmole/cell/hr)	Enzyme activity (μmole/g of protein/hr)
L_2 cell	298.7 ± 59.8	95.5 ± 30.2
L_3 cell	151.5	27.3
L_4 cell	170.3 ± 42.2	61.2 ± 17.4
L_5 cell	342.6	117.2
L_6 cell	368.9 ± 86.5	74.8 ± 9.6
L_7 cell	135.8 ± 22.7	40.7 ± 6.0
L_{10} cell	217.7 ± 43.6	76.5 ± 16.2
L_{11} cell	279.0 ± 32.5	64.6 ± 8.0
LPGc cell	180.2 ± 28.6	25.5 ± 4.0
R_2 cell	440 ± 57.5	54.2 ± 6.7
R_{14} cell	144.7 ± 23.3	34.8 ± 3.6
R_{15} cell	520.7 ± 93.5	211.1 ± 49.3
Bag cells		83.9 ± 12.6
Neuropile		66.4 ± 7.8
Branchial nerves		227.3

3. Ions

The action of many excitatory and inhibitory substances may be explained by supposing that they open additional channels or pathways through the membrane for one or several of the environmental ions. For excitatory substances, the channels are available to Na and at least one other ion species; for inhibitory substances they are available to K or Cl ions or both. It is therefore important for researchers concerned with the problem of the physiology of nervous tissue to have data relating the ion content of a cell with its physiological conditions.

(a) METHODS

Several different techniques are available for the determination of ions at the cellular level. Techniques such as activation analysis, X-ray fluorescence or microprobe analysis necessitate the use of fairly expensive equipment and the time-consuming preparation of the samples. Ordinary emission flame photometry or atomic absorption spectrophotometry do not yet have a sensitivity sufficient for work with single cells. Cation-sensitive glass electrodes have been used to measure certain ions, though it is questionable whether their sensitivity is specific for ions alone. Probably the best approach for analysing ions at the cellular level is through the use of integrating microflame photometers. For a discussion on the problems peculiar to micro- and ultramicroflame photometric instruments, and for the application of this technique when analysing ions at the cellular level, the reader is referred to the article by Haljamäe and Waldman (1972).

Since the first integrating ultramicroflame photometer designed by Ramsay *et al.* (1951), several improved instruments have been described, e.g. by Müller (1958), Malnic *et al.* (1964), Carlsson *et al.* (1967), Haljamäe and Larsson (1968) and Katz (1968). All these instruments are used mainly for Na^+ and K^+ analysis. Because of the relatively low flame temperature, the emission of Na^+ and K^+, both of which have low excitation potentials, is sufficient for analytic work, allowing as little as 10^{-14} mole of ion to be measured. The basic principle of these low-temperature flame photometers is that the sample is introduced into the flame on the tip of a 50 μ or 100 μ thick platinum–iridium alloy wire. A constant sector of the light emitted in the flame is focused via collector lens systems through transmission filters or a monochromator onto photomultipliers or other types of detectors such as selenium photoelements. The output of the detector is amplified and fed into an integrator and the integral value is registered. The usual set-up is a dual-channel instrument which enables simultaneous determinations of both Na^+ and K^+.

The use of low-temperature flame essentially reduces interference problems, due to the content of cations with higher excitation potentials in the biological material. Self-absorption in the flame is prevented by keeping the concentration of the element low in the flame. Cross-interference between Na^+ and K^+ has been reported to occur with some of the instrumental set-ups, but can be prevented by using calibration solutions with Na^+/K^+ ratios similar to those expected in the biological material. Generally the presence of phosphate markedly enhances the emission, and therefore excess ammonium phosphate is usually added to the samples and the standard solutions.

The analysis of other cations of interest in biological systems, such as Ca^{++} or Mg^{++}, is not possible with the above-mentioned ultramicroflame photometric techniques, due to the much higher excitation potentials of these bivalent cations. As much higher flame temperatures are needed, the introduction of the sample on to the tip of a wire is not possible, as the wire would melt.

Two different approaches to ultramicrodeterminations of Ca^{++} and Mg^{++} have been described. An emission microphotometric instrument has been developed by Vurek and Bowman (1965) and further improved by Vurek (1967). The principle is that through the electrical excitation of helium, very energetic, metastable helium atoms are generated, which can transfer their energy to impurity atoms and excite them. The sample is placed on an iridium loop and introduced into a helium-containing chamber. It is then volatilised by electrical heating of the wire into the region containing the excited helium atoms, and the emission is detected and integrated. An ultramicroflame spectroscopic technique has been described by Haljamäe and Wood (1971), where a special burner is mounted inside the standard lamp housing of an ordinary Beckmann DU spectroscope. The sample is electrically heated from a thin iridium loop and carried by a stream of argon into a hot H_2/O_2 flame. For emission detection, the ordinary slits and optics of the spectroscope are used with the addition of suitable photodetectors, amplifiers and integrators. The recording of non-integrated and integrated emission values has neatly dealt with interference problems due to the content of elements with different excitation potentials in biological material. This procedure of Haljamäe and Wood (1971) also enables as little as 10^{-13} moles of Cl^- to be analysed.

(b) STUDIES

The analysis of ions in tissue or cell samples using micro- or ultramicroflame photometry has been approached in two ways. The first is to place the sample directly in the flame of the instrument and then to analyse the excitation (e.g. Müller, 1958; Katzman *et al.*, 1969; Giacobini *et al.*, 1967b). The other approach is to extract the ions from the sample and then analyse them by micro- or ultramicroflame photometry (e.g. Kessey, 1968; Haljamäe and Lange, 1972; Haljamäe and Hamberger, 1971; Haljamäe and Waldman, 1973). The second approach seems more desirable as it eliminates any interference caused by organic materials.

Using the first technique, Giacobini *et al.* (1967b) followed the changes in Na^+ and K^+ content in single invertebrate neurons (crustacea stretch receptor) during the initial period after isolation and during incubation periods under conditions of reduced movement of sodium and potassium ions, by employing an integrating flame photometry method. After isolation of tissues, ionic events could be divided into two periods. During the first period a continuous recovery from the initial disturbance of the ionic content takes place. This recovery does not occur if the active transport mechanism is inhibited (e.g. with ouabain or by emitting Na^+ or K^+). The second period starts about 1 hr later, continues for at least the next 5 hr, and is characterised by stationary levels of Na^+ and K^+, which, in normal saline, approach physiological levels. It is of interest to note that tetradotoxin, which inhibits sodium ion transport across membranes, and electrical stimulation had no effect upon the sodium and potassium ion content of the stretch receptor.

Müller (1958), also using the first approach, analysed the Na^+ and K^+ concentrations in single myelinated nerve fibres of frogs. He found that the potassium and sodium values varied respectively between 50–145 mM and 50–80 mM per litre of nerve fibre. The same author was also able to correlate the movements of potassium and sodium ions within the nerve fibres with the flow of current through the nerve by using microflame photometry. Studies on intracellular ion concentrations in the crayfish axons have also been described (Wallin, 1967). In these experiments a simultaneous determination of membrane potential and intracellular concentrations of Na^+, K^+ and Cl^- ions in single axons were studied. The normal ion concentrations on crayfish giant axons were subsequently found to be 17.4 ± 0.4 mM sodium, 265 ± 1.8 mM potassium and 12.7 ± 0.7 mM chloride. In the experiments of Katzman *et al.* (1969), supramedullary neurons of the puffer fish (*Spheroides maculatus*) were frozen *in situ*, sectioned at a low temperature, and freeze-dried. Small fragments of the cells were dissected and analysed directly by ultramicroflame photometry. The results showed that the K^+ concentration in the cytoplasm averaged 134 mM per litre tissue volume and in nuclei, 113 mM, whilst the Na^+ concentration in cytoplasm fragments varied between 56 and 138 mM per litre tissue volume and in nuclei, between 40 and 135 mM.

By using the second approach where the ions are extracted from the tissue sample before analysing in a microflame photometer, Keesey (1968) studied the sodium and potassium ion concentration in nanogram samples of mammalian nervous tissue. Average weights of Na^+ and K^+ in microamounts of rabbit thoracic spinal ganglia were found to be similar to those obtained for the whole ganglia. However, the K^+/Na^+ ration from micro-amounts of rabbit thoracic ganglia 10 min *post mortem* and 20 min *post mortem* were different, being 2.64 and 3.23 respectively. In another study, Haljamäe and Lange (1972) followed changes in ion content in microquantities (CA3 region) of hippocampus tissue of rats before and after behavioural experiments (transfer of handedness). The results showed that training of rats increased Ca^{++} content in the CA3 region without changing the Na^+ or K^+ content.

Haljamäe and collaborators (Henn *et al.*, 1972; Haljamäe and Waldman, 1973; Haljamäe and Hamberger, 1971) have also studied the accumulation of specific ions into bulk prepared neuronal, glial and nuclei preparations from rabbit and rat brains, using ultramicroflame photometry. The results showed that both neuronal and glial cells from the rabbit were capable of active potassium transport and that calcium inhibited potassium uptake by glial but stimulated the uptake into neurons. It was also found that the glial cell Na^+–K^+–ATPase was always more active, approximately 2.5–4-fold, and very much more sensitive to variations in the potassium ion concentration than the neuronal ATPase. In the nuclei of rat cerebral cortex, the accumulation of potassium and sodium was very low, whereas divalent cations of magnesium, manganese and calcium accumulated to a higher concentration.

4. Transmitter Substances

A knowledge of the transmitter compound liberated from a particular nerve terminal is essential for the understanding of the interaction between two cells at their synaptic junctions. From the data of a number of studies it is now generally accepted that acetylcholine, noradrenaline, GABA and glutamate are transmitter compounds at certain synapses. Other substances, including serotonin (5-HT), dopamine, adrenaline and glycine, are likely transmitter substances, though the evidence is less compelling. The following account outlines some recently developed microprocedures which allow analysis of transmitter substances at the cellular level. It should be borne in mind, however, that detection of the enzymes (see page 31) involved in the metabolism of a transmitter often indicates that the transmitter is also present. Full details of the dansyl procedure, which is potentially one of the best methods for detecting and analysing transmitter substances, is described on page 58.

(a) SCREENING OF TRANSMITTERS

A rapid screening method for identifying possible transmitter substances at the cellular level has been described by Hildebrand *et al.* (1971). The method is based on the selective synthesis and localisation of transmitters in those cells which are presumed to release them. The procedural steps involve: (a) incubation of nervous tissue in a physiological medium with one or more radiochemical precursors or related transmitters (choline for ACh, glutamate for GABA, tyrosine for catecholamines and tyramine, tryptophan for serotonin); (b) isolation of nervous tissue components (e.g. cell bodies, parts of cells, neuropile) chemically extracted under conditions that preserve transmitter substances; (c) high voltage paper electrophoresis of the extract under conditions which separate most of the compounds of interest (see Fig. 26); (d) quantification of the isotope in each transmitter candidate. This whole procedure is very sensitive, allowing 10^{-12} moles of material to be detected. Experiments with rat superior cervical ganglia showed them to synthesise noradrenaline and ACh readily, which are known to function as transmitters in the sympathetic nervous system. Leech (*Hirudo medicinalis*) Retzius cells were also shown to produce serotonin, while experiments on identified neurons of the lobster (*Homerus*) showed inhibitory neurons to take up glutamate and convert it to GABA, while excitatory cells, though being able to accumulate glutamate, generally lacked the ability to transform it. There is much evidence suggesting that serotonin is a transmitter substance in the Retzius cells of the leech, while in the lobster GABA and glutamate are thought to be inhibitory and excitatory transmitters respectively.

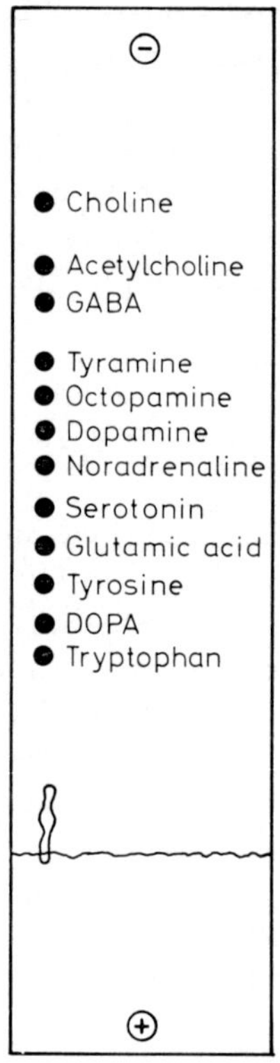

FIG. 26. Diagram showing the relative positions of some possible transmitter substances, precursors and neutral metabolites after high-voltage electrophoresis. (After Hildebrand *et al.*, 1971.)

(b) MICROCHROMATOGRAPHY OF AMINES

A microchromatographic procedure (based on the method of Bell and Somerville, 1966) for the detection of monoamines in the nanogram has been described by Osborne (1971). Standard amounts of amines (1–1000 ng), dissolved in 50% acetone in 0.01 N HCl, and applied with an ultrathin capillary to the bottom edge of a 3 × 3 cm sheet of polyamide (see page 69) can be fractionated by ascending chromatography in a 50-ml beaker. The beaker is covered to prevent evaporation of the solvent, and just enough solvent is placed in the beaker to cover the base. A single separation of the different amines can be effected by one to two solvent systems (see Fig. 27), viz. methyl acetate/isopropanol/ammonia 25% (9:7:5 by volume) or butanol/chloroform/acetic acid (4:1:1 by volume). When the solvent reaches the upper edge of the chromatogram, the plate is dried and placed in a sealed jar (500 ml) with about 3 g of paraformaldehyde. The paraformaldehyde must have been stored at a relative humidity of 60% for at least 5 days to produce optimum fluorescence. After heating the jar in an oven for 3 hr at 80°C, the amines are located as fluorescent spots by viewing the chromatogram under ultraviolet light. From Table 18, which records the minimum detectable amounts of biogenic amines by this technique, it can be seen that the procedure is particularly sensitive for the detection of the transmitter substances dopamine, noradrenaline and serotonin.

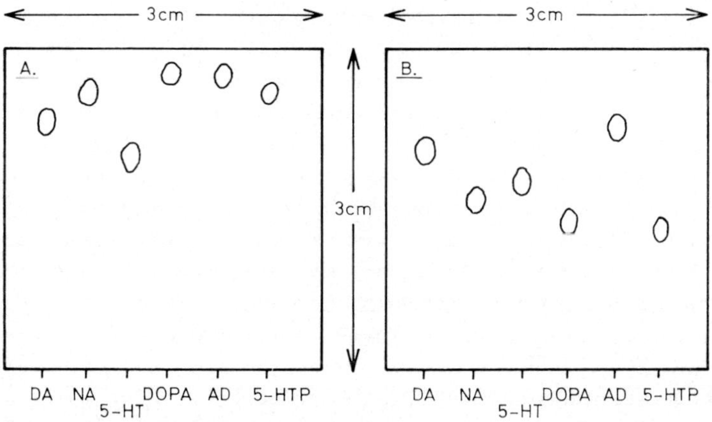

FIG. 27. Diagram of microchromatograms (polyamide layers) using the solvent systems (A) methy-acetate/isopropanol/ammonia and (B) butanol/chloroform/acetic acid for the fractionation of dopamine (DA), noradrenaline (NA), serotonin (5-HT), 3,4-dihydroxyphenylalanine (DOPA), adrenaline (AD) and 5-hydroxytryptophan (5-HTP). (After Osborne, 1971.)

TABLE 18. MINIMUM DETECTABLE AMOUNTS OF BIOGENIC AMINES AND OTHER COMPOUNDS AFTER FORMALDEHYDE TREATMENT (from Osborne, 1971)

Compound	Detectable amount (ng)
Adrenaline	100
Bufotenine	80
n,n-Dimethyltryptamine	1.000
3,4-Dimethoxyphenethylamine	15
3,4-Dihydroxyphenylalanine	50
Dopamine	6
5-Hydroxytryptophan	50
5-Hydroxytryptamine	5
3-Hydroxykynurenine	80
3-Hydroxy-4-methoxyphenylethylamine	20
Kynurenine	100
Kynurenic acid	100
Kynuramine	1.000
Melatonin	90
Metadrenaline	1.000
Mescaline	90
α-Methyl-m-tyrosine	1.000
3-Methoxy-4-hydroxyphenylethylamine	50
Noradrenaline	7
Normetadrenaline	100
Octopamine	1.000

This very simple procedure for the analysis of amines can be applied to a number of neurochemical problems. The applicability of the method is demonstrated by the following experiment. A number (6–10) of giant serotonin-containing neurons (GSCs) from the snail *Helix pomatia* are dissected (see page 79) and placed in a 5-μl capillary containing saline. The neurons are sedimented by centrifugation, and the saline is removed and replaced with 1 μl 70% acetone 0.01 NHCl. After homogenisation the capillary is placed in a freezer (−30°C) for 30 min to allow the neuronal proteins to be precipitated by the acetone. The capillary and its contents are then thoroughly centrifuged and the supernatant chromatographed on 3 × 3 cm polyamide layers using the solvent system methylacetate/isopropanol/25% ammonia. Standard amines are chromatographed separately or added to the extract to assist in the identification process. Analysis of chromatograms from extracts of as few as six neurons show a fluorescent spot to correspond both in colour and position to pure serotonin. Moreover, examination of the excitation and emission spectra reveal the substance to be serotonin. A semi-quantitative estimation of the serotonin content in the neurons can be made by simply comparing the intensity of fluorescence in the neuronal extract with that in standard amounts of amine. From twelve different experiments it was estimated that a GSC contains 0.9 ng of serotonin. This figure is slightly higher than that obtained by other procedures (see Cottrell and Osborne, 1970), where the serotonin content was calculated to be 0.7 ng serotonin per cell.

The described procedure is especially suitable for analysing the amine content in neurons thought to contain more than one type of amine and has also been successfully adapted to measure the rate of incorporation of radioactivity into specific neurons in order to measure the biosynthesis and half-lives of certain monoamines. The method has the disadvantage that amounts of proteins which would not interfere with the separation of dansyl derivatives impair the separation of the amines.

(c) FLUOROMETRY OF AMINES

Quantitative fluorometric procedures for the analysis of serotonin and dopamine at the cellular level have been developed by McCaman *et al.* (1973). The procedure involves extraction of the amines from the tissues with a special liquid cation exchange system, so that the amines free of their precursors are recovered and measured by micromodification of the methods of Maickel and Miller (1966) for serotonin, and of Shellenberger and Gordon (1971) for dopamine. The sensitivity limit of the methods is 2 pmoles for serotonin and 4 pmoles for dopamine.

The applicability of the above method has been demonstrated by analysis of the serotonin content in individual Retzius cells of the leech *Hirudo medicinalis*, where the content was shown to be 2.5 pmoles per cell. This value agrees with that found by Rude *et al.* (1969), who analysed groups of Retzius cells. An analysis of various nervous components from the molluscs *Aplysia* and *Tritonia* was also made, though none of the cells analysed had measurable levels of dopamine (see McCaman *et al.*, 1973).

References

BAHR, G. F., CARLSSON, L. and ZEITLER, E. (1961) Determination of dry weight in population of submicroscopic particles by means of quantitative electron microscopy. 1st International Biophysics Congress, Stockholm, p. 327.

BAHR, G. F. and ZEITLER, E. (1965) The determination of dry mass in populations of isolated particles. In: *Symposium on Quantitative Electron Microscopy* (Eds. G. F. Bahr and E. Zeitler), pp. 217–239. Williams & Wilkins, Baltimore.

BAHRER, R. (1952) Interference microscopy and mass determination. *Nature*, **169**, 366.

BELL, C. E. and SOMERVILLE, A. R. (1966) A new fluororescence method for detection and possible quantitative assay for some catecholamine and tryptamine derivatives on paper. *Biochem. J.* **98**, 1c–3c.

BERRY, R. W. (1969) Ribonucleic acid metabolism of a single neuron: correlation with electrical activity. *Science*, **166**, 1021–1023.

BOCHAROVA, L. S., BOROVYAGIN, V. L., DYAKONOVA, T. L., WARTON, S. S. and VEPRINTSEV, B. N. (1972) Ultrastructure and RNA synthesis in a molluscan giant neuron under electrical stimulation. *Brain Res.* **36**, 371–384.

BRIEL, G., NEUHOFF, V. and MAIER, M. (1972) Microanalysis of amino acids and their determination in biological material using dansyl-chloride. *Hoppe-Seyler's Z. Physiol. Chem.* **353**, 540–553.

BRATTGÅRD, S. O., EDSTRÖM, J. E. and HYDÉN, H. (1957) The chemical changes in regenerating neurons. *J. Neurochem.* **1**, 316–325.

BUCKLEY, G., CONSOLO, S., GIACOBINI, E. and McCAMAN, R. E. (1967) A micromethod for the determination of choline acetylase in individual cells. *Acta Physiol Scand.* **71**, 341–347.

CARLSON, L. (1965) Constants, necessary for the interferometric mass measurement of fixed material, determined by microradiography. *Acta Histochemica*, Suppl. **6**, 397–402.

CARLSSON, B., GIACOBINI, E. and HOVMARK, S. (1967) An instrument for simultaneous determination of sodium and potassium in microsamples of biological material. *Acta Physiol. Scand.* **71**, 379–390.

CEDAR, H. and SCHWARTZ, J. H. (1972) Cyclic adenosine monophosphate in the nervous system of *Aplysia californica*, I. *J. Gen. Physiol.* **60**, 570–589.

CEDAR, H., KANDEL, E. R. and SCHWARTZ, J. H. (1972) Cyclic adenosine monophosphate in the nervous system of *Aplysia californica*, II. *Gen. Physiol.* **60**, 558–569.

CONSOLO, S., GIACOBINI, E. and KARJALAINEN, K. (1968) Monoamine oxidase in sympathetic ganglia of the cat. *Acta Physiol. Scand.* **74**, 513–520.

COTTRELL, G. A. and OSBORNE, N. N. (1970) Subcellular localisation of serotonin in an identified serotonin-containing neuron. *Nature*, **225**, 470–472.

COTTRELL, G. A. and POWELL, B. (1971) Formation of serotonin by isolated serotonin-containing neurons and by isolated non-amine-containing neurons. *J. Neurochem.* **18**, 1695–1697.

CREMER, Th., DAMES, W. and NEUHOFF, V. (1972) Micro-disc electrophoresis and quantitative assay of glucose-6-phosphate dehydrogenase at the cellular level. *Hoppe-Seyler's Z. Physiol. Chem.* **353**, 1317–1329.

DANEHOLT, B. and BRATTGÅRD, S. O. (1966) A comparison between RNA metabolism of nerve cells and glia in the hypoglossal nucleus of the rabbit. *J. Neurochem.* **13**, 913–921.

DEWHURST, S. A. (1972) Choline phosphokinase activities in ganglia and neurons of *Aplysia*. *J. Neurochem.* **19**, 2217–2219.

EDSTRÖM, J. E. (1953) Ribonucleic acid mass and concentration in individual nerve cells. A new method for quantitative determinations. *Biochim. Biophys. Acta*, **12**, 361–386.

EDSTRÖM, J. E. (1956) Separation and determination of purines and pyridine nucleotides in picogram amounts. *Biochim. Biophys. Acta*, **22**, 378–388.

EDSTRÖM, J. E. (1958) Quantitative determination of ribonucleic acid in the micromicrogram range. *J. Neurochem.* **3**, 100–106.

EDSTRÖM, J. E. (1964) Microextraction and microelectrophoresis for determination and analysis of nucleic acids in isolated cellular units. In: *Methods in Cell Physiology* (Ed. D. M. Prescott), Vol. I, pp. 417–447. Academic Press, New York.

EDSTRÖM, J. E. and GRAMPP, W. (1965) Nervous activity and metabolism of ribonucleic acid of the crustacean receptor neuron. *J. Neurochem.* **10**, 725–731.

EDSTRÖM, J. E. and KAWIAK, J. (1961) Microchemical deoxyribonucleic acid determination in individual cells. *J. Biophys. Biochem. Cytol.* **9**, 619–626.

EDSTRÖM, J. E. and NEUHOFF, V. (1973) Micro-electrophoresis for RNA and DNA base analysis. In: *Micromethods in Molecular Biology* (Ed. V. Neuhoff), pp. 215–256. Springer Verlag, Berlin.

EDSTRÖM, J. E., EICHNER, D. and EDSTRÖM, A. (1962) The ribonucleic acid of axons and myelin sheaths from Mauthner neurons. *Biochem. Biophys. Acta*, **61**, 178–184.

EGYHAZI, E., RINGBORG, U., DANEHOLT, B. and LAMBERT, B. (1968) Extraction and fractionation of low molecular weight RNA on the microscale. *Nature*, **220**, 1036–1037.

EGYHAZI, E., DANEHOLT, B., EDSTRÖM, J. E., LAMBERT, B. and RINGBORG, U. (1969) Low molecular weight RNA in cell components of *Chironomus tetans* salivary glands. *J. Mol. Biol.* **44**, 517–532.

GAINER, H. (1971) Micro disc electrophoresis in sodium dodecyl sulphate: an application to the study of protein synthesis in individual identified neurons. *Analyt. Biochem.* **44**, 589–60.

GIACOBINI, E. (1957) Quantitative determination of cholinesterase in individual sympathetic cells. *J. Neurochem.* **1**, 234–244.

GIACOBINI, E. (1959) The distribution and localisation of cholinesterase in nerve cells. *Acta Physiol. Scand.* Suppl. 156, pp. 1–54.

GIACOBINI, E. (1968) *Chemical Studies on Individual Neurons*. Part I. *Neurosciences Research* (Eds. S. Ehrenpreis and O. C. Solnitzky), Vol. 1, pp. 1–66. Academic Press, New York.

GIACOBINI, E. (1969) *Chemical Studies on Individual Neurons.* Part II. *Neurosciences Research* (Eds. S. Ehrenpreis and O. C. Solnitzky), Vol. 2, pp. 112–198. Academic Press, New York.

GIACOBINI, E. and GRASSO, A. (1966) Variations of glycolytic intermediates, phosphate compounds and pyridine nucleotides after prolonged stimulation of an isolated crustacean neuron. *Acta Physiol. Scand.* **66**, 49–57.

GIACOBINI, E., HOVMARK, S. and KOMETIANI, Z. (1967b) Intracellular variations of Na and K in isolated nerve cells. *Acta Physiol. Scand.* **71**, 391–400.

GIACOBINI, E., PALMBORG, B. and SJÖQVIST, F. (1967a) Cholinesterase activity in innervated and denervated sympathetic ganglion cells of the cat. *Acta Physiol. Scand.* **69**, 355–361.

GIACOBINI, E., KARJALAINEN, K., KERPEL-FRONIUS, S. and RITZEN, M. (1970) Monoamines and monoamine oxidase in denervated sympathetic ganglia of the cat. *Neuropharmacology*, **9**, 59–64.

GILLER, E. and SCHWARTZ, J. H. (1969) Distribution of choline acetyltransferase in single neurons of the abdominal ganglion of *Aplysia. Fed. Proc.* **28**, 734.

GILLER, E. and SCHWARTZ, J. H. (1971a) Choline acetyl transferase in identified neurons of abdominal ganglion of *Aplysia californica. J. Neurophysiol.* **34**, 93–107.

GILLER, E. and SCHWARTZ, J. H. (1971b) Acetyl cholinesterase in identified neurons of abdominal ganglion of *Aplysia californica. J. Neurophysiol.* **34**, 108–115.

HAMBERGER, A. (1961) Oxidation of tricarboxylic acid cycle intermediates by nerve cell bodies and glial cells. *J. Neurochem.* **8**, 31–35.

HALJAMÄE, H. and HAMBERGER, A. (1971) Potassium accumulation by bulk prepared neuronal and glial cells. *J. Neurochem.* **18**, 1903–1912.

HALJAMÄE, H. and LARSSON, S. (1968) An ultramicroflame photometer for K and Na analysis of single cells and nanoliter quantities of biological fluids. *Chem. Instr.* **1**, 131–144.

HALJAMÄE, H. and LANGE, P. W. (1972) Calcium content and conformation changes of S-100 protein in the hippocampus during training. *Brain Res.* **38**, 131–142.

HALJAMÄE, H. and WALDMAN, A. A. (1972) Flame photometry at the cell level. In: *Techniques of Biochemical and Biophysical Morphology* (Eds. D. Glick and R. M. Rosenbaum), Vol. 1, pp. 233–268. John Wiley & Sons, New York.

HALJAMÄE, H. and WALDMAN, A. A. (1973) Cation uptake by intact nuclei from rat cerebral cortex. *J. Neurochem.* **20**, 1747–1751.

HALJAMÄE, H. and WOOD, D. C. (1971) Analysis of picomole quantities of Ca, Mg, K, Na and Cl in biological samples by ultra-microflame spectrophotometry. *Analyt. Biochem.* **42**, 155–170.

HENN, F. A., HALJAMÄE, H. and HAMBERGER, A. (1972) Glial cell function: active control of extracellular K^+ concentration. *Brain Res.* **43**, 437–443.

HILDEBRAND, J. G., BARKER, D. L., HERBERT, E. and KRAVITZ, E. A. (1971) Screening for neurotransmitters: a rapid radiochemical procedure. *J. Neurobiol.* **2**, 231–246.

HOLTER, H. (1961) The Cartesian diver. In: *General Cytochemical Methods* (Ed. J. Danielli), Vol. 2, pp. 93–128. Academic Press, New York.

HYDÉN, H. (1943) Protein metabolism in the nerve cell during growth and function. *Acta Physiol. Scand.* Suppl 17.

HYDÉN, H. (1960) The neuron. In: *The Cell* (Eds. J. Brachet and A. Mirsky), Vol. IX, pp. 215–323. Academic Press, New York.

HYDÉN. H. (1964) Biochemical and functional interplay between neuron and glia. In: *Recent Advances in Biological Psychiatry* (Ed. J. Wortis), Vol. VI, pp. 31–52. Plenum Press, New York.

HYDÉN, H. (1967) Biochemical changes accompanying learning. In: *The Neurosciences* (Eds. G. C. Quarton, T. Melnechuk and F. O. Schmitt), pp. 765–771. Rockefeller Univ. Press, New York.

HYDÉN, H., BJURSTAM, K. and McEWEN, B. (1966) Protein separation at the cellular level by microdisc electrophoresis. *Analytical Biochem.* **17**, 1–15.

HYDÉN, H. (1972) Macromolecules and behaviour. In: *Arthur Thomson Lectures* (Eds. G. B. Ansell and P. B. Bradley), pp. 3–75. Macmillan, London.

HYDÉN, H. and PIGON, A. (1960) A cytophysiological study of the functional relationship between oligodendroglial cells and nerve cells of Deiters' nucleus. *J. Neurochem.* **6**, 57–72.

HYDÉN, H. and EGYHAZI, E. (1962) Nuclear RNA changes during a learning experiment in rats. *Proc. Nat. Acad. Sci.* **48**, 1366–1373.

HYDÉN, H. and EGYHAZI, E. (1963) Glial RNA changes during learning experiment in rats. *Proc. Nat. Acad. Sci.* **49**, 618–624.

HYDÉN, H. and EGYHAZI, E. (1964) Changes in RNA content and base composition in cortical neurons of rats in a learning experiment involving transfer of handedness. *Proc. Nat. Acad. Sci.* **52**, 1030–1035.

JARLSTEDT, J. (1963) Functional localisation in the cerebellar cortex studied by quantitative determinations of Purkinje cell RNA. *Acta Physiol. Scand.* **67**, 243–254.

KATZ, G. M. (1968) Another look at ultramicro integrative flame photometry. *Analyt. Biochem.* **26**, 381–397.

KATZMAN, R., LEHRER, G. M. and WILSON, C. E. (1969) Sodium and potassium distribution in puffer fish supramedullary nerve cell bodies. *J. Gen. Physiol.* **54**, 232–249.

KEESEY, J. C. (1968) Flame photometric analysis of sodium and potassium in nanogram samples of mammalian nervous tissue. *J. Neurochem.* **15**, 547–562.

KERNELL, D. and PETERSON, R. P. (1970) The effect of spike activity versus synaptic activation on the metabolism of ribonucleic acid in a molluscan giant neuron. *J. Neurochem.* **17**, 1087–1094.

KOENIG, E. (1965a) Synthetic mechanisms in the axons. I. *J. Neurochem.* **12**, 343–356.

KOENIG, E. (1965b) Synthetic mechanisms in the axons. II. *J. Neurochem.* **12**, 357–361.

KOENIG, E. (1967a) Synthetic mechanisms in the axons. III. *J. Neurochem.* **14**, 429–435.

KOENIG, E. (1967b) Synthetic mechanisms in the axons. IV. *J. Neurochem.* **14**, 437–446.

KOENIG, E. and BRATTGÅRD, S. O. (1963) A quantitative micromethod for determination of specific radioactivity of ^{3}H-purines and ^{3}H-pyrimidines. *Analyt. Biochem.* **6**, 424–434.

KOSLOW, S. and GIACOBINI, E. (1969) An isotopic micromethod for the measurement of cholinesterase activity in individual cells. *J. Neurochem.* **16**, 1523–1528.

LEHRER, G. M., WEISS, C., SILIDES, D. J., LICHTMAN, C., FUMAN, M. and MATHEWSON, R. F. (1968) The quantitative histochemistry of supramedullary neurons of puffer fishes. I. Enzymes of glucose metabolism. *J. Cell Biol.* **37**, 575–578.

LINDERSTRÖM-LANG, K. (1937) Principle of Cartesian diver applied to gasometric technique. *Nature*, **140**, 108.

LOWRY, O. H. (1953) The quantitative histochemistry of the brain. *J. Histochem. Cytochem.* **1**, 420–428.

LOWRY, O. H. (1963) The chemical study of single neurons. *Harvey Lecture*, **58**, 1–19.

LOWRY, O. H., ROBERTS, N. R. and KAPPHAHN, J. I. (1957) The fluorometric measurement of pyridine nucleotides. *J. Biol. Chem.* **224**, 1047–1064.

LOWRY, O. H., PASSONNEAU, J. V., SCHULZ, D. W. and ROCK, M. K. (1961) The measurement of pyridine nucleotides by enzymatic cycling. *J. Biol. Chem.* **236**, 2746–2755.

LOWRY, O. H., PASSONNEAU, J. V., HASSELBERGER, F. X. and SCHULZ, D. W. (1964) Effect of ischemia on known substrates and cofactors of the glycolytic pathway of the brain. *J. Biol. Chem.* **239**, 18–30.

LOWRY, O. H. and PASSONNEAU, J. V. (1972) *A Flexible System in Enzymatic Analysis.* Academic Press, New York.

MAICKEL, R. P. and MILLER, F. P. (1966) Fluorescent products formed by reaction of indole derivatives with *o*-phthaldehyde. *Analytical Chem.* **38**, 1937–1938.

MALNIC, G., KLOSE, R. M. and GIEBISCH, G. (1964) Micropuncture study of renal potassium excretion in the rat. *Am. J. Physiol.* **206**, 674–686.

MATSCHINSKY, F. M., ELLERMAN, J. E., LANDGRAF, R., KRZANOWSKI, J., KOTLER-BRAJTBURG, J. and FERTEL, R. (1970) Quantitative histochemistry of glucose metabolism in the islets of Langerhans. In: *Recent Advances in Quantitative Histo- and Cytochemistry* (Eds. H. C. Dubach and U. Schmidt), pp. 143–182. Hans Huber, Bern.

McCAMAN, R. E. (1968) Application of tracers to quantitative histochemical and cytochemical studies. In: *Advances in Tracer Methodology* (Ed. S. Rothchild), Vol. 4, pp. 137–202. Plenum Press, New York.

McCAMAN, R. E. and DEWHURST, S. A. (1970) Choline acetyltransferase in individual neurons of *Aplysia californica. J. Neurochem.* **17**, 1421–1426.

McCAMAN, R. E. and DEWHURST, S. A. (1971) Metabolism of putative transmitters in individual neurons of *Aplysia californica. J. Neurochem.* **18**, 1329–1335.

McCAMAN, M. W., WEINREICH, D. and McCAMAN, R. E. (1973) The determination of picomole levels of 5-hydroxytryptamine and dopamine in *Aplysia*, *Tritonia*, and leech nervous tissues. *Brain Research* **53**, 129–137.

MÜLLER, P. (1958) Experiments on current flow and ionic movements in single myelinated nerve fibres. *Exp. Cell Res.* Suppl. 5, 118–152.

NEUHOFF, V. (1968) Micro-disc-electrophorese von Hirnproteinen. *Arzneimittel-Forsch.* **18**, 35–38.

NEUHOFF, V. (1973) In: *Micromethods in Molecular Biology* (Ed. V. Neuhoff). Springer Verlag, Berlin.

NEUHOFF, V. and LEZIUS, A. (1967) Nachweis der Substruktur von DNA-Polymerasen, der enzymatisch aktiven Proteinkomponente und ihre Enzym-Substrat-Komplexe mit der Micro-Disc-Electrophorese. *Hoppe-Seyler's Z. Physiol. Chem.* **348**, 1239.

NEUHOFF, V. and WEISE, M. (1970) Determination of picomole quantities of γ-aminobutyric acid (GABA) and serotonin. *Arzneim. Forsch.* (*Drug Research*), **21**, 104–107.

NEUHOFF, V., van der HAAR, F., SCHLIMME, E. and WEISE, M. (1969) Zweidimensionale Chromatographie von Dansyl-Aminosäuren im Pico-Mol-Bereich, angewandt zur direkten Charakterisierung von Transfer-Ribonucleinsäuren. *Hoppe-Seyler's Z. Physiol. Chem.* **350**, 121–128.

NEUHOFF, V., SCHILL, W. B. and STERNBACH, H. (1970) Microanalysis of pure deoxyribonucleic acid-dependent ribonucleic acid polymerase from *Escherichia coli. Biochem. J.* **117**, 623–631.

OSBORNE, N. N. (1971) A micro-chromatographic method for the detection of biologically active monoamines from isolated neurons. *Experientia*, **27**, 1502–1513.

OSBORNE, N. N. (1972) The *in vivo* synthesis of serotonin in an identified serotonin-containing neuron of *Helix pomatia. Int. J. Neuroscience*, **3**, 215–219.

OSBORNE, N. N. (1973) The analysis of amines and amino acids in microquantities of tissue. In: *Progress in Neurobiology* (Eds. G. A. Kerkut and J. W. Phillis), Vol. 1, part 4, pp. 299–329. Pergamon Press, Oxford.

OSBORNE, N. N., BRIEL, G. and NEUHOFF, V. (1971) Distribution of GABA and other amino acids in different tissues of the gastropod *Helix pomatia*, including *in vitro* experiments with ^{14}C-glucose and ^{14}C-glutamic acid. *Int. J. Neuroscience*, **1**, 265–272.

PETERSON, R. P. (1970) RNA in single identified neurons of *Aplysia. J. Neurochem.* **17**, 325–338.

PETERSON, R. P. and KERNELL, D. (1970) Effects of nerve stimulation on the metabolism of ribonucleic acid in a molluscan giant neuron. *J. Neurochem.* **17**, 1075–1085.

PIGON, A. and EDSTRÖM, J. E. (1959) Nucleic changes during starvation and encystment in a ciliate (*Urostyla*). *Exptl. Cell Res.* **16**, 648–656.

RAMSAY, J. A., FALLOON, S. W. H. W. and MACHIN, K. E. (1951) An integrating flame photometer for small quantities. *J. Sci. Instrum.* **28**, 75–80.

RÜCHEL, R., MESECKE, S., WOLFRUM, D. I. and NEUHOFF, V. (1973) Mikroelectrophorese an kontinuierlichen Polyamid Gradient Gelen. I. Herstellung und Eigenschaften von Gelgradienten in Kapillaren; ihre Anwendung zur Proteinfraktionierung und Molgewichtsbestimmung. *Hoppe-Seyler's Z. Physiol. Chem.* **354,**1351–1368.

RUDE, S., COGGESHALL, R. E. and van ORDEN III, L. S. (1969) Chemical and ultrastructural identification of 5-hydroxytryptamine in an identified neuron. *J. Cell Biol.* **41**, 832–854.

SHELLENBERGER, M. K. and GORDON, J. H. (1971) A rapid, simplified procedure for simultaneous assay of norepinephrine, dopamine and 5-hydroxytryptamine from discrete brain areas. *Analyt. Biochem.* **39**, 356–372.

SCHIEFER, H. G. and NEUHOFF, V. (1971) Fluorometric microdetermination of phospholipids at the cellular level. *Hoppe-Seyler's Z. Physiol. Chem.* **352**, 913–926.

TOSCHI, G. and GIACOBINI, E. (1965) Puromycin and the impulse activity of crayfish stretch-receptor neuron. *Life Sci.* **4**, 1831–1834.

VUREK, G. C. (1967) Emission photometry of picometer of calcium, magnesium and other metals. *Analyt. Chem.* **39**, 1599–1601.

VUREK, G. C. and BOWMAN, R. L. (1965) Helium-glow photometer for picomole analysis of alkali metals. *Science*, **149**, 448–450.

WALLIN, B. G. (1967) Intracellular ion concentrations in single crayfish axons. *Acta Physiol. Scand.* **70**, 419–430.

WEINREICH, D., DEWHURST, S. A. and McCAMAN, R. E. (1972) Metabolism of putative transmitters in individual neurons of *Aplysia californica*: aromatic amino acid decarboxylase. *J. Neurochem.* **19**, 1125–1130.

WILSON, D. L. (1971) Molecular weight distribution of proteins synthesised in single identified neurons of *Aplysia. J. Gen. Physiol.* **57**, 26–40.

WOLFRUM, D. I., RÜCHEL, R., MESECKE, S., NEUHOFF, V. (1974) Mikroelektrophorese an kontinuierlichen Polyacrylaminen Gradienten Gelen. III. Extraktion und Elektrophorese von Ribonucleinsäuren (in preparation).

ZEUTHEN, E. (1961) The Cartesian diver balance. In: *General Cytochemical Methods* (Ed. J. Danielli), Vol. 2, pp. 61–90. Academic Press, New York.

CHAPTER 6

General Techniques used in Microprocedures

1. Weight Determination

A number of procedures exist for accurate dry weight determination of nervous tissue components, e.g. X-ray absorption, interferometry and quartz fibre balance. Most widely used is the quartz fibre balance, originally developed by Lowry (see Lowry and Passonneau, 1972) and since modified in various ways (see e.g. Moss, 1972). Essentially, it is a type of spring balance and its construction is based on mounting a quartz fibre of suitable dimensions horizontally in a case to protect it from air currents and static charges. Its displacement in micrograms or smaller loads is measured with a micrometer ocular in a wide-field horizontally mounted microscope, which is also used for the loading and unloading of the balance. The balance case can be made from a large syringe with the plunger reversed. The fibre is sealed with epoxy cement to a piece of copper wire previously sealed to the plunger with the same kind of cement. To hold samples weighing more than 0.1 μg, there is a pan of very thin glass mounted at the end of the fibre. A diagram of a Lowry-type balance is shown in Fig. 29. In practice, a number of fibres of different lengths (e.g. 5–80 mm long

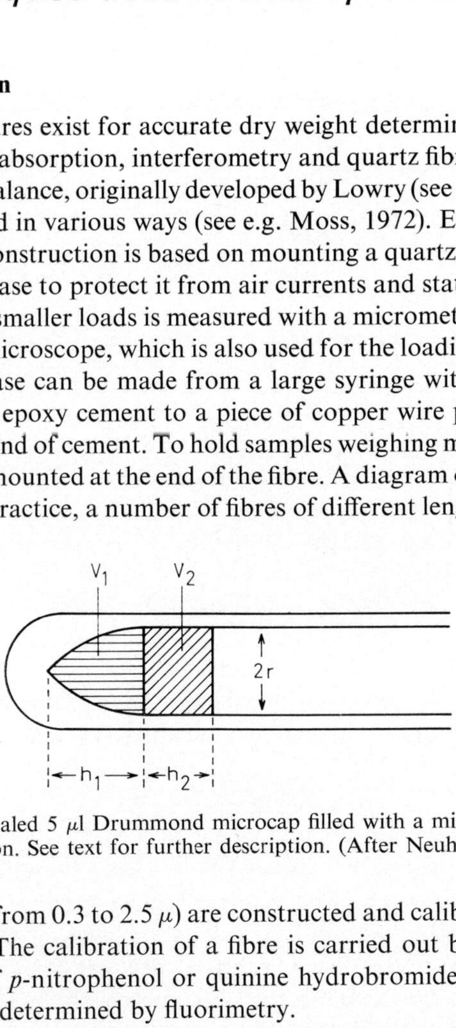

FIG. 28. Sketch of a heat-sealed 5 μl Drummond microcap filled with a minute amount of fresh tissue for wet weight determination. See text for further description. (After Neuhoff, 1971.)

with diameters ranging from 0.3 to 2.5 μ) are constructed and calibrated so that a range of loads can be weighed. The calibration of a fibre is carried out by noting the deflection produced by crystals of *p*-nitrophenol or quinine hydrobromide, and the weight of the crystals is subsequently determined by fluorimetry.

Probably the most useful weight determination of single neurons or parts of neurons is dry weight. Wet weight determinations at the cellular level are generally inaccurate due to loss of water when the sample is being transferred to the balance. However, it is often convenient and necessary to weigh fresh tissue. A suitable microbalance (e.g. Cahn microbalance, Scientific Glass Apparatus Co., Bloomfield, New Jersey, USA) can be used for weighing freshly dissected invertebrate giant neurons, though possible water loss must be taken into consideration. A more accurate way of weighing fresh tissue (0.01 mg–1 mg) is that described by Neuhoff (1971), which can be carried out in the following way.

A Drummond microcap (1–10 μl or larger) is carefully heat-sealed at one end by rotating it on the edge of a spirit burner flame so that the seal is paraboloid in shape (see Fig. 28).

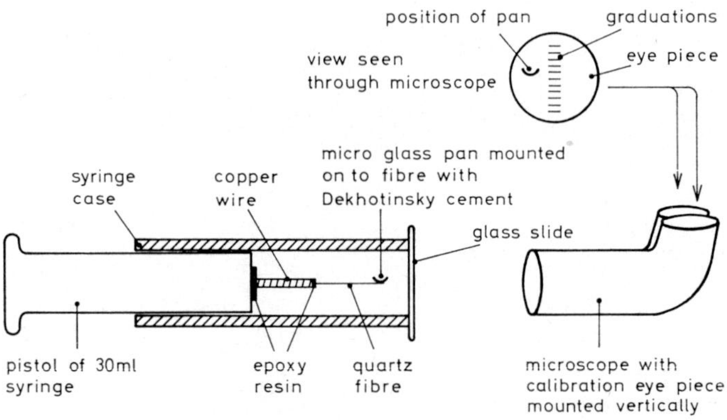

FIG. 29A shows a diagrammatic set-up which can be used to construct a quartz fibre balance. A syringe is particularly useful as the syringe case with a glass slide over the end of it provides a draught-free area for the weighing procedure. The actual quartz fibre measures about 5 cm and the pan about 1 mm square (though these parameters can be altered to suit the weights required). The balance is graduated by noting the deflection made through the eye-piece with a crystal of e.g. *p*-dinitrophenol, the weight of which is determined fluorometrically. The graduated balance can then be used to weigh parts of dried neurons.

The paraboloid shape is important because the calculation of the volume is based on such a form. The capillary (microcap) is then filled with a suitable buffer, into which the minute piece of fresh tissue is placed, and the capillary is centrifuged at 20,000 *g* for 15 min. It is possible to buy centrifuges (e.g. Heraeus-Christ Haematocrit Centrifuge, Osterode, Germany) with special adaptors for capillaries, though other types of centrifuges can be adapted easily. The capillary is embedded in water, observed through a microscope, and the image of the part of the capillary containing the sample projected onto graph paper (see Fig. 28). The volume is then calculated from a sketch of the sample using the formula for a paraboloid of revolution, corrected for the magnification:

$$V_1 = \frac{\pi}{2}\left(\frac{r}{\text{mag}}\right)^2 \frac{h_1}{\text{mag}}.$$

If the specific weight of the tissue is known (for practical purposes it is often assumed to be 1), the exact wet weight of minute amounts of tissue can be calculated. If only the paraboloid part of a heat-sealed 5 μl capillary is filled, the amount of nervous tissue is about 0.01–0.03 mg. Should more than the paraboloid part of the capillary be filled with the sample, it is necessary to calculate this part separately with the formula:

$$V_2 = \pi\left(\frac{r}{\text{mag}}\right)^2 \frac{h_2}{\text{mag}}.$$

V_1 and V_2 are then added together to give the total amount of tissue.

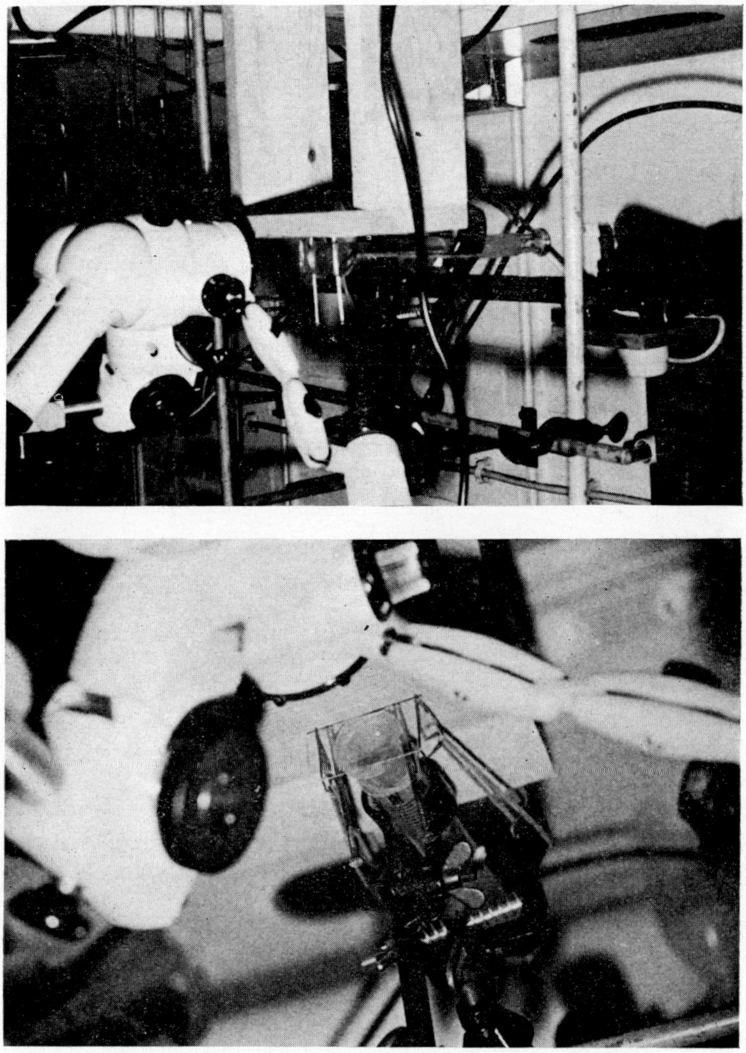

Fig. 29B shows two photographs of a constructed quartz fibre balance taken from two different viewpoints.

2. Disruption of Nervous Tissue, e.g. Homogenisation

The disruption of nervous tissue is often a prerequisite in microbiochemical analysis. Isolated cells or very small cone cylinders of tissue samples can often be effectively disrupted by means of repeated freezing and thawing. Another procedure is to add minute splinters of razor blade to the sample (more than 0.1 mg) and solution in a microtube and shake vigorously on a whirlmixer. This completely disrupts the nervous tissue, though the process can be intensified by first freezing and thawing the tissue. The splinters can then be removed by drawing a magnet along the glass tube. Yet another method of effective disruption is homogenisation. Eichner (1966) developed a very simple but effective way of homogenising microgram quantities of tissue, particularly when the homogenisation is

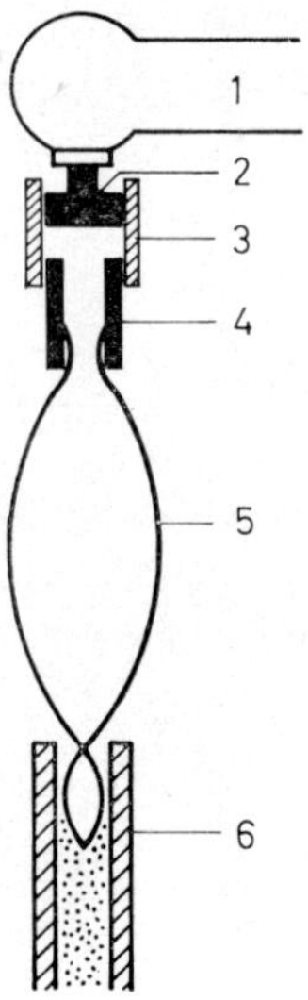

FIG. 30. Schematic diagram of microhomogeniser developed by Eichner (1966). 1, Handpiece of a dentist's drill. 2, Small milling head. 3, Polythene tubing. 4, Wire loop. 5, Capillary containing the contents to be homogenised.

done in 1-μl or 2-μl capillaries. The homogeniser, as shown in Fig. 30, consists of a loop made from Mikrothal wire (AB Kanthal, Halsthammer, Sweden), the two ends of which are connected to the motor of a dentist's drill by two polythene tubes. Actual homogenisation is carried out under microscopic vision; the high elasticity of the wire allows for quite a lot of manoeuvrability. The heat generated is unimportant as too high a speed is not attained.

When homogenising tissue in 5-μl or 10-μl Drummond capillaries, a technically simpler homogenisation is possible with dentist's nerve canal drills (Neuhoff, 1968). These drills (e.g. Beutelrock drills, diameter 0.25 mm for 5-μl capillaries, diameter 0.35 mm for 10-μl capillaries) are fitted into the hand-piece of a dentist's drill which is held in a fixed position, and worked with an infinitely variable foot treadle (see Fig. 31). It is important for the drill's motion to be anticlockwise to its thread, to prevent any solution from being drawn out of the capillary. Because of the large surface area of the drill, the contents of the capillary can heat up while being homogenised. It is therefore advisable, especially when analysing temperature-sensitive substances, to cool the capillary in a small beaker of iced water throughout the procedure.

The 'dentist's drill method' can also be adapted for the homogenisation of greater amounts of nervous tissue in specially made microtubes (see page 24). Instead of employing drills, pestles fitted with an appropriately sized head of Dynal are used (see Fig. 32). Throughout the homogenisation the contents are kept cool in a beaker of iced water. Allowance must be made for volume displacement by the pestle head, and care must be taken to have a maximum grinding surface between the bottom of the pestle head and the base of the microtube. The latter is achieved by the use of flat-based microtubes (see page 24) with pestle heads made to fit.

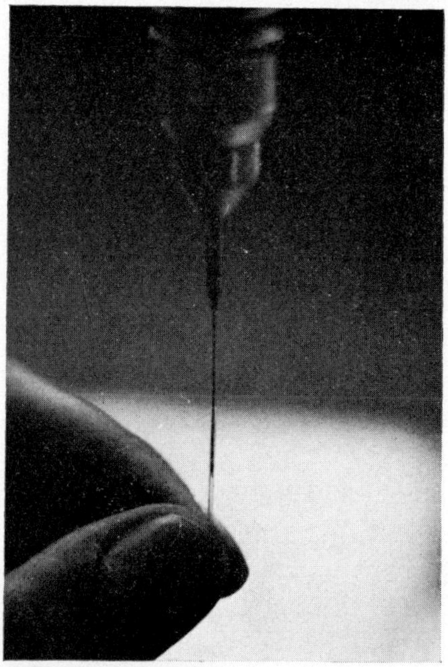

FIG. 31. This figure shows the procedure used to homogenise a sample in a 5-μl capillary with a nerve canal drill. The nerve canal drill is connected to a dentist's drill, the speed of which can be adjusted by means of a foot treadle.

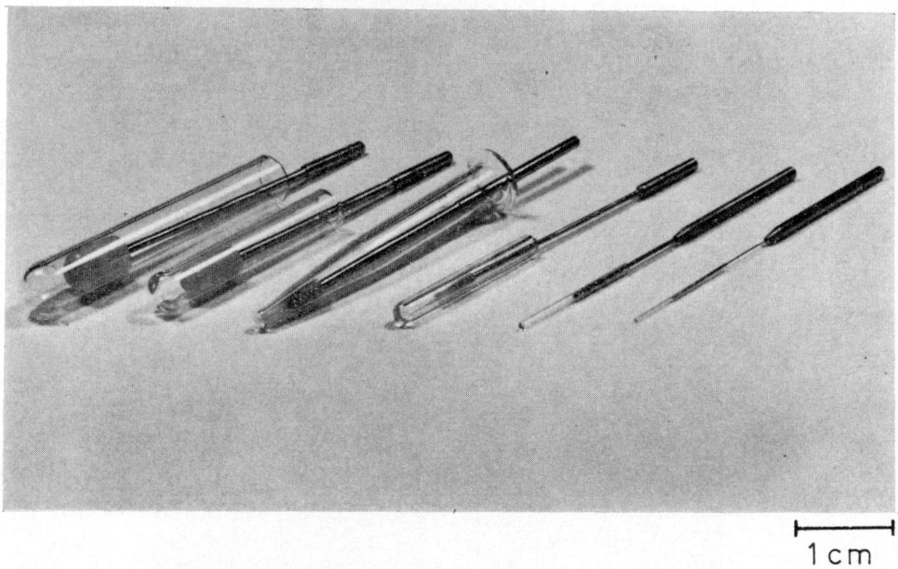

FIG. 32. A photograph showing a number of pestles of different sizes which can be attached to a dentist's drill to homogenise appropriate samples.

3. Dialysis

Often small quantities of protein need to be purified by dialysis. Small volumes of protein solution (15, 50, 100, 250, 500 μl) can easily be dialysed in the Dynal chambers described by

Neuhoff and Kiehl (1969). A schematic diagram of such chambers for volumes of 15 μl and 100 μl (obtainable from E. Schütt jr., Göttingen, Germany) is shown in Fig. 33. In practice a wet, semi-permeable membrane, a similarly sized, bored-out rubber washer, and a bored-out sealing cap (which holds the other two structures in position) are first fixed onto

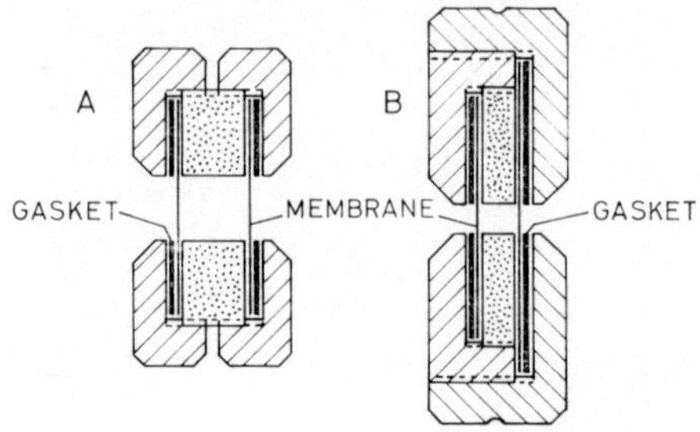

FIG. 33. Schematic diagram of microdialysis chambers for volumes of solutions of 100 μl (A) and 15 μl (B). (After Neuhoff and Kiel, 1969.)

FIG. 34. The set-up for microdialysis, showing the dialysis chamber attached to a bar magnet within the beaker.

a thick disc (4 or 2 mm thick with an external diameter of about 20 mm). The diameter of the central hole measures between 1 and 6 mm, depending on the volume to be dialysed. The solution is then placed in the central cavity and the opening is closed by means of a wet disc of semi-permeable membrane, a bored-out washer and a bored-out sealing cap. The dialysed chamber is then fitted into a holder made of rustless steel and containing a Plexiglass enclosed magnet. Figure 34 shows a complete dialysis apparatus (for 15 μl of solution) which is placed in a beaker containing a suitable external medium and rotated during dialysis by means of a magnetic stirrer. Following dialysis, a glass capillary is used to pierce one of the membranes and the purified protein solution is removed from the chamber.

References

EICHNER, D. (1966) Ein Mikro-Homogenisator. *Experientia*, **22**, 620.

LOWRY, O. H. and PASSONNEAU, J. V. (1972) *A Flexible System in Enzymatic Analysis*. Academic Press, New York.

MOSS, M. L. (1972) Quartz fibre ultramicrobalance. *J. Histochem.* **20**, 545–547.

NEUHOFF, V. (1968) Micro-Disc-Electrophorese von Hirnproteinen. *Arzneim.-Forsch.* (*Drug Research*), **18**, 35–39.

NEUHOFF, V. (1971) Wet weight determination in the lower milligram range. *Analyt. Biochem.* **41**, 270–271.

NEUHOFF, V. and KIEHL, F. (1969) Dialysiergeräte für Volumen zwischen 10 und 500 μl. *Arzneim.-Forsch.* (*Drug Research*), **19**, 1898–1899.

CHAPTER 7

Microdetermination of Amines and Amino Acids as Dansyl Derivatives

1. Introduction

Appreciable amounts of amines and amino acids occur in the nervous system of animals. It is generally accepted that the amino acids are predominantly concerned with general metabolic processes and the maintenance of water and ion distribution across cellular membranes. In contrast, ever since the discovery that amines such as adrenaline, noradrenaline, dopamine, histamine and serotonin are biologically very active, a number of observations have been made implicating them in the basic functioning of the nervous system. Initially the strongest reasons for presuming the amines to have transmitter-like functions were the connection between amine metabolism and some mental disorders, together with the fact that several amines (mescaline, dimethyltryptamine, psilocybine, amphetamine derivatives, lysergic acid diethylamide, etc.) produce in humans hallucinations and psychotic states which resemble those in acute schizophrenic cases. Comparatively recent studies have shown that certain amino acids, e.g. glycine, alanine, glutamic acid, aspartic acid, GABA and taurine, also probably have transmitter roles. In the instances of GABA (Krnjević, 1970; Hebb, 1970; Curtis and Johnston, 1970), glutamic acid (Krnjević, 1970; Kravitz *et al.*, 1970; Usherwood, 1971) and to some extent glycine (Hebb, 1970; Aprison *et al.*, 1970), the evidence available is almost conclusive. There is no reason why individual amino acids cannot play important metabolic roles as well as function as transmitter substances. All that is required is for the cell to regulate the amino acid level so that it adequately serves both purposes. Therefore the occurrence of individual amino acids and, to some extent, amines, may indicate this additional function. It would, however, be much more meaningful when making a study to examine the nature of the mechanism in each instance for maintenance of relative, constant, steady-state concentration of the amino acid or amine contents, rather than to interpret the significance of the variation in the actual amounts of substance.

The aim of this chapter is to describe one method for analysing the functional roles of amines and amino acids at the cellular level. Essentially the technique involves the reaction of –OH or $-NH_2$ groups of amino acids, amines and related substances with dansyl-chloride (1-dimethylaminonaphthalene-5-sulphonyl-chloride) at alkaline pH to form intensely fluorescent dansyl derivatives. The dansyl derivatives are then separated by chromatography on 3×3 cm polyamide layers using certain solvent systems. In practice, ^{14}C–dansyl-chloride is normally used, so that autoradiograms of the microchromatograms containing the separated ^{14}C–dansyl derivatives can be prepared. Moreover, individual radioactive substances can be removed from the microchromatograms and counted in a scintillation spectrometer, so that data can be obtained on the content of substances which react with ^{14}C–dansyl-chloride. This process detects as little as 1 pmole of amine and amino acid in minute quantities of tissue (Neuhoff *et al.*, 1969; Neuhoff and Weise, 1970; Osborne, 1973a).

The first significant use of dansyl-chloride was by Weber (1952). He used the substance to prepare highly fluorescent conjugates of albumen to estimate the size and shape of the albumen molecule. Later on, dansyl-chloride was useful in particular for the histochemical localisation of antigens in immunochemistry, virology and bacteriology, and for the study of active centres of enzymes. Perhaps one of the most important applications for the substance was that of end-group determinations of proteins and peptides (Gray and Hartley, 1963), which made it possible to use dansyl-chloride in a way analogous to 2,4-dinitrofluorobenzene (Sanger, 1945).

It is now clear that the usefulness of dansyl-chloride in biochemical analysis stems from its ability to react with amino acids, amines, phenols, proteins, peptides, imidozoles and sulphydryl groups under favourable conditions to form highly fluorescent derivatives (see Seiler and Wiechmann, 1970; Seiler, 1970). These fluorescent products can easily be separated with suitable chromatographic systems, and their intensities are so great that, under good conditions, 0.005 nmole of an amine can be determined by direct fluorometry, making the sensitivity comparable to methods where isotopes are used. For the analysis of amines and amino acids, both thin-layer chromatography and paper chromatography can be used to separate their dansyl derivatives (see Seiler and Wiechmann, 1970; Seiler, 1970), though Woods and Wang (1967) showed polyamide layers to be particularly appropriate.

The actual reaction between amines, amino acids and dansyl-chloride is not a simple process, and with amino acids it is often accompanied by fragmentation (see Neadle and Pollitt, 1965; Seiler, 1970). This, together with the fact that dansyl derivatives localised on chromatograms can often only be identified by elaborate means, e.g. mass spectrometry, provides a number of difficulties in the analysis of amines and amino acids, especially from tissue samples.

2. Theoretical and Practical Considerations

(a) THE REACTION

Dansyl-chloride reacts easily with aliphatic amino (–NH_2) or hydroxy (–OH) groups at alkaline pH to form intensely fluorescent derivatives (Seiler, 1970; Gray, 1967). However, the reaction (Neadle and Pollitt, 1965; Seiler, 1970) is not a simple one as shown in Fig. 35. Since water is present in a reaction mixture, the dansylation of an amino acid (AA–NH_2) at alkaline pH can be assumed to involve at least three reactions:

(1) AA–NH_2 + dansyl-Cl $\rightarrow$ AA–NH-dansyl + HCl

(2) H_2O + dansyl-Cl $\rightarrow$ HO-dansyl + HCl

(3) AA–NH-dansyl + dansyl-Cl $\rightarrow$ dansyl-NH_2 + other products

If only reactions (1) and (2) were to occur, a 100% yield of dansylated amino acids should be possible. An optimal pH has obviously to be found, since an increase in pH will push the equilibrium of both reaction (1) and the competing reaction (2) to the right. In addition, reaction (2) will have a measurable rate dependent on the dansyl-chloride concentration and independent of the concentration or type of amino acid present. However, reaction (3)

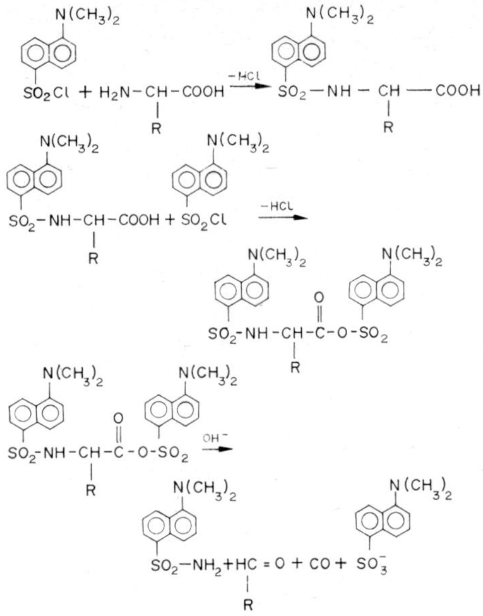

FIG. 35. The dansyl reaction according to Seiler (1970).

is consecutive to reaction (1), resulting in a decomposition of the dansylated amino acid, so that a 100% yield of the desired product cannot be achieved. Therefore, the yield of reaction (1) is the sum of the yields of dansylated amino acid and dansyl-NH_2. Thus it is very important to minimise the extent of unwanted side reactions by variation of the reaction parameters.

In order to achieve maximal concentrations of the formed dansyl product it is essential to consider (a) the concentration of dansyl-chloride used, (b) the reaction time, (c) the temperature of the reaction mixture, and (d) the pH of the reaction mixture. The optimal dansylation conditions for any group of substances are best achieved by fixing one of the conditions, e.g. reaction temperature of 37°C, and then determining the others by trial and error experiments. For the microdetermination of amino acids, the optimal conditions required were worked out by Briel *et al.* (1972). A reaction temperature of 37°C was chosen as a standard so that the other conditions could be determined.

(b) DETERMINATION OF OPTIMUM pH

The following experiment was done by Briel *et al.* (1972) in order to determine the optimum pH required for the dansylation of small amounts of amino acids at 37°C. A mixture of twenty amino acids (lysine, histidine, ornithine, tyrosine, tryptophan, methionine, asparagine, glutamic acid, aspartic acid, alanine, proline, leucine, glycine, phenylalanine, threonine, isoleucine, valine, glutamine, taurine and serine) was dissolved in 0.05 M $NaHCO_3$ pH 10, each at a concentration of 5×10^{-4} moles. The solution was divided into a number of 2-μl portions and each was adjusted to give pH values of 8.5, 9.0, 10.0, 10.5, 11.0 and 11.5. To each 2.5-μl sample 2 μl acetone and 4 μl ^{14}C–dansyl-chloride (10.63 μmole/ml) were added and incubated in the dark for 30 min at 37°C. A defined portion of each sample (0.2–0.5 μl) was then subjected to microchromatography and the spots containing the individual ^{14}C–dansyl amino acids were removed from the chromatogram and counted in a liquid scintillation counter. From the amino acids tested, including the bifunctional amino acids, viz. lysine, ornithine and histidine (see Fig. 36), it was clearly shown that for optimum dansylation conditions a pH of 10.0 is best.

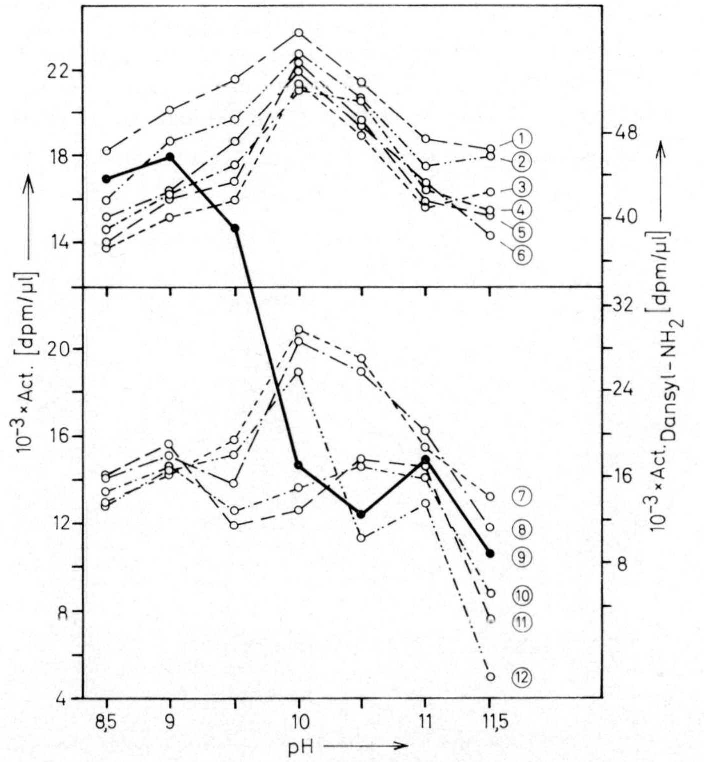

FIG. 36. The influence of pH on the dansylation of some amino acids. 1, dansyl-glycine; 2, dansyl-proline; 3, dansyl-tryptophan; 4, dansyl-valine; 5, dansyl-leucine; 6, dansyl-phenylalanine; 7, dansyl-bis-lysine; 8, dansyl-bis-ornithine; 9, dansyl-NH_2; 10, dansyl-glutamic acid; 11, dansyl-aspartic acid; 12, dansyl-bis histidine. (After Briel *et al.*, 1972.)

This pH (i.e. 10.0) for the dansylation of amino acids in the microscale is higher than that used by other workers. According to Seiler and Wiechmann (1970), primary and secondary amines and amino acids are best dansylated at a pH of 8.0, though a higher pH of 10.0 was required for the reaction of phenols and diphenols. Similarly, Gros and Labouesse (1969), Zanetta *et al.* (1970) and Spivak *et al.* (1971) found pHs ranging between 8.0 and 9.0 best for the dansylation of amino acids. It must, however, be remembered that the optimum pH determined for the microprocedure is very much dependent on the reaction time of 30 min and temperature of 37°C.

(c) INCUBATION TIME

In a series of experiments by Briel *et al.* (1972), a mixture of twenty-four substances (leucine, alanine, phenylalanine, tyrosine, tryptophan, serine, threonine, methionine, 5-hydroxytryptophan, 5-hydroxyindole, isoleucine, glutamic acid, GABA, taurine, aspartic acid, arginine, glycine, valine, proline, ornithine, lysine, histidine, asparagine and glutamine), each at a concentration of 4×10^{-4} molar in 0.05 M $NaHCO_3$ pH 10, was mixed with acetone and ^{14}C–dansyl-chloride (10.63 μmole/ml) in the same proportions as for the determination of optimum pH experiments, and incubated at 37°C. Definite portions were then removed after 5, 15, 30 and 120 min, and subjected to microchromatography. The spots corresponding to dansyl-ornithine, dansyl-histidine, dansyl-GABA, dansyl-phenylalanine, dansyl-glutamic acid, dansyl-isoleucine, dansyl-glycine and dansyl-NH_2 were then removed from the chromatograms and counted for radioactivity. From the results as shown in Fig. 37, it can be seen that the reaction reached its maximum between

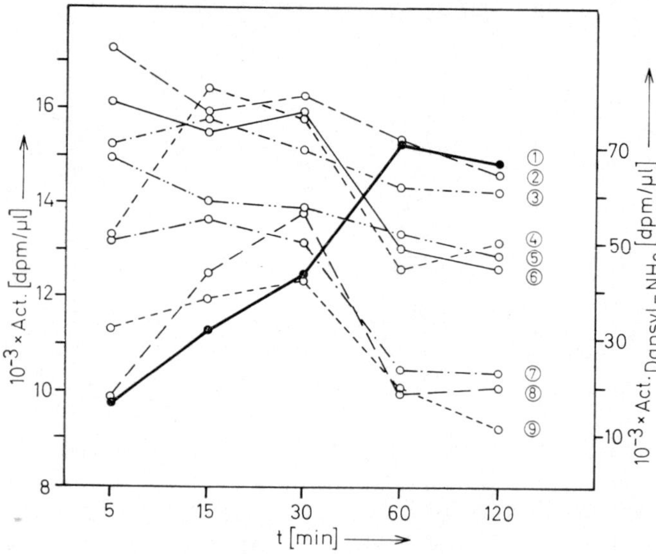

FIG. 37. The influence of incubation time on the formation of optimum amounts of some amino acids 1, dansyl-NH_2; 2, dansyl-proline; 3, dansyl-GABA; 4, dansyl-glycine; 5, dansyl-glutamic acid; 6, dansyl phenylalanine; 7, dansyl-bis-ornithine; 8, dansyl-isoleucine; 9, dansyl-bis-histidine. (After Briel *et al.*, 1972)

5 and 30 min incubation time, and then for most amino acids decreased. A reaction time of 30 min at 37°C would appear to be the most suitable condition for dansylation, since maximal dansylation of substances like histidine and ornithine is observed after 30 min, while the maximal dansylation of phenylalanine, GABA, glutamic acid, glycine and proline, though occurring earlier, remains more or less constant, even after an incubation time of 30 min.

For other procedures various incubation times have been reported for the reaction between dansyl-chloride and the substances to be dansylated, though the incubation temperatures were different. Gray and Hartley (1963) and Gray (1967) used an incubation time of 3 hr at room temperature, or 1 hr at 37°C for end-group determinations of proteins and peptides. An incubation time of 3 hr at room temperature was also used by Crowshaw *et al.* (1967) for the dansyl procedure, though Spivak *et al.* (1971) found an incubation time of 40 min at room temperature sufficient for the dansylation. This contrasts with the experiments of Neadle and Pollitt (1965), who did not state the period required for the dansylation of amino acids at room temperature, while Seiler and Wiechmann (1966) found it necessary to incubate the reagents for 16 hr at room temperature. Gros and Labouesse

(1969) found an incubation time of 3 hr at 20°C best, although 95 % of the dansylation occurred within the first 30 min. This was confirmed by Zanetta *et al.* (1970), but they preferred a longer incubation period (2 hr at 20°C) since the monoaminodicarboxylic acids need a longer time for dansylation.

(d) CONCENTRATION OF DANSYL-CHLORIDE

In the experiments of Briel *et al.* (1972), a constant amount of ^{14}C–dansyl-chloride (10.63 μmole/ml) was incubated with various amounts of a pool of amino acids in the ratios 1:2, 1:9, 1:18, and 1:89. In all experiments the reaction volume was kept constant (2 μl amino acid sample + 2 μl acetone + 4 μl ^{14}C–dansyl-chloride) and an incubation time 30 min at 37°C was used. The yield of individual dansyl amino acids as a percentage of the possible (100 %) yield extracted from the individual chromatograms is shown in Table 19. It can be seen very clearly that the yield of the various dansylated amino acids varies with the amount of dansyl-chloride available. With most of the amino acids the degree of dansylation decreases as the dansyl-chloride increases, except for proline, glycine and alanine, where the dansyl-chloride content is not very important.

TABLE 19. RECOVERY IN % COMPARED WITH THE THEORETICAL VALUE (= 100%) WHEN REACTING DIFFERENT CONCENTRATIONS OF AMINO ACIDS WITH A CONSTANT CONCENTRATION OF ^{14}C–DANSYL-CHLORIDE (10.63 × 10^{-3} M)
Reaction mixture: 2 μl amino acid solution + 2 μl acetone + 4 μl ^{14}C–dansyl-chloride. Incubation time 30 min at 37°C (from Briel *et al.*, 1972)

	Dansylation (%) at amino acid concentrations of			
Dansyl substance	4 × 10^{-4} M	10^{-4} M	5 × 10^{-5} M	10^{-5} M
Dansyl-tryptophan	96.6	74.6	61.1	49.6
Dansyl-bis-lysine	95.1	71.8	52.4	30.6
Dansyl-phenylalanine	105.2	81.5	64.9	38.3
Dansyl-bis-histidine	85.8	61.5	51.2	28.6
Dansyl-isoleucine	92.1	69.0	56.7	28.0
Dansyl-proline	118.5	107.6	101.6	—
Dansyl-glycine	116.5	110.4	103.7	—
Dansyl-glutamine	98.6	83.5	66.7	—
Dansyl-aspartic acid	113.1	93.5	85.2	—
Dansyl-GABA	102.3	90.7	68.2	—
Dansyl-methionine	38.4	28.3	22.2	—
Dansyl-taurine	—	111.7	118.7	—
Dansyl-bis-serotonin	52.0	64.0	67.4	49.3
Dansyl-N-serotonin	26.1	12.4	12.6	—
Dansyl-alanine	102.7	111.2	96.8	—
Dansyl-leucine	88.6	73.3	55.1	37.2
Dansyl-bis-tyrosine	87.2	76.3	58.7	40.8
Dansyl-5-hydroxyindole	57.3	79.9	96.3	102.6
Dansyl-bis-ornithine	92.0	75.3	57.5	30.7
Dansyl-valine	97.8	81.6	67.9	63.2
	88.7	72.4	97.9	—
$\frac{^{14}\text{C–Dansyl-Cl}}{\text{pool of amino acids}}$	2	9	18	89
$\frac{^{14}\text{C–Dansyl-Cl}}{\text{each amino acid}}$	53	212	425	2125

The importance of dansyl-chloride concentration is also demonstrated in the following experiment. Rat brain tissue was homogenised in 0.05 M $NaHCO_3$ pH 10, so that the concentration was 1 mg nervous tissue in 10 μl $NaHCO_3$. After centrifugation the supernatant was mixed with an equal volume of acetone to precipitate the proteins and then centrifuged again. 3 μl of the supernatant was reacted with either 1 μl, 2 μl, 3 μl or 4 μl of ^{14}C–dansyl-chloride (10.63 μmole/ml) and incubated in the dark for 30 min at 37°C. The contents of dansyl derivatives were then evaporated to dryness, each was resuspended in 4 μl acetone/acetic acid (3:2 by volume), and 0.3 μl of each sample chromatographed. It can be seen from Fig. 38, which shows the autoradiograms of the chromatograms, that the amine and amino acid patterns vary depending on the content of dansyl-chloride. Not only do the intensities of the individual spots differ, but also the amounts of derivatives. The variation in the amount of dansyl derivatives is probably the result of fragmentation (see Seiler, 1970).

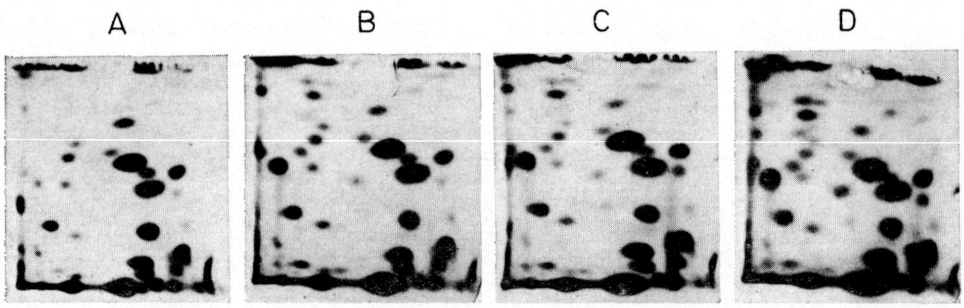

FIG. 38. In this experiment amines and amino acids were extracted from a rat's brain and a certain amount of the extract was reacted with 1 μl (A), 2 μl (B), 3 μl (C) and 4 μl (D) ^{14}C–dansyl-chloride under identical conditions (for further explanation see text). The above microchromatograms from the ^{14}C–products were then prepared. It is clear that as the concentration of dansyl-chloride used is increased, so more spots are produced. Moreover, the intensity of the various spots on the different chromatograms also reveals their dependence on the concentration of dansyl-chloride. This experiment clearly demonstrates the importance of using optimal dansyl-chloride concentrations.

(e) DEDUCTIONS

The optimal conditions for the dansylation of amino acids at 37°C require the following:

1. The reaction conditions should be at a pH of 10.0.
2. The reaction time should be 30 min.
3. The concentration of dansyl-chloride required must be critically assessed.

Should the optimum conditions of dansylation not exist, some of the formed dansyl derivatives can undergo fragmentation to form corresponding aldehydes, keytones and dansyl-NH_2. The actual extent of fragmentation is also dependent on the physiochemical properties of the individual substances to be dansylated. However, if there are optimum dansylation conditions for a group of known substances, it is possible, on the basis of the specific activity of the used dansyl-chloride, to calculate the content of each substance, since it is known that 1 mole of reactive group reacts with 1 mole of dansyl-chloride under these conditions (Casola and Matteo, 1972). It would, however, be necessary first to obtain calibration curves of the individual substances under optimum conditions and the recovery of each substance used in the estimation in the analysed mixture.

When analysing amines and amino acids in small quantities of tissue, the difficulty in practice is to determine the concentration of dansyl-chloride required, and to a lesser extent to make certain that the pH of the extracted substances in 0.05 M $NaHCO_3$ is 10. Often, the pH of the small quantity (as little as 1–2 μl in the case of the analysis of neurons) of extracted substances in 0.05 M $NaHCO_3$ can be estimated precisely by trial experiments. It should be emphasised here that the pH has to be measured before the addition of acetone (see scheme on page 67), otherwise this would result in an apparent decrease in pH.

In reality the problem in the analysis of tissue samples by the microprocedure is to determine the concentration of dansyl-chloride required. For this to be done exactly, one would have to know precisely how many reactive groups (this presupposes knowledge of all substances present) occur in the tissue sample that react with dansyl-chloride. This cannot be achieved in practice unless one extracts known substances from the tissue sample and then reacts them with dansyl-chloride. However, such an extraction process would result in a decrease in sensitivity in the dansyl procedure by a factor of at least 1000, incompatible with the whole concept of microprocedure. In any case, the identification of unknown substances by the dansyl method is often an exceedingly advantageous part of the method as shown in the example on page 86, where electrical stimulation of a nerve cell resulted in the neuron metabolising glucose to form substances which the unstimulated cell did not produce.

When analysing tissue samples with the microprocedure, it has been found from experience that it is best to determine the content of dansyl-chloride required by initial trial-and-error experiments. Once the concentration has been determined, the same concentration should be used in any one series of experiments. It should, however, always be borne in mind that it is only an approximate concentration and that any attempt to analyse the content of the individual substances in the tissue samples can only be at best semi-quantitative. The problems arising in quantification of dansyl substances are discussed on page 76.

3. Requirements for the Analysis of Tissue Samples

The following is a list of some of the apparatus, glassware, chemicals and sundries which are necessary:

(a) APPARATUS AND INSTRUMENTS

- Dissecting microscope
- Dissecting instruments
- Ultraviolet lamp (e.g. Minerallight UVSL.25)
- Scintillation spectrometer
- Whirlmixer (Mixomat)
- Centrifuge, adapted for capillaries and microtubes (e.g. Heraeus-Christ Zeta 20, Germany)
- Microhomogenizer (see page 53)
- Suitable freeze-dryer that can also be used as a vacuum evaporator
- Polyamide layers cut to measure 3 × 3 cm (Carl Schleicher & Schüll, F170 Micro-Polyamide layers, Germany)
- Microknives (see Fig. 44)

Micropipettes (see page 18)
Pasteur pipettes
Special 'cross-action' forceps (see Fig. 59), for handling polyamide plates
Deep-freezer (−20°C)
Suitable forge for making tubes, capillaries, etc.
Hair-dryer
Incubation oven

(b) GLASSWARE

50-ml beakers (at least three)
Small Petri dishes to cover 50-ml beakers
Drummond capillaries, 1, 2, 5 and 10 μl
Measuring cylinders
Microtubes (see page 24)

(c) CHEMICALS AND SOLUTIONS

Purified samples of amino acids
Purified samples of amines
^{14}C–dansyl-chloride (see page 67)
Unlabelled dansyl-chloride
Developing solution 1, water/formic acid (100:3 by volume)
Developing solution 2, benzene/acetic acid (9:1 by volume)
Developing solution 3, ethyl acetate/methanol/acetic acid (20:1:1 by volume)
Acetone/acetic acid (3:2 by volume)
Acetone
0.05 M $NaHCO_3$ adjusted to pH 10 with 0.1 N NaOH
Scintillation fluid (4 g PPO and 0.1 g POPOP per litre)

(d) SUNDRIES

Rubber tubing with mouthpiece for micropipettes (see Fig. 17)
Scintillation vials
Ice bucket
Plasticine to hold capillaries or microtubes
Spirit flame for sealing Drummond capillaries
Film for autoradiography (see page 75)
Developer for autoradiograms
Wooden tongs (e.g. a clothes-peg) for holding microtubes
Parafilm
Stop clock

4. General Procedure

A summary of the method used to obtain microchromatograms for the analysis of nerve tissue from a terrestrial vertebrate or invertebrate follows:

1 mg tissue 20 μl 0.05 M $NaHCO_3$ pH 10

↓

homogenised thoroughly

↓

centrifuged for 15 min at 20,000 *g*

↓

supernatant transferred to a clean tube and an equal volume of acetone added

↓

kept at -20°C for 30 min

↓

centrifuged for 15 min at 20,000 *g*

↓

4 μl portion + 4 μl 5 μM/ml of ^{14}C–dansyl-chloride

↓

incubated in the dark for 30 min at 37°C

↓

dried *in vacuo*

↓

redissolved in 4 μl acetone/acetic acid (3:2 v/v)

↓

0.4 μl applied to a single 3 × 3 cm microchromatogram.

The microchromatogram is developed in an ascending way in two or three solvent systems and viewed under ultraviolet light. Since each dansylated substance is radioactive, the content can be quantified or autoradiograms prepared.

(a) DANSYL-CHLORIDE SOLUTION

^{14}C–dansyl-chloride solution in acetone (2.7 mg/ml, specific activity 49 mCi/mM or 98 mCi/mM) can be obtained from CEA, Gif-Sur-Yvette, France, or Schwarz/Mann, Orangeburg, New York. ^{3}H–dansyl-chloride solution in benzene (1000 Ci/mM) can be obtained from Amersham/Buchler, Braunschweig, Germany. Since the dansylation reaction is dependent to a great extent upon the concentration of radioactive dansyl-chloride (Table 19) the concentrations of the purchased solutions should be checked thoroughly. Known concentrations of radioactive dansyl-chloride can then be prepared and stored at -20°C until required. Any evaporation of acetone or benzene while handling dansyl-chloride stock solutions must be avoided. One way of doing this is to handle the tube containing the dansyl-chloride with a pair of wooden tongs, thus eliminating much of the heat transfer which is inevitable when fingers are used. The dansyl-chloride stock solution should always be surrounded by ice while a sample is being removed.

The actual concentration of the purchased ^{14}C–dansyl-chloride can be checked in the following way. The vial is opened, the walls washed down with acetone and the whole of the contents, still in the vial, dried *in vacuo*. The residue in the vial is dissolved by the addition of 200 μl acetone, and 1 μl of this solution is then removed (with a 1 μl Drummond capillary) and placed in a cuvette containing 100 μl absolute ethanol. The concentration of the 1 μl sample can subsequently be determined spectrophotometrically, using an extinction coefficient of 3.67×10^3 at 369 nm (Gray, 1964) and applying the formula

$$C = \frac{A}{B \times D}$$

where C = concentration (moles/litre), A = optical density, determined spectrophotometrically, of 1 μl sample diluted one-hundredfold, B = the extinction coefficient used, viz. 3.67×10^3, and D = the optical path length, viz. 1.

For a more rapid way of ascertaining the concentration, the following procedure can be used. The purchased radioactive material is made up to exactly 200 μl by volume as described above, and 1 μl is removed with a Drummond capillary and transferred directly to a counting vial previously filled with scintillation solution. The capillary is then broken up in the vial with a glass rod in order to release its contents and the radioactivity is counted. Since the total activity of the 200 μl ^{14}C–dansyl-chloride solution is stated by the manufacturers (e.g. CEA Schwarz/Mann, Orangeburg, New York) to be 0.1 mCi and the specific activity (= Ci/mole) to be 49 mCi/mM, and since by definition 1 Ci contains 2.22×10^{12} dpm the actual concentration of the solution can easily be determined, provided the efficiency of the spectrophotometer is taken into account.

(b) EXTRACTION OF AMINES AND AMINO ACIDS

A small tissue sample (5–0.05 μg wet weight) is thoroughly homogenized in 0.05 M $NaHCO_3$, pH 10. A concentration of 1 mg nerve tissue/20 μl $NaHCO_2$ is most suitable. After centrifugation at 45,000 *g* for 15 min, the supernatant is transferred to another tube, its proteins are precipitated by the addition of an equal quantity of cold acetone and the process is completed by placing the solution in a freezer (−20°C) for 30 min. The solution is allowed to thaw, centrifuged once more (20,000 *g* for 30 min), and the supernatant is transferred to a clean tube ready for dansylation.

The 'buffer' used for the extraction procedure (0.05 M $NaHCO_3$ adjusted to pH 10 by 1 M NaOH) is similar to the system used by Gray and Hartley (1963) and Hartley (1970) for end-group determination of peptides and proteins, though their molarity of $NaHCO_3$ was higher (0.1 M). Other 'buffer' systems have also been used by other workers, and these include 0.2 M sodium phosphate buffer or 0.2 M $KHCO_3$ (Zanetta *et al.*, 1970; Spivak *et al.*, 1971). It is important to use the lowest practical molarity of 'buffer' for the microdetermination of amines and amino acids to avoid the accumulation of salts on the very small application point on microchromatograms, otherwise the quality of separation could be affected. The usual buffer systems used for extracting amino acids in the macroscale, e.g. picric acid, perchloric acid, sulphosalicylic acid, ethanol/HCl, are not suitable as they all affect the final dansyl reactions.

In order to obtain reproducible results it is advisable to homogenize identical quantities of tissue for the same length of time. For a tissue sample of 1 mg a homogenizing time of 5 min (using a dentist's drill, about 24,000 rpm) is sufficient. About 85% of all extractable amines and amino acids are released and recovered in the supernatant under these conditions (Briel, 1972).

(c) DANSYLATION

A specific quantity of solution containing the amines and amino acids is reacted with a known amount of dansyl-chloride in a microtube, mixed thoroughly on a Whirlmix and incubated in the dark for 30 min at 37°C. After incubation the dansylated products are evaporated to dryness. In the case of nervous tissue, normally 3–5 μl of solution containing the substances (corresponds to 1 mg wet weight tissue per 20 μl 0.05 M $NaHCO_3$ pH 10 + 20 μl acetone) is reacted with an equal volume of 5 μM/ml of either ^{14}C–dansyl-chloride or unlabelled dansyl-chloride (see scheme of general procedure, page 67). However, when specific amines or amino acids are being analysed, other convenient concentrations of these substances can be used.

Small amounts of solution (3–5 μl) are transferred to the microtubes with capillaries of suitable size (see page 66). Care must be taken to deposit the solutions at the bottom of the microtube as any of the acetone solutions which touch the microtube walls will tend to evaporate and thus cause inaccuracies. It is necessary to have the proportion of acetone high in the final reaction mixture so that the content of water (approximately 25%) can be kept as low as possible. This is because dansyl-chloride reacts with water, and since only a specific amount of dansyl substances can be fractionated on a single microchromatogram, the dansyl-OH content should be kept as low as possible to observe the other dansyl substances.

When the solutions are in the microtubes, it is advisable to seal the tubes with parafilm to avoid evaporation. The contents are then thoroughly mixed by touching the base of the microtube on a Whirlmix, and the tube is jerked in a downwards movement, so that any of the contents sprayed by the mixing are once again deposited on the bottom. The mixture is then incubated in the dark for 30 min at 37°C and evaporated to dryness. The actual forming of the vacuum used for drying should be a gradual process so that none of the solution can be lost by ejection out of the tube. The dried dansylated products may be stored at −20°C until required for further analysis.

(d) MICROCHROMATOGRAPHY

Microchromatography of the dansyl products is carried out on polyamide sheets cut to measure 3 × 3 cm and obtained from Schleicher & Schüll (TLC Ready-Plastic sheets F 1700, Micro-Polyamide). These have proved more suitable for chromatography than the polyamide film obtained from the Cheng-Chin Trading Co., Taiwan, used originally by Woods and Wang (1967) to separate dansyl amino acids in the normal scale. This is because the Schleicher & Schüll polyamide layers are more homogeneous, thus producing sharper resolution of the dansyl derivatives.

The dried dansylated substances are resuspended in 2–5 μl of acetone/acetic acid (3:2 v/v), depending upon the concentration of the dansylated amines and amino acids (see scheme of general procedure, page 67). A precise volume (0.1–0.5 μl) of this solution is transferred by suction into the ultra-thin tip of a capillary connected to the mouth by rubber tubing (see Fig. 17) and applied to one corner of a 3 × 3 cm polyamide layer. The application of the dansylated substances to the polyamide layer is normally done under a stereomicroscope to ensure that the starting point is sharp (diameter 0.5–1.0 mm).

The application of the sample to the microchromatogram is of critical importance in that it often determines the final quality of separation. The capillary used for the application should be fine, the end of the shaft having a diameter wide enough to allow the dansylated substances in acetone/acetic acid to be released without too much pressure being applied via the mouthpiece. When the sample is being applied, the capillary tip is brought carefully into contact with the polyamide layer so that on application the sample is drawn from the capillary on to the layer. Should the capillary tip not touch the polyamide layer, the solution released from the tip has the tendency to migrate slightly up the outside of the capillary walls, drying and thus affecting the accuracy of the procedure. It must be stressed that the capillary tip should barely touch the polyamide layer, for any damage caused to the layer will have drastic effects on the quality of separation. The actual point of application should be at one corner of the sheet, 3–4 cm from the edge, and should be no greater than 0.5–1.0 mm. It is helpful to mark the point with a soft pencil and to use the warm air from a hair-dryer to speed up the process. To prevent the polyamide from being blown away by the stream of warm air, it is best to fix the layer to the base plate of the stereomicroscope with the springs of the microscope, which are normally used for holding the microscope slides (see Fig. 39).

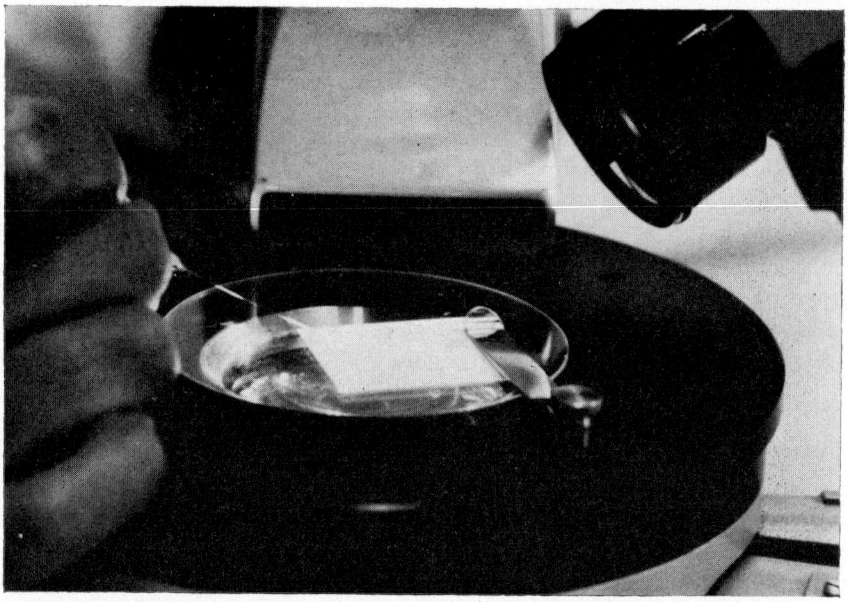

FIG. 39. This figure shows the procedure used to apply a small amount of dansyl substances by means of a capillary pipette to one corner of a micropolyamide chromatogram. It can be seen that the clip used on the microscope to hold glass slides is here employed to secure the microchromatogram. Also visible is the end of a hair-dryer resting on the microscope stage to provide a continuous stream of warm air.

Microchromatograms are developed in an ascending fashion in 50-ml beakers, which need not be covered, though this is recommended for the second and third solvent systems. Chromatography in the first dimension is with water/formic acid (100:3 by volume), and in the second dimension benzene/acetic acid (9:1 by volume) (Woods and Wang, 1967). In order to separate certain groups of substances, e.g. dansyl-alanine from dansyl-NH_2, dansyl-glutamic acid from dansyl-aspartic acid, dansyl-threonine, dansyl-glutamine, dansyl-serine and dansyl-asparagine from dansyl-arginine, dansyl-ϵ-lysine, dansyl-α-amino-histidine and dansyl-cystine, a third chromatography in the second dimension is required (see Figs. 40 and 43). In this instance the solvent system employed is ethyl acetate/methanol/acetic acid (20:1:1 by volume) (Woods and Wang, 1967) and often it is best to control the separation under ultraviolet light. The time to complete the chromatography in the first solvent system is about 3 min, the second 5 min and the third

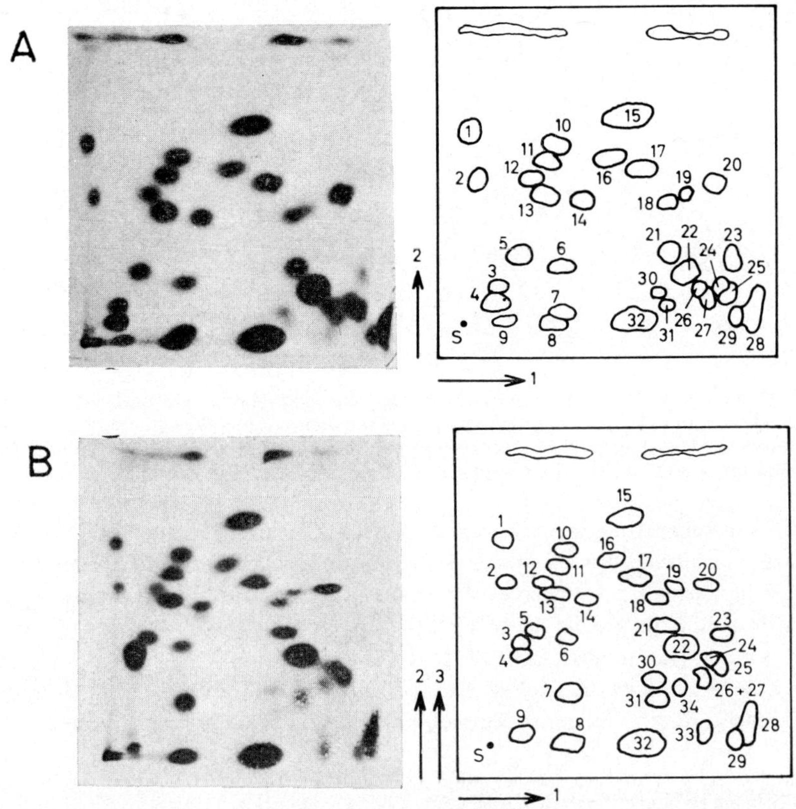

FIG. 40. Autoradiograms of 3 × 3 cm microchromatograms and corresponding maps to show the positions of various amines and amino acids which had been reacted with ^{14}C–dansyl-chloride and separated in two (A) and three (B) solvent systems. The directions of chromatography are indicated by the arrows and the solvent systems used were water/formic acid (10:3 by vol) in the 1st direction, benzene/acetic acid (9:1 by vol) in the 2nd direction, ethylacetate/methanol/acetic acid (20:1:1 by vol) in the 3rd direction. S, starting point; 1, dansyl-bis-serotonin; 2, dansyl-bis-tyrosine; 3, dansyl-tryptophan; 4, dansyl-N-serotonin; 5, dansyl-bis-lysine; 6, dansyl-bis-ornithine; 7, dansyl-N-tyrosine; 8, dansyl-taurine; 9, dansyl-cystein; 10, dansyl-isoleucine; 11, dansyl-leucine; 12, dansyl-bis-histidine; 13, dansyl-phenylalanine; 14, dansyl-methionine; 15, dansyl-proline; 16, dansyl-valine; 17, dansyl-GABA; 18, dansyl-alanine; 19, dansyl-NH_2; 20, dansyl-ethanolamine; 21, dansyl-glycine; 22, dansyl-hydroxyproline; 23, not identified; 24, dansyl-glutamine; 25, dansyl-asparagine; 26, dansyl-threonine; 27, dansyl-serine; 28, dansyl-arginine, dansyl-α-amino-histidine and dansyl-α-lysine; 29, dansyl-cystine; 30, dansyl-glutamic acid; 31, dansyl-aspartic acid; 32, dansyl-OH; 33 and 34, not identified.

about the same. The layer must be dried thoroughly with hot air and then cooled before each chromatography.

A number of other chromatographic systems can be used in order to resolve specific substances. For example, in order to separate many of the substances in a second chromatography which is normally carried out with a third chromatographic system, the mixture n-heptane/n-butanol/acetic acid (3:3:1 by volume) can be used (Roberts *et al.*, 1973). However, this results in many other substances running closely together (cf. Figs. 40 and 41), so making it difficult to remove the individual spots from the microchromatogram for radioactive counting. As an alternative to solvent system 3, the mixture 0.5 M tri-sodium phosphate/ethanol (3:1 by volume) is particularly useful (Tang and Hartley, 1970),

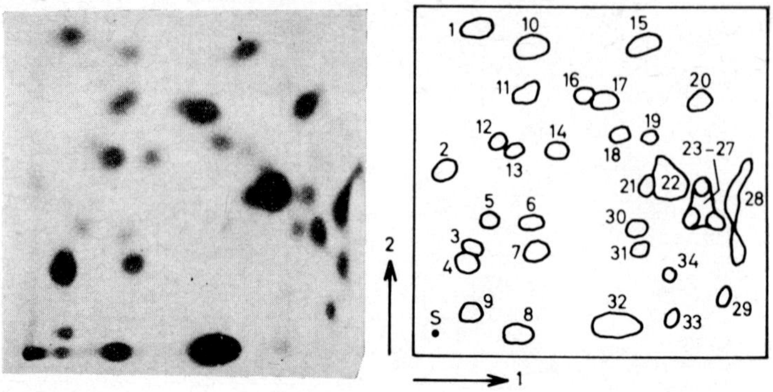

FIG. 41. Autoradiograms of a two-dimensional microchromatogram (directions of chromatography indicated by arrows) and a corresponding map to show the separation of the same substances as indicated in Fig. 40. The solvent system used for the first chromatography was water/formic acid (9:1 by vol) and for the second chromatography n-heptane/n-butanol/acetic acid (3:3:1 by vol).

especially when separating dansyl-ϵ-lysine, dansyl-α-amino-histidine and dansyl-arginine. The following combinations of solvent systems can also be applied for the separation of particular substances, e.g. hydroxyproline from glutamic and aspartic acid:

Carbon tetrachloride/glacial acetic acid (4:1 by volume)
n-Butyl acetate/glacial acetic acid (9:1 by volume)
Diethyl ether/glacial acetic acid (9:1 by volume)
95% Ethanol/dimethylformamide/glacial acetic acid (4:1:2:4 by volume)

(e) IDENTIFICATION OF AMINES AND AMINO ACIDS ON MICROCHROMATOGRAMS

As little as 1 pmole of an individual amino acid can be detected on a single microchromatogram viewed under ultraviolet light (about 248 mμ). Dansyl-OH, the reaction product of dansyl-chloride with water, appears as an intensive blue spot which does not migrate with the second or third solvent. When the water content is high, additional minor dansyl-OH spots can appear (see Fig. 43). In contrast the dansyl amines and dansyl-amino

---

FIG. 42. Autoradiograms of microchromatograms (original size 3 × 3 cm) to show the positions of substances reacted with ^{14}C–dansyl-chloride. Each microchromatogram was developed in an ascending way using two solvent systems: 1st development (horizontal direction) water/formic acid (100:3 v/v) and 2nd development (vertical direction) benzene/acetic acid (9:1 v/v). The starting points (S) and the positions of dansyl-OH (W) are indicated in all the chromatograms while the positions of dansyl-NH_2 (N) and dansyl-glycine (G) are only indicated on some of the chromatograms. Dansyl-NH_2, the reaction product of dansyl-chloride and ammonium, appears on all the chromatograms, although its concentration on some is too low for identification. Glycine was added to some samples so that the position of the dansyl products of a particular substance could be seen in relation to glycine. Substances dansylated are (A) putreanine (2 products), (B) putrescine (2 products), (C) spermine (4 products), (D) spermidine (3 products) (E) noradrenaline (5 products), (F) adrenaline (5 products), (G) octopamine (4 products), (H) bufotenine, (2 products), (I) dihydroxyphenylalanine (2 products), (J) tyramine (4 products), (K) 4-hydroxy-methoxy-mandelic acid (2 products), (L) 4-hydroxy-3-methoxy-phenylacetic acid (3 products), (M) α-methyl-m-tyrosine (2 products), (N) homoserine (1 product), (O) α-amino butyric acid (spot 1) and γ-amino butyric acid (spot 2), (P) homocystein (1 product), (Q) γ-amino-β-hydroxybutyric acid (3 products), (R) citrullines (2 products), (S) allo-cystathionine (2 products), (T) p-chlorophenylalanine (2 products). All substances were obtained from manufacturers and not purified before dansylation. Some minor spots may therefore be products of impurities.

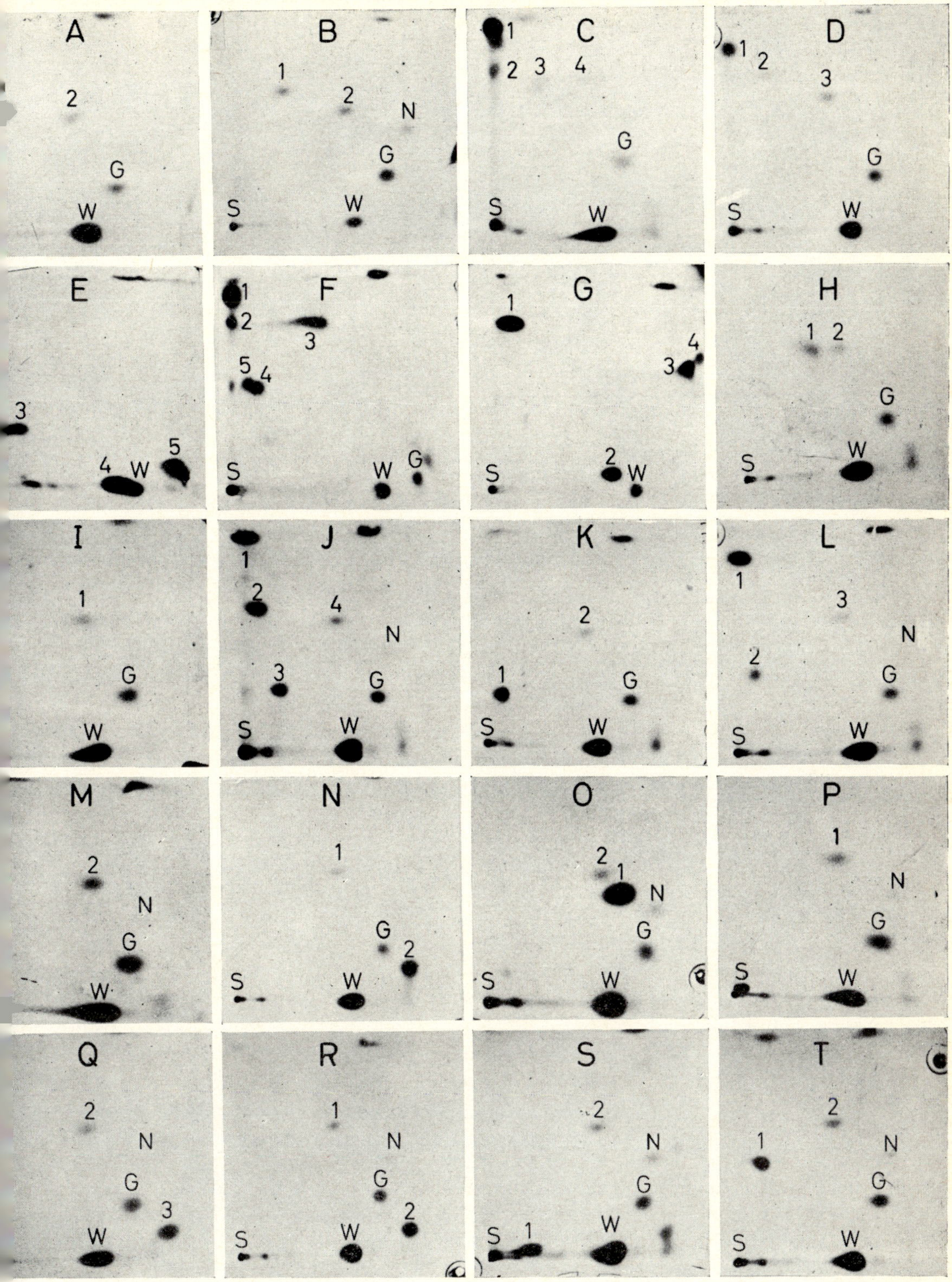
A
2
G
W
B
1
2
N
G
S
W
C
1
2
3
4
G
S
W
D
1
2
3
G
S
W
E
3
4
W
5
F
1
2
3
5
4
S
W
G
G
1
4
3
2
S
W
H
1 2
G
S
W
I
1
G
W
J
1
2
4
N
3
G
S
W
K
2
1
G
S
W
L
1
3
N
2
G
S
W
M
2
N
G
W
N
1
G
2
S
W
O
2
1
N
G
S
W
P
1
N
G
S
W
Q
2
N
G
3
W
R
1
N
G
2
S
W
S
2
N
G
S
1
W
T
2
1
N
G
S
W

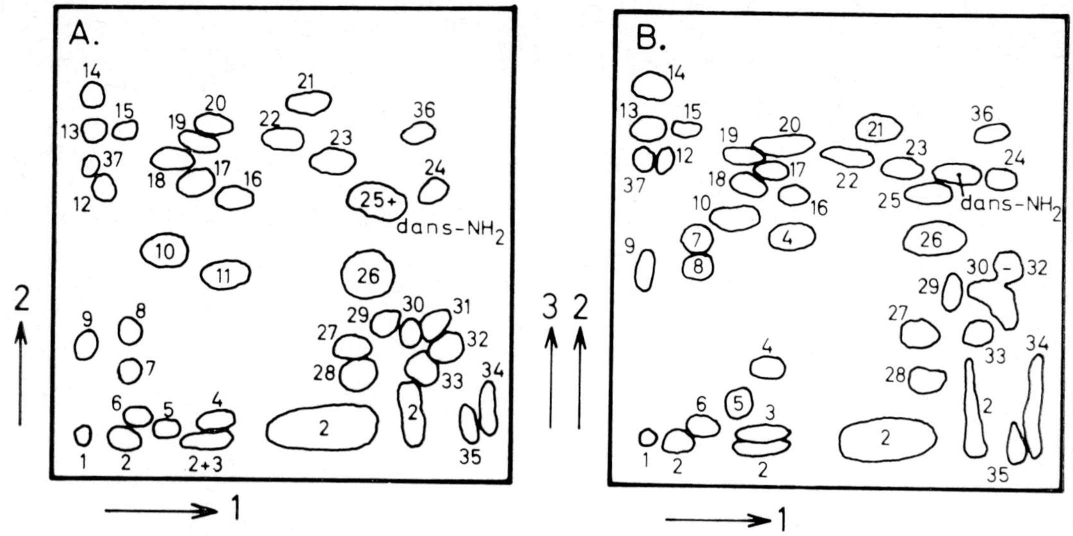

FIG. 43. Positions of some dansylated amines and amino acid after chromatography on 3 × 3 cm polyamide layers using either two (A) or three (B) solvent systems. 1, starting point; 2, dansyl-OH (reaction of dansyl-chloride with water); 3, dansyl-taurine; 4, dansyl-N-tyrosine; 5, dansyl-bis-5-hydroxytryptophan; 6, dansyl-cystein; 7, dansyl-N-serotonin; 8, dansyl-tryptophan; 9, dansyl-N-5-hydroxytryptophan; 10, dansyl-bis-lysine; 11, dansyl-bis-ornithine; 12, dansyl-bis-tyrosine; 13, dansyl-bis-serotonin; 14, dansyl-5-hydroxyindole; 15, dansyl-tryptamine; 16, dansyl-methionine; 17, dansyl-phenylalanine; 18, dansyl-bis-histidine; 19, dansyl-leucine; 20, dansyl-isoleucine; 21, dansyl-proline; 22, dansyl-valine; 23, dansyl-GABA; 24, dansyl-ethanolamine; 25, dansyl-alanine; 26, dansyl-glycine; 27, dansyl-glutamic acid; 28, dansyl-aspartic acid; 29, dansyl-hydroxyproline; 30, dansyl-threonine; 31, dansyl-glutamine; 32, dansyl-asparagine; 33, dansyl-serine; 34, dansyl-arginine, dansyl-ε-lysine, dansyl-α-amino-histidine; 35, dansyl-cystine; 36, dansyl-tyramine; 37, dansyl-N-acetyl-serotonin.

acids appear either yellow or yellow-green when seen under ultraviolet light. The distribution of various dansylated amines and amino acids on microchromatograms after two-dimensional chromatography using two or three solvent systems is shown in Figs. 40, 41, 42 and 43. Certain amino acids, e.g. tyrosine, methionine, lysine and serine, form mono- and disubstituted derivatives as does serotonin, whilst histidine appears in three forms. Individual catecholamines, in contrast, form a number of dansyl derivatives.

Standard dansylated substances can be added to a tissue extract or chromatographed separately to assist in the identification process. Alternatively, the technique of finger-printing can be employed by applying standard dansylated substances onto the reverse side of a polyamide layer. The coincidence of the starting points of sample and standards can be checked under ultraviolet light (the layer is transparent for the fluorescent spots). Although dansyl-chloride reacts theoretically with all aliphatic –OH and $-NH_2$ groups, the actual number of dansyl substances formed from substances containing two or more reactive sites varies and depends upon a number of criteria, viz. reaction conditions, structure of the molecule, etc. This complicates the already difficult process of characterising each dansyl derivative formed from a substance containing two or more reactive groups. Analysis of the emission and excitation spectra of individual dansyl derivatives formed from a single substance has proved a way to help in the identification of, for example, serotonin (Neuhoff and Weise, 1970) and histidine (Neuhoff *et al.*, 1971). However, the characterisation of individual substances on microchromatograms formed from the catecholamines (e.g. dopamine and noradrenaline can each theoretically form seven dansylated substances) has so far proved very complicated, and for this reason they have not yet been analysed by the microprocedure.

It is possible to characterise the dansyl derivatives of a substance like serotonin in the following manner. Serotonin has two groups (an –OH and $-NH_2$) which can react with dansyl-chloride, so that theoretically three dansyl-serotonin derivatives (i.e. dansyl-H-, dansyl-O- and dansyl-bis-serotonin) can be found. However, under the conditions used in microchromatography, viz. pH 10 reaction time of 30 min at 37°C, only two dansyl derivatives occur on chromatograms. From the chromatographic behaviour of these two dansyl-serotonin derivatives a number of predictions can be made. Since neither migrates very far in the aqueous phase, it is probable that the very much faster migrating derivative in the organic phase will in all probability be the dansyl-bis- form. This prediction concerning the positions of the mono- and bis- derivatives can be confirmed by eluting the individual spots with ethanol and redansylating each. Following chromatography it can be deduced that the dansyl-N-derivative was converted to dansyl-bis-serotonin and both compounds now run in the same chromatographic position. Another way of showing that one of the spots is a mono- and the other a bis- derivative of serotonin is to dansylate ^{14}C–serotonin with ^{3}H–dansyl-chloride and count the radioactivity associated with each dansylated derivative. These predictions were proved by Neuhoff and Weise (1970) by producing a number of microchromatograms and removing the individual spots corresponding to the two dansyl derivatives. The individual samples were then pooled and eluted with absolute ethanol. After checking the purity of each sample by rechromatographing a small amount, their emission and excitation spectra were critically recorded by use of a Turner Model 210 Spectrophotometer. By comparing the spectra with the structure of the molecules corresponding to dansyl-N-, dansyl-bis- and dansyl-O-serotonin with serotonin, it was concluded that the two derivatives are dansyl-N-serotonin and dansyl-bis-serotonin.

(f) AUTORADIOGRAPHY AND MEASUREMENT OF RADIOACTIVITY

Autoradiograms of microchromatograms can be obtained in the dark by using either Agfa-Gevaert Denture Ultra Rapid L film or Dupont Dosimeter film 046/545, exposure time 2–6 days for ^{14}C-labelled dansyl derivatives. When very high specific activity tritiated labelled substances are used, the exposure time is very much longer.

For quantitative measurements of radioactive fluorescent spots on the microchromatograms, the spots are first clearly marked under ultraviolet light with a sharp soft pencil. Care must be taken not to damage the polyamide layer in any way, which can often occur if too hard a pencil is used. The pencil markings should always be larger than the fluorescent spots. The actual removal of the polyamide layer from the carrier foil is carried out under microscopic vision using a special razor-splinter knife (see Neuhoff, 1973) with a knife-edge of between 0.5 and 1.00 mm wide (see Fig. 44). The razor-splinter knife is not only used for removing the area of polyamide marked out, but also for transferring the polyamide directly into counting vials. The polyamide layer is best removed from the foil by beginning at one edge, scraping parallel strips and then transferring them directly to a counting vial. It is advisable to inspect the knife under the stereomicroscope, and also the area scraped from the microchromatogram, to see whether any pieces of polyamide have been missed accidentally.

In practice, most of the spots on the left-hand side of the microchromatogram are

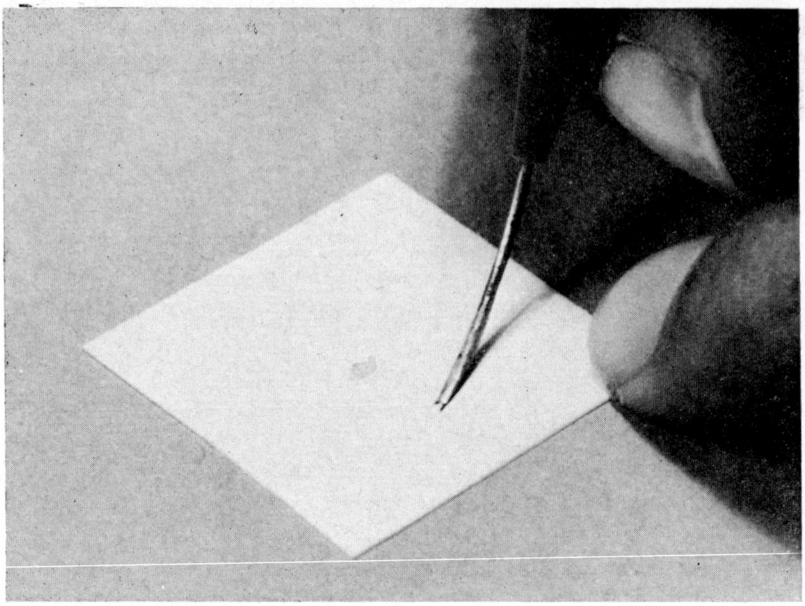

FIG. 44. This photograph shows the type of knife used for removing individual areas from a 3 × 3 cm polyamide layer. Each area removed corresponds to the position of a single dansyl-derivative which was identified under ultraviolet light and marked around its perimeter with a soft pencil.

removed after chromatography in two solvent systems. The chromatogram is then developed in a third chromatography solvent system in order to separate in particular dansyl-alanine from dansyl-NH_2, dansyl-glutamic acid from dansyl-aspartic acid, dansyl-threonine, dansyl-glutamine, dansyl-serine and dansyl-asparagine from dansyl-arginine, dansyl-ϵ-lysine, dansyl-α-amino-histidine and dansyl-cystine. These and some other spots are then marked under ultraviolet light and removed. Each sample is suspended in 10 ml scintillation solution (4 g PPO and 0.1 g POPOP in 1 litre toluene) and the radioactivity counted. The quenching effects produced by the pieces of polyamide layers are negligible (Neuhoff and Weise, 1969).

(g) QUANTIFICATION

A serious difficulty encountered in using dansyl-chloride lies in the quantification of the results. This is because of the undesirable side reactions, variations in the degree to which different amino acids are dansylated and differences due to changes in the rate of reagent to substrate. All these points have already been discussed. Indeed, unless the identity and structure of all substances dansylated are known, it is not possible to determine the components of a tissue extract quantitatively. For example, in extracts from snail tissues, over thirty unknown substances are present, thereby making quantification of the substances impossible. Problems also arise in obtaining even a semi-quantitative assessment of the individual substances in extracts containing so many unknowns. It is generally unsatisfactory to use internal standards because the inclusion of exogenous substances alters the optimal dansylation of all the substances dansylated, leading to the possibility of false results. Where the amino acid and amine content of a series of tissue extracts is not significantly different, then internal standards (e.g. norvaline) can be used. However, in circumstances where one or more amino acid in the extract may be considered altered, for example when in studying a drug effect, the inclusion of an exogenous amino acid will significantly affect the result. External standards can be used; in this case different amounts of standard amino acid mixtures (0.01–1.00 M) are dansylated and chromatographically separated. Standard curves can then be made of the relationship between the concentration of the substance and the radioactivity and these can be referred to, in order to determine the concentration of substances in a biological extract.

From experience it has been found useful to express the quantity of each individual substance as a percentage of the total radioactivity of the dansylated substances which are removed from the chromatogram. While this method allows relative concentrations of any amine and amino acid to be determined it is only related to the absolute concentration of all the substances which are detectable on the chromatogram and removed. This way of expressing the results is, however, unsatisfactory for detecting changes in amino acids or amines, since a change in any one substance of the mixture will obviously affect the apparent concentration of the other substances when expressed in this fashion. Nevertheless, this method of expressing the results for different cells from the same tissue does enable the assessment of the proportions of one substance to the total present in the mixture (Osborne *et al.*, 1973b). As it is possible to detect many unknown dansyl substances on a single chromatogram, there are clearly many disadvantages in expressing the results in this way.

When radioactivity is first incorporated into tissue amines and amino acids (e.g. by incubating brain slices in a medium containing ^{14}C–glucose), quantification can be obtained by reacting tissue extracts with ^{3}H–dansyl-chloride. The ^{3}H/^{14}C ratio for the individual substances can therefore be analysed and corrected, so that a very reliable method for quantification becomes available. This is particularly useful when studying turn-over rates of a few substances in a mixture where it is often thought that the absolute content of the substances does not alter. However if the absolute concentration of substances in the mixture is altered, this method, like the others, is unreliable for quantitative analysing.

Weise and Eisenach (1972) have evaluated autoradiograms by densitometry in an attempt to quantify the procedure, and they were able to detect as little as 10^{-14} moles of amino acid. However, this procedure is unsuitable because the problem occurring in the quantification of tissue substances lies in the actual dansyl procedure (as described above). Moreover, a number of erroneous results can occur when analysing the radioactivity on autoradiograms by densitometry. For example, the background and extinction depend on the sensitivity of the film, the exposure time and the conditions of developing and fixing the film. These factors make comparisons between films difficult unless the conditions are rigidly controlled. Neither can spots which are close together be determined with any degree of accuracy. Moreover, the assumption is made that the film is homogeneous, which it is probably not.

In summary, the complexity of the reactions between dansyl-chloride and substances in tissue extracts makes it very difficult to determine the absolute content of the individual substances in an extract. It is, however, possible to determine the absolute content when the nature of all substances which react with dansyl-chloride is known. This can be achieved when analysing tissue extracts simply by selectively extracting certain substances before dansylation. However, this greatly decreases the sensitivity of the method, which is fundamentally in opposition to the concept of the microprocedure. When trying to determine the contents of only a few substances which react with dansyl-chloride, internal and external standards can be used. However, this becomes a difficult and laborious task when trying to analyse the contents of, for example, all the free amino acids.

5. Procedure for Analysing Neurons

The principle for analysing isolated cells is similar to that used for the analysis of small amounts of nervous tissue. However, instead of using small reaction tubes for homogenization, centrifugation, etc., the whole procedure is carried out in 10-μl Drummond capillaries with an internal diameter of 560 μ. This complicates the analysis and necessitates more careful handling in all the steps involved.

The problem in analysing neurons lies essentially in keeping the volume of solution as small as practically possible, since the whole of the contents will finally be applied to a single microchromatogram. The amount of 0.05 M $NaHCO_3$ used is particularly important, when one bears in mind that dansyl-chloride reacts with water and also that the salt content has to be kept low in order to achieve a good separation of substances by chromatography. The procedure used is as follows.

A 10-μl Drummond capillary (microcap) is carefully heat-sealed at one end and all but 0.05 cm of the tubing removed. The minute tube, with a volume of about 2 μl and measuring 0.5 cm long with a diameter of 560 μ, is then filled with 0.05 M $NaHCO_3$ pH 10 using a pipette capillary, and kept upright in a piece of plasticine. The cells to be analysed are placed on the meniscus of the 'buffer' with a pipette capillary and a piece of 10-μl Drummond glass tubing then attached to the open end with polythene tubing (see Fig. 45).

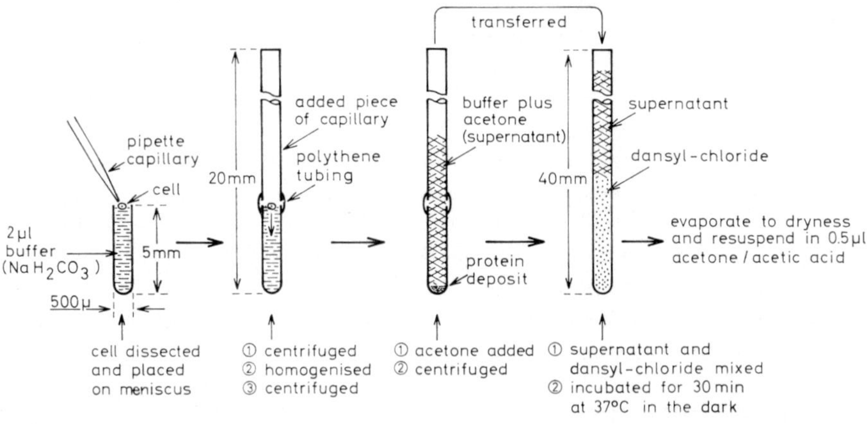

FIG. 45. Schematic diagram of procedure used to analyse single cells for amine and amino acid content using the microdansyl procedure.

Any attempt to place the cells in the buffer must be avoided since this often results in the loss of cells or in the loss of buffer, so producing inaccuracies. The capillary is then centrifuged to sediment the cells and the contents are homogenized with a nerve canal drill (see page 53). The capillary is centrifuged again and 2 μl acetone are added to precipitate the proteins. In order to increase the precipitating effect, the capillary and contents are placed in a freezer (−20°C) for 30 min. Once more the capillary is centrifuged and the supernatant carefully transferred to a 10-μl microcap, heat-sealed at one end. This transference is critical and depends solely on the kind of pipette capillary used and the way in which it is manipulated (see page 23). A specific amount of dansyl-chloride (found out by trial experiments) is added to the 10-μl microcap and the contents are incubated in the dark at 37°C for 30 min. After the incubation period, the microcap contents are vacuum-dried and 0.5 μl acetone/acetic acid (3:2 by volume) is used, not only to wash down the walls of the microcap, but also to dissolve the dansylated contents. The whole of the solution (0.5 μl) is then applied to the corner of a single 3 × 3 cm polyamide chromatogram (see Fig. 45).

The procedure described above can be employed successfully to obtain good chromatograms from single *Aplysia* neurons. However, the likelihood of experimental error is very high in this method. Moreover, the nerve canal drill used for the homogenization process has to be scrupulously clean. An alternative method is to dissect a large number of neurons and put them directly into a microglass tube (inner diameter 1.5 mm) containing distilled water. The tube is then frozen in liquid nitrogen and thawed. This 'shocking' process is repeated three times before the sample is finally frozen and freeze-dried. A suitable amount of 0.05 M $NaHCO_3$ pH 10 and acetone is used to resuspend the sample and the contents are analysed as described earlier. The disadvantage of this procedure is that it is impossible to apply it to the single-cell level.

6. Biological Application of Procedure

(a) AMINES AND AMINO ACIDS IN CHARACTERISED NEURONS

Although the present method has made it possible to study the amine and amino acid content of specific cell types and consistent results have been obtained, it is still difficult to know the exact chemical content of each cell type. This is probably because of the isolation procedure of the cells and also the neuron's physiological activity. Free-hand dissection of living neurons can incur a number of errors, particularly from contamination of surrounding tissue and blood. Furthermore, the cell's membrane can be damaged during the dissection, thus allowing leakage of chemicals from the cells and the destruction of the integrity of the neuron. For these and other reasons, very many experiments are required before trying to decide upon the exact chemical content of each cell type. In practice this might prove impossible, should the chemical content of each neuron depend on the activity of the individual animal. In addition the microprocedure involving the use of dansyl-chloride produces difficulties in quantifying all the individual contents in a cell, since this would require precise knowledge of the amount of substances present which react with dansyl-chloride (see page 64).

(i) *Two* Helix *neurons which make synaptic contact*

In each cerebral ganglion of the snail *Helix pomatia* (and also a number of other gastropods) (Osborne and Cottrell, 1971; Cottrell and Osborne, 1970) there is a large or giant neuron located in the anterior–ventral position. Conclusive evidence has shown that these giant neurons contain serotonin (Cottrell and Osborne, 1970; Osborne, 1972a, 1973a) and the cells have consequently been termed the giant serotonin cells or GSCs. The membrane properties of the GSCs in *Helix* have been studied by Kandel and Tauc (1966), who discovered that the cells show a high degree of anomalous rectification, and that each GSC sends an axon to the ipsilateral exterior lip nerve and one to each cerebral buccal connective. Cottrell (1970a, 1971), studying these cells in greater detail, has demonstrated that the two GSCs form synaptic excitatory links with other neurons in the buccal ganglia. These neurons in the buccal ganglia are equally as large as the GSCs and lack biogenic amines (Osborne, 1972a; Osborne and Cottrell, 1972a). From the diagram (Cottrell, 1971) in Fig. 46, which shows the position and axonal processes of the GSCs and the non-amine-

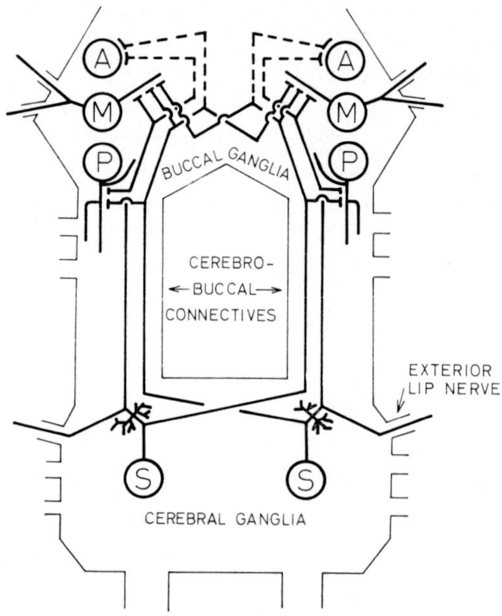

FIG. 46. Diagrammatic representation of the cerebral and buccal ganglia of *Helix pomatia* showing the position and axonal processes of the GSCs and the neurons with which they make contact (redrawn after Cottrell, 1971). The neurons in the buccal ganglia which receive synaptic input from the GSCs are called anterior (A), middle (M) and posterior (P) cells. None of the three giant buccal cells contains monoamines as observed by fluorescence microscopy. The processes from the different cells were traced either by dye injection or by electrophysiological methods.

containing buccal neurons with which they make contact, it can be seen that a number of biochemical studies can be made on these cells before and after electrophysiological experiments.

Extracts prepared from four cells of either GSCs or posterior buccal cells (non-amine-containing buccals) of *Helix pomatia* were sufficient to produce a single satisfactory micro-chromatogram (Briel *et al.*, 1971; Osborne and Cottrell, 1972a). Examination of more than thirty pairs of chromatograms showed that it is indeed possible to obtain consistent results. Analysis of ^{14}C–dansyl derivatives in the GSCs (Fig. 47) and buccal cells (Fig. 48) shows clearly that only the GSCs contain any serotonin. The amine occurs in two forms: dansyl-N-serotonin (spot 25) and dansyl-bis-serotonin (spot 26). Moreover, the GSCs alone contain 5-hydroxyindole (spot 14), 5-hydroxytryptophan (5-HTP) (spot ?) and an unknown substance (spot 5), which may be a metabolite of serotonin (Briel *et al.*, 1971).

Generally, the distribution of substances which react with ^{14}C–dansyl-chloride is dis-

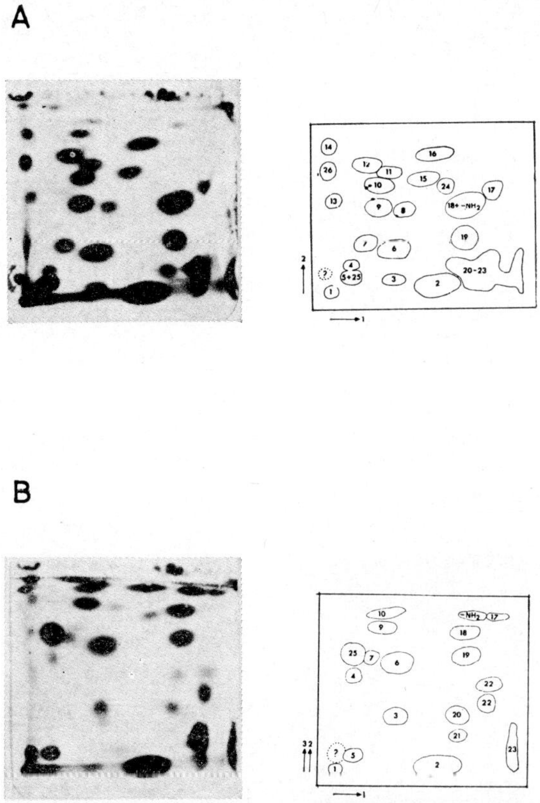

FIG. 47. Autoradiograms of microchromatograms of extracts of GSCs from *Helix pomatia* which had been reacted with ^{14}C–dansyl-chloride, developed in either two (A) or three (B) solvent systems in order to separate most dansyl derivatives. The direction of chromatography is indicated by the arrows. 1st direction, water/formic acid (100:3), 2nd direction, benzene/acetic acid (9:1), 3rd direction, ethylacetate/methanol/acetic acid (20:1:1). Identification keys are presented to assist in the interpretation of dansyl derivatives on the microchromatograms: i.e. 1, starting point; 2, dansyl-OH; 3, N-tyrosine; 4, tryptophan; 5, unknown; 6, bis-ornithine; 7, bis-lysine; 8, methionine; 9, phenylalanine; 10, leucine; 11, isoleucine; 12, unknown; 13, bis-tyrosine; 14, 5-hydroxyindole; 15, valine; 16, proline; 17, ethanolamine; 18, alanine; 19, glycine; 20, glutamic acid; 21, aspartic acid; 22, glutamine, serine and threonine; 23, arginine, ϵ-lysine, cystine and α-amino-histidine; 24, GABA; 25, N-serotonin; 26, bis-serotonin; ?, 5-hydroxytryptophan.

similar in each cell type, although the buccal cells contain more methionine (spot 8) and ornithine (spot 6) but less glycine (spot 19) than the GSCs. These results could only be obtained by counting the radioactivity associated with each spot on a number of microchromatograms of cell extracts. The higher levels of methionine in the buccal cells (Osborne and Cottrell, 1972a) are of particular interest in view of the recent report by Malyuk and Maiskii (1970), who showed by autoradiography that certain neurons in the buccal ganglia of *Planorbis corneus*, which may be analogous to the buccal cells of *Helix pomatia*, specifically take up radioactive methionine.

If GSCs were stored at -20°C in an alkaline buffer for 3–5 days before dansylation and chromatography, the intensity of the serotonin spots was reduced; none was visible at all in extracts which had been stored for 2 weeks (Osborne and Cottrell, 1972a). This loss of serotonin spot is consistent with the known lability of the amine at an alkaline pH. GSCs

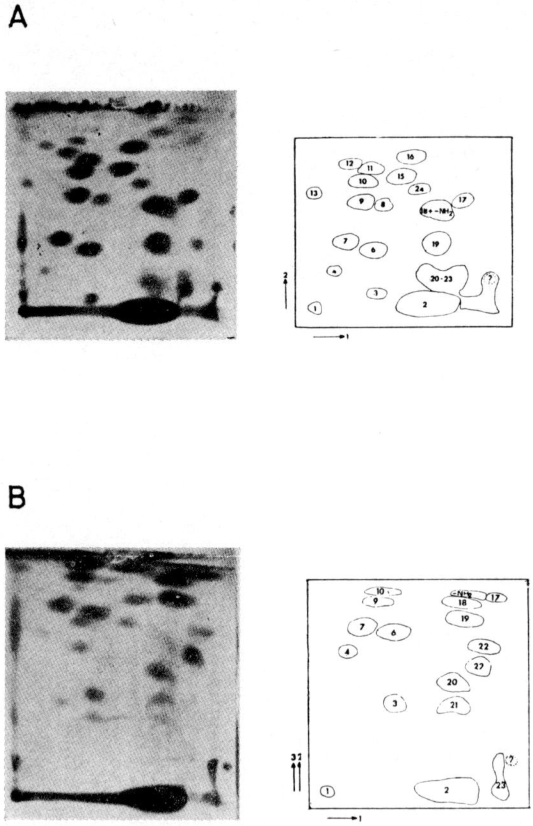

FIG. 48. Autoradiograms of microchromatograms from extracts of buccal neurons (posterior giant buccal cells) from *Helix pomatia* after having been reacted with ^{14}C–dansyl-chloride and separated in two (A) and three (B) solvent systems. The spot numbers on the identification keys and the conditions of chromatography are as explained in Fig. 47.

dissected after staining with methylene blue in order to obtain a purer sample also indicated a reduction of the serotonin content (Briel *et al.*, 1971). It seems therefore possible that methylene blue might indeed cause a 'release' of chemicals from cells.

(ii) *Serotonin in the GSCs,* in vivo *studies*

The content of endogenous serotonin in the GSCs was measured in the following way: extracts of the cells were reacted with ^{14}C–dansyl-chloride and the radioactive derivatives separated by microchromatography. The radioactivity associated with the serotonin spots was scraped from the chromatograms and related to standard chromatographed concentrations of ^{14}C–dansyl-serotonin. From a number of experiments like this it was estimated that a single GSC contains 1.1 ng serotonin (Osborne, 1972a). Another type of microchromatographic procedure, not involving the use of dansyl-chloride, has substantiated this (see page 44).

In order to study the *in vivo* synthesis of serotonin in the GSC, the central nervous system of the snail (CNS) can be perfused with ^{14}C–5-HTP for varying periods of time (Osborne,

1972a). Extracts from the GSCs and the buccal cells are then both reacted with unlabelled dansyl-chloride, and the dansyl derivatives chromatographed. The positions of dansyl-N–^{14}C–serotonin and dansyl-bis–^{14}C–serotonin in each sample are analysed for radioactivity, and the serotonin formed from its precursor is calculated from the specific activity of the perfused ^{14}C–5-HTP. It is clear from the results as shown in Table 20 that only the GSCs form ^{14}C–serotonin from ^{14}C–5-HTP. In no instance is ^{14}C–serotonin found to exist in the non-amine containing cells. The formation of ^{14}C–serotonin in the GSCs at 20°C reaches a peak value after 2 hr perfusion with ^{14}C–5-HTP through the snail's CNS. From a combination of a number of experiments it was calculated that a single GSC formed 0.8 ng/2 hr ^{14}C–serotonin.

TABLE 20. CONTENT OF SEROTONIN PRODUCED IN GSCS AND BUCCAL CELLS OF *Helix Pomatia* AFTER PERFUSING THE SNAILS' CNS WITH ^{14}C–5–HTP

In a single experiment, four cells were used in each instance. The figure shown in parentheses represents the number of experiments performed (from Osborne, 1972a)

Perfusion of ^{14}C–5–HTP (hr)	GSCs ^{14}C–serotonin formed/cell	Buccal cell ^{14}C–serotonin formed/cell
0.5	0.2 ± 0.1 ng (6)	0 ng (6)
1	0.4 ± 0.2 ng (8)	0 ng (8)
2	0.8 ± 0.2 ng (5)	0 ng (8)
3	0.8 ± 0.3 ng (8)	0 ng (8)
4	0.8 ± 0.1 ng (6)	0 ng (6)

Since the volume of the cytoplasm of each GSC is approximately 1 nl, the rate of ^{14}C–serotonin production under experimental conditions was about 2.3 pmole/nl of cell cytoplasm/hr. This is very much lower than the values in the results obtained by Cottrell and Powell (1971) who studied the *in vitro* formation of serotonin in the GSCs, showing it to be 60 pmole/nl of cell cytoplasm/hr. These authors (Cottrell and Powell, 1971) incubated cell extracts in 5-HTP for varying periods of time and the amount of serotonin formed was separated by paper electrophoresis and assayed biologically on an isolated heart preparation. The results showed the buccal cells to lack enzymes capable of producing serotonin from 5-HTP. In contrast, the GSCs produced serotonin at a rate of 60 pmole/nl of cytoplasm/hr. There are at least two possible reasons why the rate of serotonin synthesis *in vitro* is higher than *in vivo*. Firstly, in the *in vivo* perfusion experiments the GSCs do not immediately come into contact with ^{14}C–5-HTP as they do in the *in vitro* studies. A second possible reason is that in the *in vivo* studies the integrity of the neuron is maintained during the synthesis of serotonin, so making available only precise amounts of enzymes for serotonin production.

The *in vivo* metabolism of ^{14}C–tryptophan in the GSCs and buccal cells was also investigated (Osborne, 1973b). In these experiments the central nervous system of individual animals was perfused with snail saline (Meng, 1960) containing ^{14}C–tryptophan (Radiochemical Centre, Amersham, specific activity 57 mCi/mmol, 5 μCi/ml of perfused snail saline) at a rate of 1 ml/2 hr (see Osborne, 1972b) and then rinsed in pure snail saline for 5 min. In some experiments the normally silent GSCs (Cottrell, 1971; Osborne and Cottrell, 1972b) were selectively stimulated via the exterior lip nerves during the perfusion procedure to produce a firing of spikes. Extracts of neurons (either unstimulated or stimulated GSCs or buccal cells) were then reacted with unlabelled dansyl-chloride, and the dansylated derivatives chromatographed. Examination of the autoradiograms revealed that both the buccal cells (Fig. 49A) and GSCs (Figs. 49B and 49C) incorporate radioactive

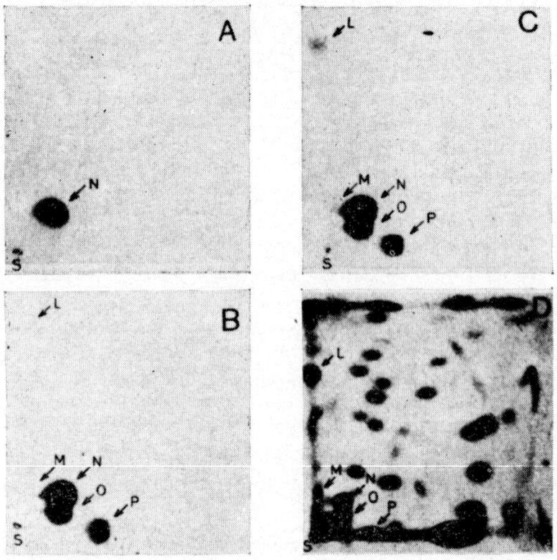

FIG. 49. Autoradiograms of microchromatograms from extracts of non-serotonin-containing buccal cells (A) and serotonin-containing cells, or GSCs (B), and stimulated GSCs (C) from snails which were perfused with ^{14}C-tryptophan and the cell extracts of which reacted with unlabelled dansyl-chloride. D is an autoradiogram of substances (to which standard amounts of tryptophan, serotonin and 5-hydroxy-tryptophan have been added) in the GSCs which react with ^{14}C-dansyl chloride. Microchromatograms were developed in two solvent systems as described for Fig. 47. It can be seen that both the buccal cells (A) and the GSCs (B) incorporate tryptophan (as dansyl-tryptophan (N)), but only the GSCs form serotonin (as dansyl-N-serotonin (O) and dansyl-bis-serotonin (L)) and some 5-hydroxytryptophan (as dansyl-N-5-hydroxytryptophan (P) and dansyl-bis-5-hydroxytryptophan (M)). Electrical stimulation of the GSCs (C) shows an increase in the serotonin content in relation to tryptophan and 5-hydroxytryptophan. S, starting points on chromatograms. (From Osborne, 1973b.)

tryptophan, but only the GSCs metabolise the amino acid to form serotonin and 5-hydroxy-tryptophan. Examination of the autoradiograms from a number of different experiments showed that the GSCs incorporate slightly more radioactivity from the tryptophan into serotonin than the 5-hydroxytryptophan (Osborne, 1973a). Electrical stimulation of the GSCs in separate experiments resulted in a slight increase in the radioactivity associated with all the substances (tryptophan, 5-hydroxytryptophan and serotonin), but greatest in the case of serotonin (Fig. 49C). In experiments where the cells were stimulated for a period of twice as long as in the previous experiments, while the central nervous system was perfused with double the amount of ^{14}C–tryptophan, a slight increase in labelled serotonin and perhaps tryptophan occurred, although the 5-hydroxytryptophan content remained stable.

That tryptophan is metabolized to form serotonin and 5-hydroxytryptophan in the GSC was clearly demonstrated in this study. Since the GSCs are known to convert 5-hy-droxytryptophan into serotonin, these data provided additional proof that 5-hydroxy-tryptophan is the intermediary product in the synthesis of serotonin from tryptophan. It was, however, surprising to find only slightly lower amounts of radioactivity associated with 5-hydroxytryptophan compared to serotonin, when it was observed from earlier experiments that the amount of endogenous 5-hydroxytryptophan in the GSC is very much lower than that of serotonin. Electrical stimulation of the GSCs strong enough to produce a firing of spikes, however, clearly produces an increase in the amount of radioactive serotonin, while the content of labelled 5-hydroxytryptophan remains more or less constant. Unfortunately the radioactivity associated with the metabolites from tryptophan on the individual microchromatograms was very slight, so that it was impossible to measure reliably the increase in the serotonin content following stimulation. It may be worthwhile noting that over a 100% increase in the synthesis of serotonin from labelled tryptophan has been observed in the mid-brain region of the rat after electrical stimulation (Shields and Eccleston, 1972). Doubling the length of stimulation and the amount of perfused tryptophan was found to increase only slightly the radioactivity associated with serotonin. This was surprising, especially since the fluorescence associated with the dansyl-serotonin derivatives on the individual chromatogram was at all times only a little less than that associated with dansyl-tryptophan. It could be that only a proportion of the GSCs' endogenous serotonin incorporates radioactivity even after prolonged stimulation, due to various factors. Firstly, serotonin is known to be localised in a number of compartments within the GSC, including synaptic-type vesicles, lysosome-like particles (Cottrell and Osborne, 1970; Osborne, 1973c) and also the cytoplasm. Each compartment may have very different turn-over rates. Secondly, stimulation of the GSC will result in an increase of the axoplasmic transport of serotonin, which occurs in snails (Osborne and Cottrell, 1970). Moreover, stimulation probably alters the tryptophan concentrations, as this amino acid is known to be reduced on stimulation of the ganglia (Osborne *et al.*, 1972a). Obviously, all these factors have to be considered should the precise rate of metabolism of serotonin within the resting or stimulated cell have to be completely analysed.

Though studies on vertebrate tissue have shown the distribution of the tryptophan hydroxylating enzyme (tryptophan-hydroxylase) to be similar to that of serotonin (Lovenberg, 1972), these studies provide the first direct proof of the localisation of such an enzyme in serotonin-containing cells (GSCs) alone. Moreover, in some experiments snails were pre-injected with *p*-chlorophenylalanine, which is known to affect the activity of tryptophan-hydroxylase in the vertebrates (Koe and Wiessmann, 1966; Jequier *et al.*, 1967), before perfusion of the central nervous system with labelled tryptophan. The results revealed that *p*-chlorophenylalanine, though not interfering with the uptake of tryptophan into the GSCs, almost completely prevented the formation of 5-hydroxytryptophan and serotonin. As well as showing that snail nervous tissue contains an enzyme very similar or identical to tryptophan-hydroxylase, they also demonstrated that the enzyme was only in cells containing serotonin.

When the central nervous system of the snail was perfused with ^{14}C–serotonin (10^{-7} M radioactive serotonin per ml: 1.5 ml per snail over a period of 4 hr), and the GSCs and buccal cells were dissected and eventually subjected to analysis, it was found that only the GSCs accumulated the amine while the buccal cells lacked this ability (Osborne *et al.*, 1973a; Osborne, 1973d). However the accumulated ^{14}C–serotonin within the GSCs was not metabolised, in contrast to what occurs in the central nervous system. 'Brain' tissue not only accumulates serotonin, but also metabolises a part of it to form 5-hydroxyindole–acetic acid. It appeared from these studies that perhaps serotonin-containing somata, e.g. GSCs, though being able to accumulate the amine, lacked the enzymes to activate the substance, while serotonin nerve endings and axons, which occur in abundance in the whole brain, possessed the ability. Furthermore, the possibility existed that one part of a serotonin-containing cell, e.g. cell body, accumulated the amine in a different manner from the others, e.g. serotonin axons and synapses. In order to test these possibilities the effect of imipramine (10^{-4} mol/ml), desipramine (10^{-4} mol/ml) and nialamide (10^{-4} mol/ml) on the accumulation of ^{14}C–serotonin into the GSCs and whole 'brain' tissue was investigated. Imipramine and desipramine are thought to act on the presynaptic membrane of vertebrate nerves by inhibiting the re-uptake of serotonin into the neuron (Glowinski and Axelrod, 1964; Segawa and Kuruma, 1968), while nialamide inhibits MAO which is involved in the breakdown of serotonin. The results of such a study are shown in Table 21, where it can be seen that imipramine, desipramine and nialamide had a greater effect upon the accumulation of serotonin in brain tissue than in the GSCs. The inhibition produced by imipramine and desipramine argues in favour of the idea that there are different accumulation mechanisms associated with the cell body and other parts of a serotonin-containing neuron. Since nialamide caused no significant increase in the accumulated serotonin in the GSCs, while very much more amine accumulated in the nervous tissue, it would appear that MAO is present in the brain but absent from its serotonin-containing neuronal somata. It may also be deduced from the data that serotonin in the snail is activated in two ways: enzymatic oxidisation and re-uptake into synaptic terminals.

TABLE 21. EFFECT OF IMIPRAMINE (10^{-4} mol/ml), DESIPRAMINE (10^{-4} mol/ml) AND NIALAMIDE (10^{-4} mol/ml) ON THE ACCUMULATION OF ^{14}C–SEROTONIN INTO THE GSCs OF *Helix pomatia*

The figure in brackets represents the number of experiments. Also indicated are the mean values ± S.E.M.

	Accumulation of ^{14}C–serotonin into GSCs	Accumulation of ^{14}C–serotonin into brain tissue
% Inhibition of ^{14}C–serotonin accumulation caused by imipramine	24 ± 6% (5)	51 ± 4% (5)
% Inhibition of ^{14}C–serotonin accumulation caused by desipramine	18 ± 3% (4)	40 ± 5% (4)
% Increase of ^{14}C–serotonin accumulation caused by nialamide	3 ± 1% (5)	48 ± 5% (5)

(iii) *Effect of electrical stimulation on the metabolism of GSCs*

In these experiments the snail brain is perfused with labelled glucose or glutamic acid (at a rate of 1 ml per 3 hr) and the GSCs are rapidly dissected to examine the metabolism of these substances (Osborne, 1972b). The effect of electrical stimulation of the GSCs on their glucose and glutamic acid metabolism was also studied. The GSCs are stimulated via the exterior lip nerves (see Fig. 48), the parameter of stimulation being such that it produces a firing of spikes in the normally silent GSCs (Osborne and Cottrell, 1972b). The dissection of a single GSC from a ganglion takes less than 5 min so that it seems unlikely that further metabolism or redistribution of the labelled substances occurs during this time. The extracts from fourteen GSCs are then reacted with unlabelled dansyl-chloride and chromatographed, after which autoradiograms are prepared to examine radioactive metabolites formed from the labelled glucose or glutamic acid. Figure 50 shows that the GSC metabolises ^{14}C–glucose to form a number of substances. Any unmetabolised

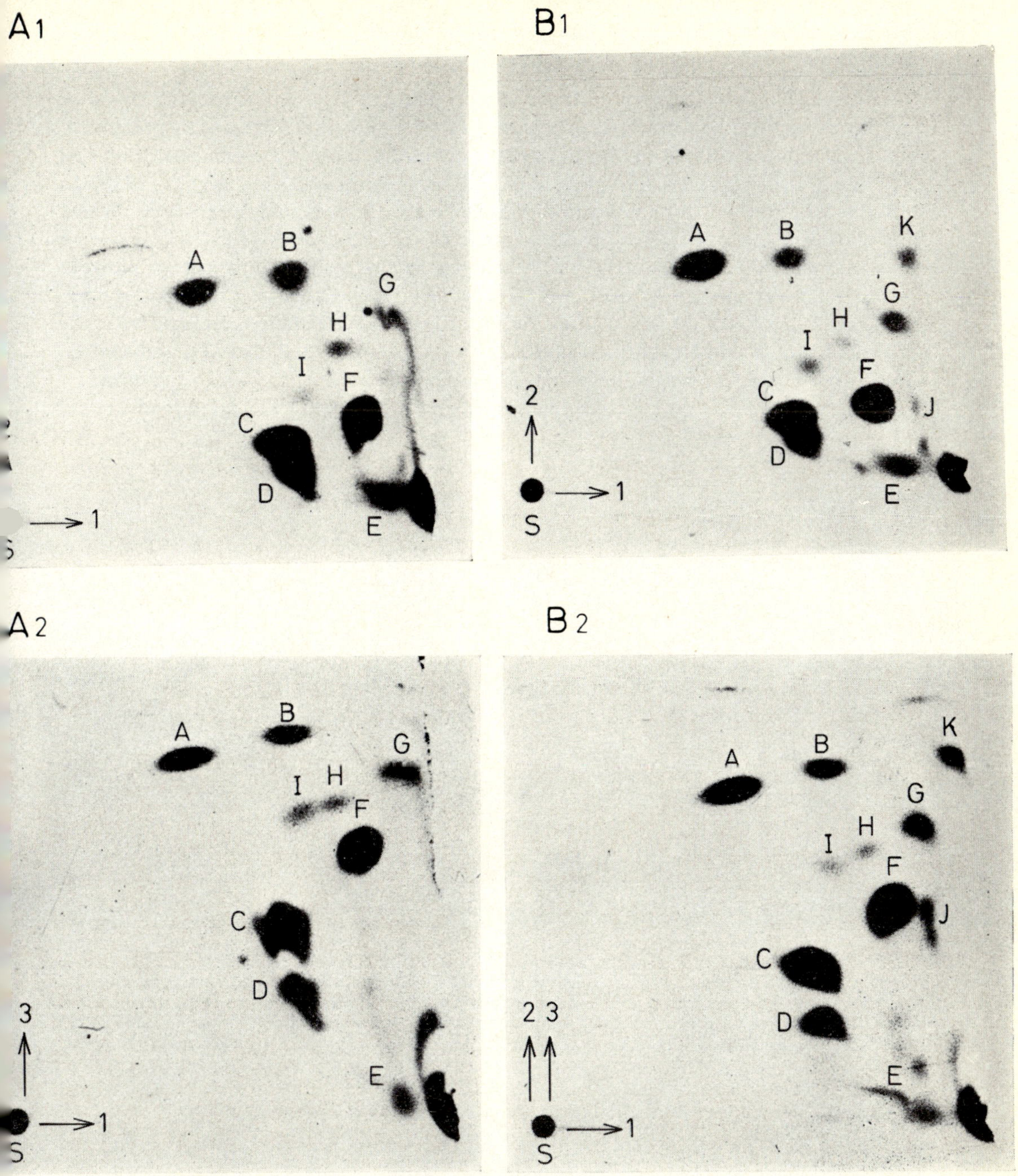

FIG. 50. Autoradiograms of microchromatograms of extracts from identified snail neurons (GSCs) dissected from animals which had been perfused with ^{14}C-glucose and reacted with unlabelled dansyl-chloride. The chromatographic conditions were as described in Fig. 47. Metabolites from ^{14}C-glucose which occur in the GSCs are shown in A1 and A2. The substances are identified as alanine (spot B), glutamic acid (C), aspartic acid (D), glutamine (F) and unknowns (spots A, H, I, G, E). The nature of separation of spot E after a third chromatography (A2) indicates the occurrence of histidine (as dansyl-α-amino-histidine) and arginine as part of this substance. B1 and B2 show the effect of electrical stimulation on the metabolism of ^{14}C-glucose in the GSCs. The concentration of spot A (unknown) has increased, while spot B (alanine) and spot H (unknown) have decreased. In addition, two unknown metabolites appear: spots K and J, which can be seen in chromatograms developed in either two (B1) or three (B2) solvent systems. (From Osborne, 1972b.)

^{14}C–glucose cannot be localised on the microchromatograms, as this substance runs with the solvent fronts. The formation of alanine (spot O), glutamine (spot S), aspartic acid (spot L), glutamic acid (spot M) and the absence of GABA is consistent with data published about the *in vitro* metabolism of glucose in snail nervous tissue (Osborne *et al.*, 1971). In addition, a number of unknown substances (spots N, Q, P, R and V) are also formed from glucose. Especially interesting is the unknown spot N, since this substance occurs in higher amounts than alanine (O), which is normally the predominant metabolite from glucose in nervous tissue (Osborne *et al.*, 1971). The effect of electrical stimulation upon the metabolism of glucose is also shown in Fig. 50. In this case there is an increase in the amount of spot N and a slight decrease in the concentration of alanine (O) and spot Q. Two further unknown metabolites (spots T and U) are also formed in high quantities.

Figure 51 shows that electrical stimulation and perfusion with ^{14}C–glutamic acid could

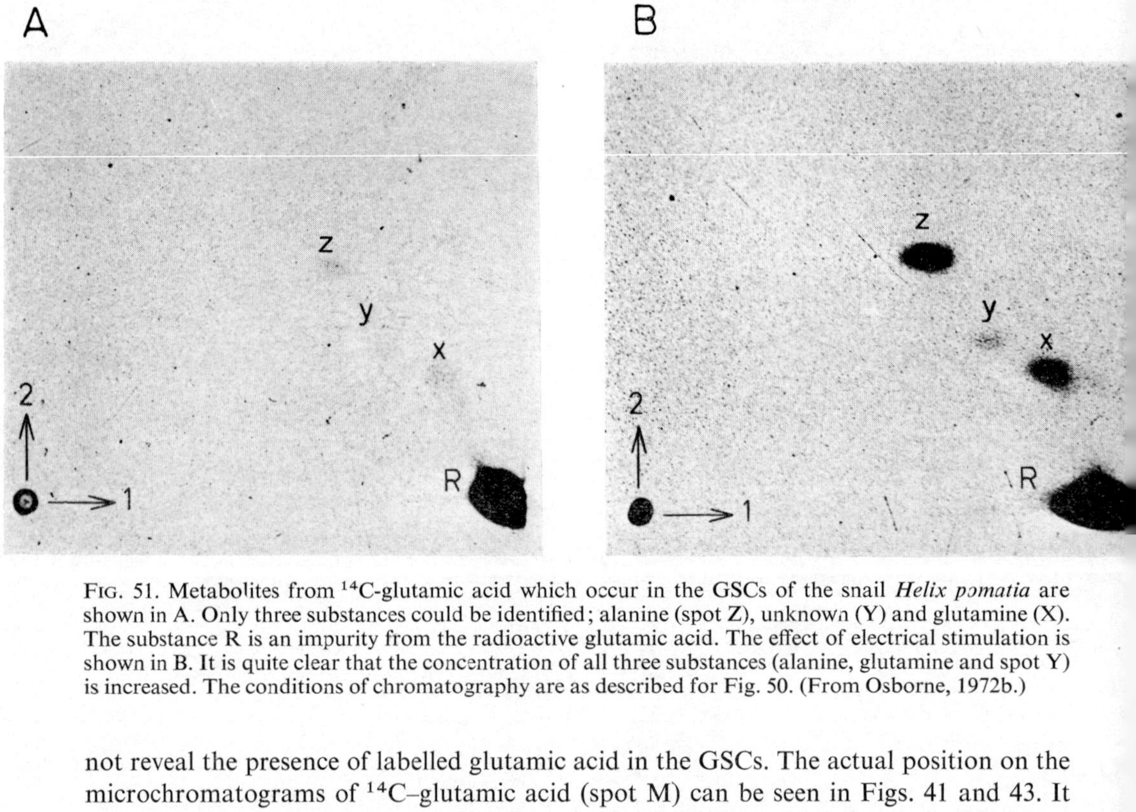

FIG. 51. Metabolites from ^{14}C-glutamic acid which occur in the GSCs of the snail *Helix pomatia* are shown in A. Only three substances could be identified; alanine (spot Z), unknown (Y) and glutamine (X). The substance R is an impurity from the radioactive glutamic acid. The effect of electrical stimulation is shown in B. It is quite clear that the concentration of all three substances (alanine, glutamine and spot Y) is increased. The conditions of chromatography are as described for Fig. 50. (From Osborne, 1972b.)

not reveal the presence of labelled glutamic acid in the GSCs. The actual position on the microchromatograms of ^{14}C–glutamic acid (spot M) can be seen in Figs. 41 and 43. It therefore appears as if the neurons lack the capacity to take up glutamic acid. However, an unknown contamination substance (spot D), which is normally present as an impurity in the used ^{14}C–glutamic acid, is found in the giant neuron. Also present are small amounts of alanine (spot B), glutamine (A) and an unknown substance (spot C). It is not clear whether these three substances are formed from glutamic acid in the surrounding glial and then taken up by the cell, or whether the rate of metabolism of the labelled glutamic acid is very much quicker than the incorporation of the substance into the neuron. If the latter is the case it would mean that the cell contains a number of pools of glutamic acid, since this amino acid exists in high amounts within the GSC (see Osborne and Neuhoff,

1973a). The mechanism of metabolism is probably complicated by the known action of glutamic acid on the cell itself (Cottrell *et al.*, 1972). From Fig. 51 it is clear that electrical stimulation of the cell results in an increased formation of alanine (B), glutamine (A) and substance C. Unlike *in vitro* studies with whole brain tissue of *Helix pomatia*, aspartic acid and GABA were not detected.

It is evident from such studies that *in vivo* metabolism of glucose and glutamic acid in the GSC is different from that observed to occur in the whole snail brain. The major metabolite from glucose is glutamic acid and not alaninc. It is possible that some of the unknown metabolites from glucose are intermediates from the citric acid cycle or from lactate, since the precise nature of all spots localised on the microchromatograms is not yet known. Citric acid cycle intermediates, lactate and other substances have been demonstrated to incorporate more radio carbon than the amino acids from ^{14}C–glucose or ^{14}C–glutamic acid in whole brain of *Helix* (Kerkut *et al.*, 1965; Bradford *et al.*, 1969). However, the presence of a single unidentified substance only in neurons perfused with ^{14}C–glutamic acid would seem to argue against the idea that the unknown substances are intermediated from the citric acid cycle and lactate. It must nevertheless be stressed that the known physiological action of glutamic acid on the cell somata probably affects the metabolism of the neuron (Cottrell *et al.*, 1972).

While it is presumed in experiments such as this that ^{14}C–glucose and ^{14}C–glutamic acid are metabolised in the GSCs, there is no direct proof that this process does not occur in the surrounding tissue and that the metabolites are not taken up by the GSCs. It is clear, however, that stimulation of the cell strong enough to elicit a firing of spikes gives rise to an increased formation of substances from labelled glucose and glutamic acid. The unknown compounds (spots T and U) formed from ^{14}C–glucose following electrical stimulation could be of special interest. These two substances may have important physiological roles in the functioning of neurons, perhaps regulating the firing frequency. Further studies are required to identify these and other unknown substances in order to analyse their functions in the nervous system.

(iv) *Giant neurons in the suboesophageal ganglia of* Helix pomatia

Six different identifiable cells in the suboesophageal ganglia of *Helix pomatia* were analysed (Osborne *et al.*, 1973b). The position of each of the neurons is indicated in Fig. 52. Three of the neurons are in the left parietal ganglion, two in the right parietal ganglion, and one in the visceral ganglion. Analysis of ^{14}C–dansyl derivatives showed that two of the six neurons (cells A and B) contain serotonin (see Table 22 and cf. Fig. 53). All six cells studied have similar amino acid patterns. It must be added that the six cells analysed are not a representative selection of neurons from the snail brain. They are all giant neurons with a diameter in excess of 100 μ. Cells A, B, D, E and G correspond to the neurons designated 5, 7, 9, 8 and 10 by Kerkut *et al.* (1970), while cell F is similar in position to neuron 9 as termed by Gainer (1972). From the reports of these authors it would appear that only cells A and B are active, while the other four neurons are inactive or silent. The possibility exists, therefore, that there is a link between the amine content and its electrophysiological activity, as only the A and B cells contain serotonin. However, electrophysiological experiments on the GSC of *Helix pomatia* have shown these serotonin-containing cells to be electrophysiologically silent (Osborne and Cottrell, 1972b; Cottrell, 1970a, b).

A comparison of the amino acids in each of the six neurons reveals a great similarity;

TABLE 22. PERCENTAGE COMPOSITION OF DANSYLATED SUBSTANCES IN SIX IDENTIFIABLE GIANT NEURONS IN THE SUBOESOPHAGEAL GANGLIONIC MASS OF *Helix pomatia*

The position of each neuron is shown in Fig. 52. ± S.E.M. for six experiments (from Osborne *et al.*, 1973b)

Spot no.	Substance	Cell A	Cell B	Cell D	Cell E	Cell F	Cell G
1	Starting point	—	—	—	—	—	—
2	Dansyl-OH	—	—	—	—	—	—
3	Dansyl-N-tyrosine	0.60 ± 0.06	0.57 ± 0.04	0.65 ± 0.03	0.79 ± 0.06	0.89 ± 0.06	0.58 ± 0.04
4	Dansyl-taurine	0.24 ± 0.03	0.34 ± 0.06	0.26 ± 0.02	0.30 ± 0.03	0.38 ± 0.06	0.38 ± 0.03
5	Dansyl-N-serotonin	0.46 ± 0.04	0.48 ± 0.07	abs.	abs.	abs.	abs.
6	Dansyl-tryptophan	1.29 ± 0.19	0.98 ± 0.17	0.86 ± 0.14	0.88 ± 0.15	0.88 ± 0.09	0.81 ± 0.11
7	Dansyl-bis-lysine	2.65 ± 0.32	3.02 ± 0.42	2.50 ± 0.45	2.61 ± 0.49	2.98 ± 0.55	2.51 ± 0.61
8	Dansyl-bis-ornithine	7.20 ± 0.66	10.58 ± 0.71	6.47 ± 0.73	6.57 ± 0.59	8.46 ± 0.61	10.12 ± 0.63
9	Dansyl-methionine	0.50 ± 0.04	1.06 ± 0.09	0.76 ± 0.05	0.81 ± 0.06	0.70 ± 0.06	1.44 ± 0.10
10	Dansyl-phenylalanine	3.10 ± 0.36	3.88 ± 0.46	5.59 ± 0.39	5.03 ± 0.41	5.82 ± 0.31	4.98 ± 0.32
11	Dansyl-bis-histidine	0.37 ± 0.04	0.41 ± 0.06	0.33 ± 0.07	0.36 ± 0.09	0.43 ± 0.05	0.33 ± 0.05
12	Dansyl-unknown (U_1)	0.69 ± 0.05	0.60 ± 0.04	0.76 ± 0.09	0.64 ± 0.11	0.85 ± 0.12	0.55 ± 0.05
13	Dansyl-bis-tyrosine	1.93 ± 0.11	2.40 ± 0.21	2.40 ± 0.17	2.27 ± 0.21	1.79 ± 0.09	2.13 ± 0.17
14	Dansyl-bis-serotonin	0.42 ± 0.05	0.47 ± 0.05	abs.	abs.	abs.	abs.
15	Dansyl-5-hydroxyindole	1.13 ± 0.09	1.83 ± 0.11	1.59 ± 0.14	1.90 ± 0.13	1.66 ± 0.17	2.27 ± 0.11
16	Dansyl-unknown (U_2)	8.90 ± 1.11	8.47 ± 1.39	7.12 ± 0.91	7.50 ± 1.12	8.87 ± 1.23	8.10 ± 0.99
17	Dansyl-isoleucine	1.75 ± 0.13	2.19 ± 0.12	2.74 ± 0.09	2.39 ± 0.09	2.20 ± 0.07	1.66 ± 0.09
18	Dansyl-leucine	3.91 ± 0.31	6.13 ± 0.29	4.98 ± 0.31	4.22 ± 0.33	3.61 ± 0.36	6.59 ± 0.32
19	Dansyl-valine	2.81 ± 0.28	2.23 ± 0.22	2.16 ± 0.36	3.01 ± 0.28	2.85 ± 0.22	2.38 ± 0.27
20	Dansyl-GABA	0.31 ± 0.05	0.29 ± 0.06	0.37 ± 0.05	0.30 ± 0.04	0.34 ± 0.07	0.27 ± 0.03
21	Dansyl-alanine	19.57 ± 1.43	18.66 ± 1.72	17.74 ± 1.33	16.88 ± 1.27	17.44 ± 1.81	16.32 ± 1.39
22	Dansyl-ethanolamine	2.47 ± 0.24	1.11 ± 0.27	1.83 ± 0.28	1.91 ± 0.31	1.62 ± 0.22	1.41 ± 0.21
23	Dansyl-glycine	10.49 ± 1.08	9.10 ± 1.12	10.95 ± 1.32	10.19 ± 0.98	10.34 ± 0.97	8.83 ± 1.21
24	Dansyl-aspartic acid	2.85 ± 0.41	2.33 ± 0.46	2.46 ± 0.51	2.66 ± 0.38	2.90 ± 0.39	2.10 ± 0.41
25	Dansyl-glutamic acid	7.90 ± 1.01	6.43 ± 1.06	9.13 ± 1.07	9.79 ± 1.1	9.20 ± 0.91	7.21 ± 0.81
26	Dansyl-glutamine, dansyl-serine, dansyl-asparagine and dansyl-threonine	5.41 ± 0.92	5.25 ± 1.10	5.95 ± 0.98	6.60 ± 0.99	5.42 ± 1.02	6.80 ± 1.11
27	Dansyl-arginine, dansyl-ϵ-lysine, dansyl-α-amino-histidine and dansyl-cystine	9.12 ± 1.01	7.58 ± 1.12	7.93 ± 1.21	7.80 ± 0.91	6.17 ± 1.01	6.80 ± 1.04
28	Dansyl-proline	3.73 ± 0.34	3.99 ± 0.42	4.57 ± 0.61	4.61 ± 0.59	4.14 ± 0.72	3.65 ± 0.61
29	Dansyl-unknown (U_3)	0.61 ± 0.05	0.32 ± 0.05	0.65 ± 0.07	0.79 ± 0.04	0.89 ± 0.06	0.58 ± 0.05
30	Excess ^{14}C–dansyl-chloride	—	—	—	—	—	—
31	Impurities of ^{14}C–dansyl-chloride	—	—	—	—	—	—

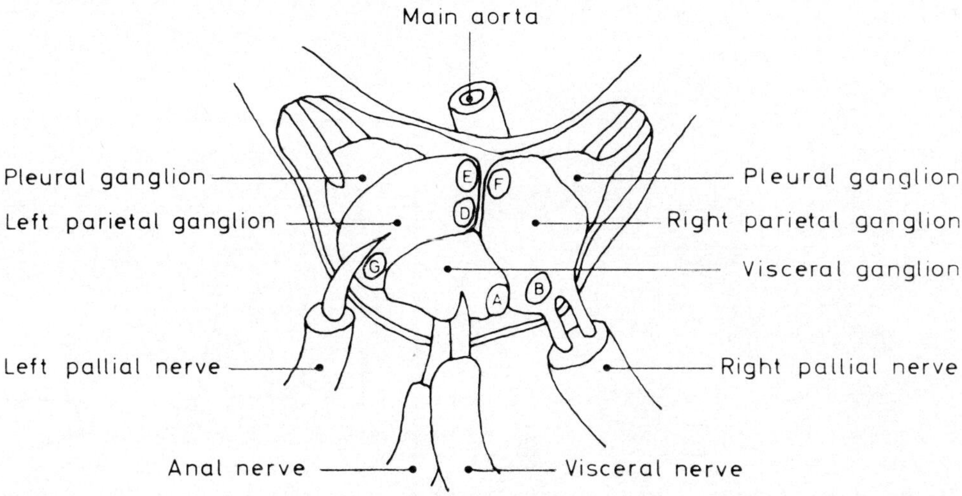

FIG. 52. A diagram of *Helix pomatia* suboesophageal ganglionic mass (dorsal view) to show the position of six identifiable giant cells, analysed for their amine and amino acid contents. (See Table 22 for actual analysis.)

the most predominant amino acids include glutamic acid, glycine, alanine and aspartic acid (see Table 22). The evidence for considering glutamic acid to have a transmitter function in the snail is quite strong and comes mainly from the work of Kerkut *et al.* (1965, 1970) and Cottrell *et al.* (1972). In contrast to this there is little proof that glycine, alanine and aspartic acid have transmitter functions in the snail. GABA, a transmitter substance in a number of situations (Krnjević, 1970; Hebb, 1970; Curtis and Johnston, 1970; Werman, 1971), but probably not in the snail, occurs in all cell types, though in low amounts. The other amino acids: tryptophan, lysine, asparagine, ornithine, phenylalanine, histidine, leucine, isoleucine, tyrosine, proline, valine, glutamine, serine, threonine, arginine, methionine, cystine, and taurine, none of which is seriously thought to have a transmitter function in gastropods, have a similar distribution in each of the cell types. Ornithine perhaps needs special mention because it occurs in high amounts, especially in cells B and G. More interesting, however, are the slightly lower amounts of phenylalanine, and the slightly higher amounts of tryptophan in the two serotonin-containing cells (cells A and B). Both tryptophan and phenylalanine are involved in the metabolism of serotonin.

Although the amino acids have a similar pattern of distribution in the cell samples, they differ in detail. For example, in two instances (cells B and G) the order of content is ornithine > glutamic acid > glycine, whilst in the other instances (cells A, D, E and F) the order is glutamic acid > glycine > ornithine. The variations in the actual amounts of individual amino acids may or may not reflect functional differences. However, before this can be decided, an examination of the nature of the mechanism in the various cells for the maintenance of relative, constant, steady-state concentrations of the individual amino acids has to be carried out.

Other substances present in all the cell samples are ethanolamine and 5-hydroxyindole. Ethanolamine is involved in the synthesis of phospholipids, and 5-hydroxyindole could be a metabolite in the synthesis of serotonin. However the occurrence of 5-hydroxyindole in all the cells analysed in the suboesophageal ganglia would seem to contradict this idea. It

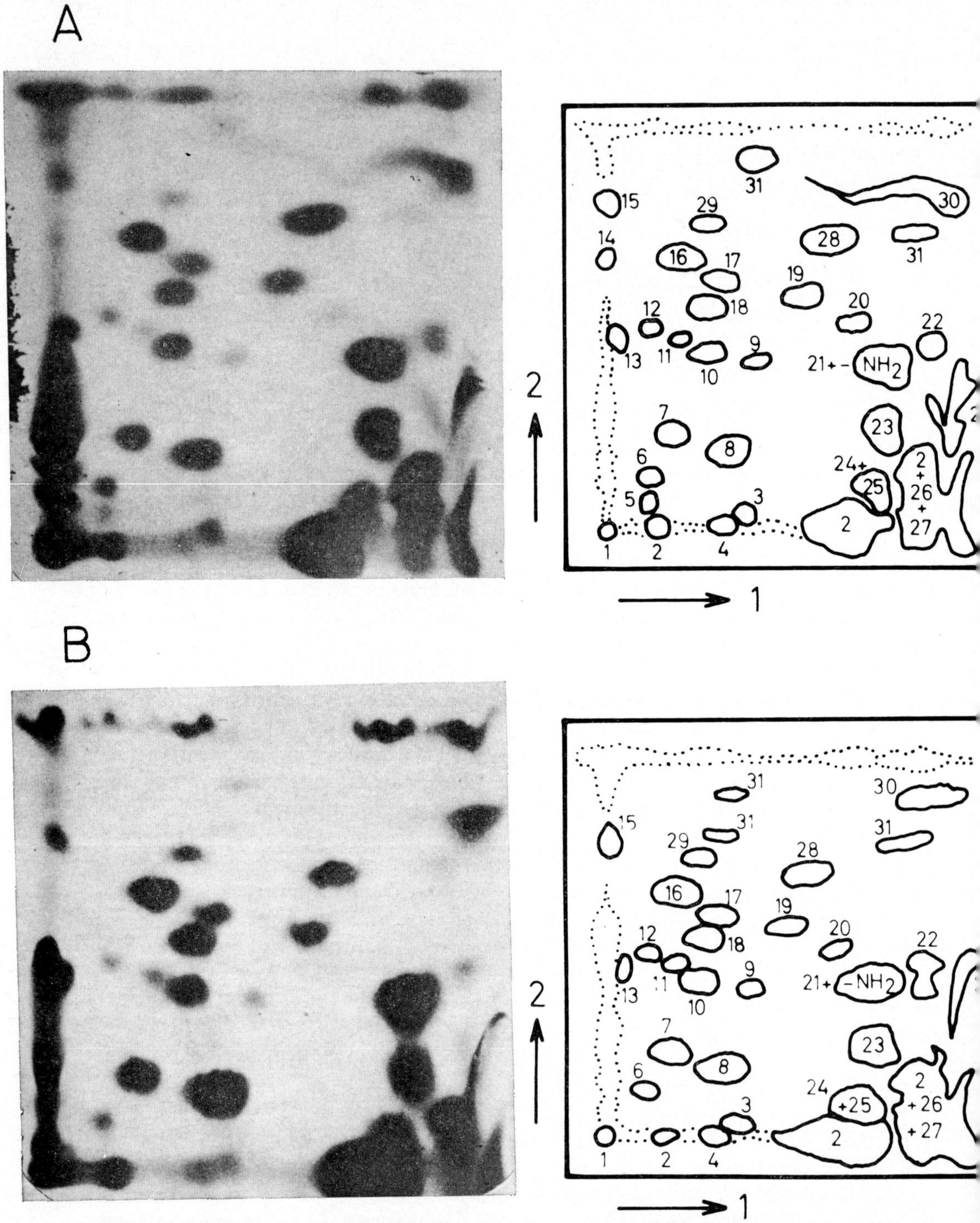

FIG. 53. Autoradiograms of dansylated substances in cell A and cell B present in the suboesophageal ganglionic mass of *Helix pomatia*. The position of each neuron is indicated in Fig. 52, and the content of individual amines and amino acids can be seen by comparing the numbers on the corresponding map with those given in Table 22. Dansyl-peptides are indicated by the dotted lines on the maps. The direction of the chromatography is shown by the arrows and is as described for Fig. 47. (From Osborne *et al.*, 1973b.)

must be pointed out that in these experiments 5-hydroxytryptophan was not analysed, since the main product of this substance formed at pH 10 (dansyl-bis-5-hydroxytryptophan) ran only in the second dimension and was contaminated with dansyl-peptides (see Osborne *et al.*, 1973b). This was because the whole procedure, including dansylation, was performed in a single microglass tube in order to minimise losses often incurred when transferring minute quantities of material, which meant, therefore, that the peptides could not be extracted.

(v) *The Retzius cells in the leech* Hirudo medicinalis

An investigation into the amine and amino acid composition of the Retzius cells of the leech *Hirudo medicinalis* has been made by Osborne *et al.* (1972b). These neurons are globe-shaped with a diameter of up to 80 μ and their constant position and superficial location allow them to be easily distinguished from the surrounding cells in living preparations. One of the most striking features of the analysis is the presence of high levels of serotonin and 5-hydroxytryptophan in the Retzius cells (see Figs. 54 and Table 23). The existence of serotonin in the neurons has previously been proved by fluorescence microscopy (Kerkut *et al.*, 1967), microspectrometry (Rude *et al.*, 1969) and chromatography (Rude *et al.*, 1969). It is worth noting that the abdominal nerve chain also has a fairly high serotonin content (see Table 23), which is not surprising since each ganglion contains at least four large serotonin-containing neurons (Kerkut, 1969).

On the whole, the distribution of amino acids in the Retzius cells and abdominal nerve chain of the leech is similar. Both have approximately the same amounts of tyrosine, lysine, ornithine, leucine, phenylalanine, methionine, GABA and arginine (plus dansyl-α-amino-histidine and dansyl-ϵ-lysine), although the abdominal nerve chain contains greater quantities of histidine, serine, proline, valine, threonine, glutamine and asparagine, alanine, glycine and lesser amounts of aspartic and glutamic acids. The occurrence of very much more tryptophan and less serine in the Retzius cells is of special interest, though the significance of the high levels of serine in the abdominal nerve chain is not yet understood.

There are some difficulties in comparing the contents of the Retzius cells with those of the serotonin-containing neurons to be found in the snail (see above), because the amine and amino acid composition of the snail nervous tissue is different from that of the leech. It does appear, however, that the Retzius cell, like the GSC in the snail, contains greater amounts of serotonin, 5-hydroxytryptophan and tryptophan than its surrounding nerve tissue (note that the 5-hydroxytryptophan levels of the serotonin-containing neurons in the suboesophageal ganglia of *Helix* were not analysed and so cannot be compared with the levels observed in the Retzius cells). In addition, the Retzius cell, again like the snail serotonin-containing neurons, contains large amounts of glycine and alanine, but differs from the snail in having lower amounts than its surrounding nerve tissue. It has also more glutamic and aspartic acid (the latter to a lesser extent) than the snail serotonin-containing neuron. So that although the amine and amino acid content of the serotonin-containing neurons in the leech and snail are similar, there is some variation in the quantities of individual substances. The significance of this variation is complicated by the fact that not only does each animal species vary in the amino acid composition of its nervous system, but that even individual animals can differ within the one species. However, it must be emphasised that studies like these on individual neurons are required to obtain precise biochemical information on the functioning of nervous systems. The comparatively high levels of the serotonin precursors, for example 5-hydroxytryptophan and tryptophan, in the serotonin-containing neurons of the leech and snail may be a characteristic of all such neurons.

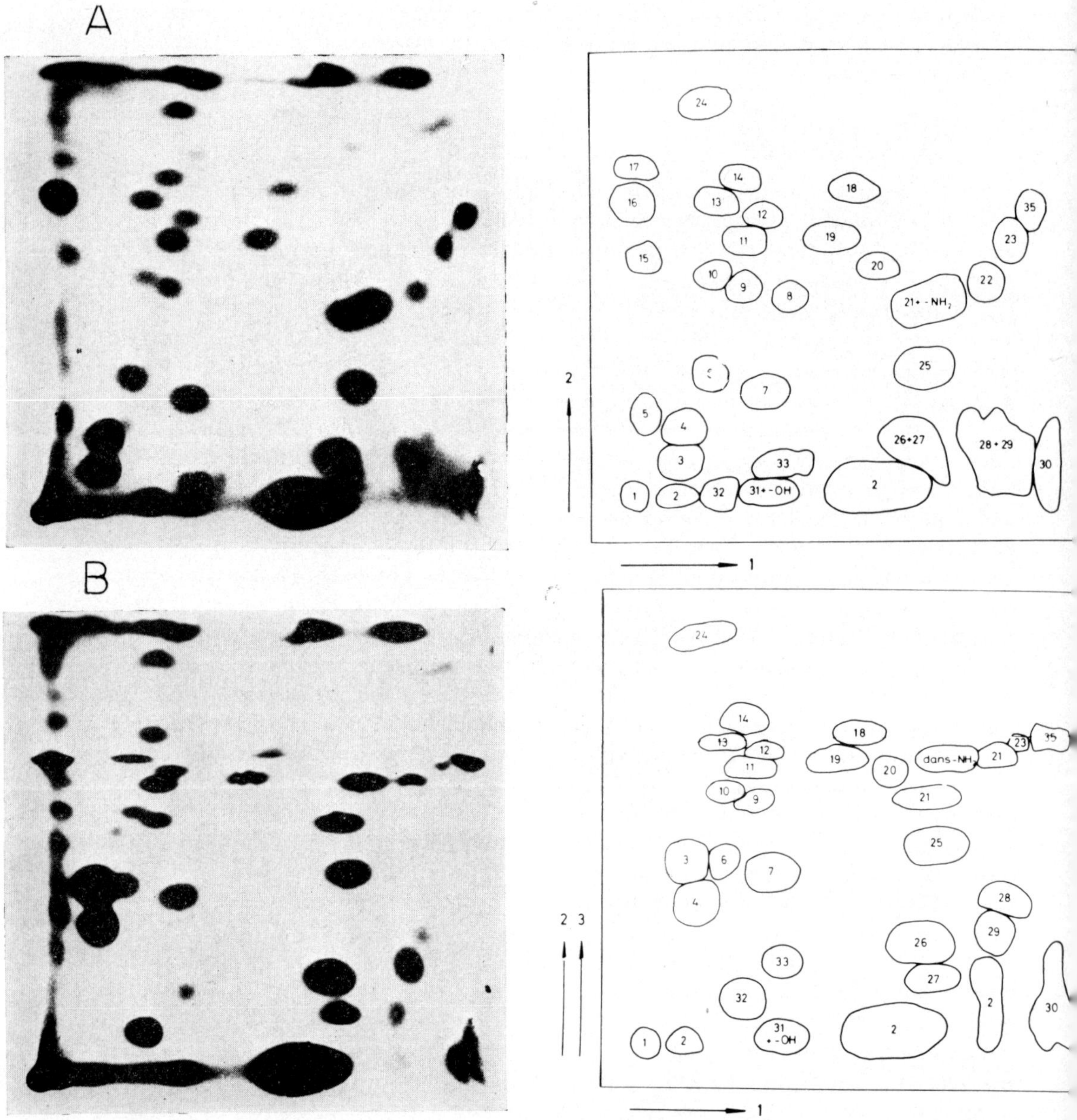

FIG. 54. Autoradiograms from extracts of Retzius cells of *Hirudo medicinalis* after having been reacted with ^{14}C-dansyl chloride and chromatographed in either two (A) or three (B) solvent systems. The conditions of chromatography are as described for Fig. 47. By comparing the spot numbers on each accompanying map with those shown in Table 23, each substance can be identified. The unmarked spots on chromatograms belong either to impurities of ^{14}C-dansyl-chloride or to other unknown compounds. (From Osborne *et al.*, 1972b.)

TABLE 23. PERCENTAGE COMPOSITION OF SUBSTANCES IN THE RETZIUS CELLS AND ABDOMINAL NERVE CHAIN (EXCLUDING THE RETZIUS CELLS) OF *Hirudo medicinalis* WHICH REACT WITH ^{14}C–DANSYL-CHLORIDE

Compare with Figs. 41 and 42 to assist in the identification of each substance on the chromatograms. ± S.E.M. for five experiments (from Osborne *et al.*, 1972b)

Spot no.	Substance	Abdominal nerve chain (free of Retzius cells)	Retzius cells
1	Starting point		
2	Dansyl-OH		
3	Dansyl-N-serotonin	1.8 ± 0.72	10.6 ± 3.52
4	Dansyl-tryptophan	1.24 ± 0.69	5.66 ± 1.04
5	Dansyl-bis-5-hydroxy-tryptophan	absent	1.74 ± 0.71
6	Dansyl-bis-lysine	3.24 ± 0.91	2.80 ± 0.88
7	Dansyl-ornithine	6.6 ± 1.2	4.21 ± 0.73
8	Dansyl-methionine	0.52 ± 0.03	0.42 ± 0.05
9	Dansyl-phenylalanine	1.47 ± 0.71	1.44 ± 0.61
10	Dansyl-bis-histidine	1.67 ± 0.72	0.96 ± 0.42
11	Dansyl-leucine	3.11 ± 0.88	2.81 ± 0.69
12	Dansyl-isoleucine	2.04 ± 0.69	1.20 ± 0.42
13	Dansyl-unknown 1	1.3 ± 0.52	1.41 ± 0.61
14	Dansyl-unknown 2	0.72 ± 0.21	1.29 ± 0.42
15	Dansyl-bis-tyrosine	1.91 ± 0.49	1.34 ± 0.52
16	Dansyl-bis-serotonin	1.39 ± 0.21	4.72 ± 0.91
17	Dansyl-5-hydroxyindole	1.81 ± 0.71	1.01 ± 0.69
18	Dansyl-proline	2.00 ± 0.22	0.99 ± 0.32
19	Dansyl-valine	3.09 ± 0.81	1.92 ± 0.72
20	Dansyl-GABA	0.44 ± 0.03	0.29 ± 0.09
21	Dansyl-alanine	10.66 ± 2.03	5.52 ± 0.88
22	Dansyl-ethanolamine	2.01 ± 0.29	1.21 ± 0.32
23	Dansyl-unknown 3	absent	1.13 ± 0.22
24	Dansyl-unknown 4	absent	1.41 ± 0.41
25	Dansyl-glycine	11.97 ± 2.10	6.61 ± 0.99
26	Dansyl-glutamic acid	9.60 ± 1.92	12.71 ± 2.16
27	Dansyl-aspartic acid	2.72 ± 0.71	4.12 ± 0.72
28	Dansyl-threonine + dansyl-glutamine + dansyl-asparagine	2.20 ± 0.91	1.01 ± 0.09
29	Dansyl-serine	7.71 ± 0.98	2.92 ± 0.51
30	Dansyl-arginine + dansyl-α-amino-histidine + dansyl-ε-lysine	4.11 ± 0.71	5.41 ± 0.88
31	Dansyl-unknown 3 + dansyl taurine	absent	4.81 ± 0.39
32	Dansyl-N-5-hydroxy-tryptophan	0.92 ± 0.09	4.52 ± 1.01
33	Dansyl-N-tyrosine	0.93 ± 0.03	0.88 ± 0.09
34	Dansyl-unknown 6	3.72 ± 0.72	absent
35	Dansyl-unknown 7	6.39 ± 1.21	2.31 ± 0.71
36	Dansyl-unknown 8	0.94 ± 0.31	absent
37	Dansyl-unknown 9	0.84 ± 0.22	absent
38	Dansyl-unknown 10	0.92 ± 0.19	absent

(b) DETERMINATION OF CONTENTS OF SPECIFIC AMINO ACIDS AND TRYPTAMINE

It is generally accepted that the amino acids in the brain are predominantly concerned with metabolic processes and the maintenance of water and ion distribution across cellular membranes. Certain amino acids are also thought to be involved in the transmission of information across specific synaptic junctions (see, for example, Werman, 1971). It is for the latter reason that the content of four amino acids (glycine, glutamic acid, GABA and taurine) in the nervous system of a variety of invertebrates and two fish species was studied (Osborne, 1971, 1972c). The occurrence of an individual amino acid in high concentrations may indicate the possibility of the substance having an additional role (i.e. transmitter substance), although it is only when the turn-over rate is known that a real indication of its function in the nervous system becomes possible. This has to be borne in mind when analysing the contents, although the quantity of an individual amino acid gives an indication of its functional role. Moreover, it has often to be measured before the turnover rate can be studied.

(i) *Glycine and glutamic acid*

Autoradiograms of microchromaograms illustrating the occurrence of ^{14}C–dansylated substances in a variety of animal nerve tissues after chromatography in three solvent systems are shown in Fig. 55A. The positions of glycine and glutamic acid on each of the microchromatograms are indicated by the arrows. The content of each substance in the different nerve tissues is summarised in Table 24. Nervous tissue from all the animals examined contained both glycine and glutamic acid, and in each instance the glycine content was greater than the glutamic acid.

Compared with the vertebrates (plaice and dogfish) and arthropods, where glutamic acid is thought to be a transmitter substance (Krnjević, 1970; Usherwood, 1971; Kravitz *et al.*, 1970), the echinoderms and annelids have very low amounts of this substance. The functional significance of the glutamic acid in the echinoderms is not clear because of a lack of experimental data. The molluscs have slightly more glutamic acid in their nervous system (3.6–6 μmoles/g), which agrees with the observation made about the brain of *Helix* by Kerkut and Cottrell (1962). The uptake and transport of glutamic acid from brain to muscle (Kerkut *et al.*, 1967) and its release following stimulation of the brain have been reported for *Helix*. Together with the known sensitivity of certain neurons in the snail brain to glutamic acid (Kerkut *et al.*, 1965), the data for considering this amino acid as a transmitter substance seem good.

In contrast to the vertebrates, where glycine is thought to function as a transmitter (Hebb, 1970; Aprison *et al.*, 1970), the echinoderms have an even higher amount of this amino acid in their nervous tissue. The significance of this is not clear, although it must be noted that Stephens and Schinske (1961) showed echinoderms to be capable of removing glycine from diluted sea water. Ferguson (1970) followed this up with autoradiographic studies showing clearly that echinoderms can incorporate amino acids from their surroundings. This may indicate that marine animals utilise glycine in a special way compared with terrestrial forms. Another neutral amino acid, taurine, exists in large amounts in nervous tissue of marine animals, while terrestrial forms have little (see below). In this connection, it is interesting to note that the terrestrial mollusc *Limax* has less glycine than its marine forms, *Buccinum* and *Eledone*, which is not the case in the annelids where *Aphrodite* (marine) has less glycine than *Lumbricus* (terrestrial). In addition the locust (terrestrial) has more glycine than the two marine arthropods (*Carcinus* and *Nephrops*) examined.

(ii) *GABA and taurine*

The content of each substance in the different nerve tissues is summarised in Table 25. All nerve tissues examined contained both taurine and GABA to varying extents. The

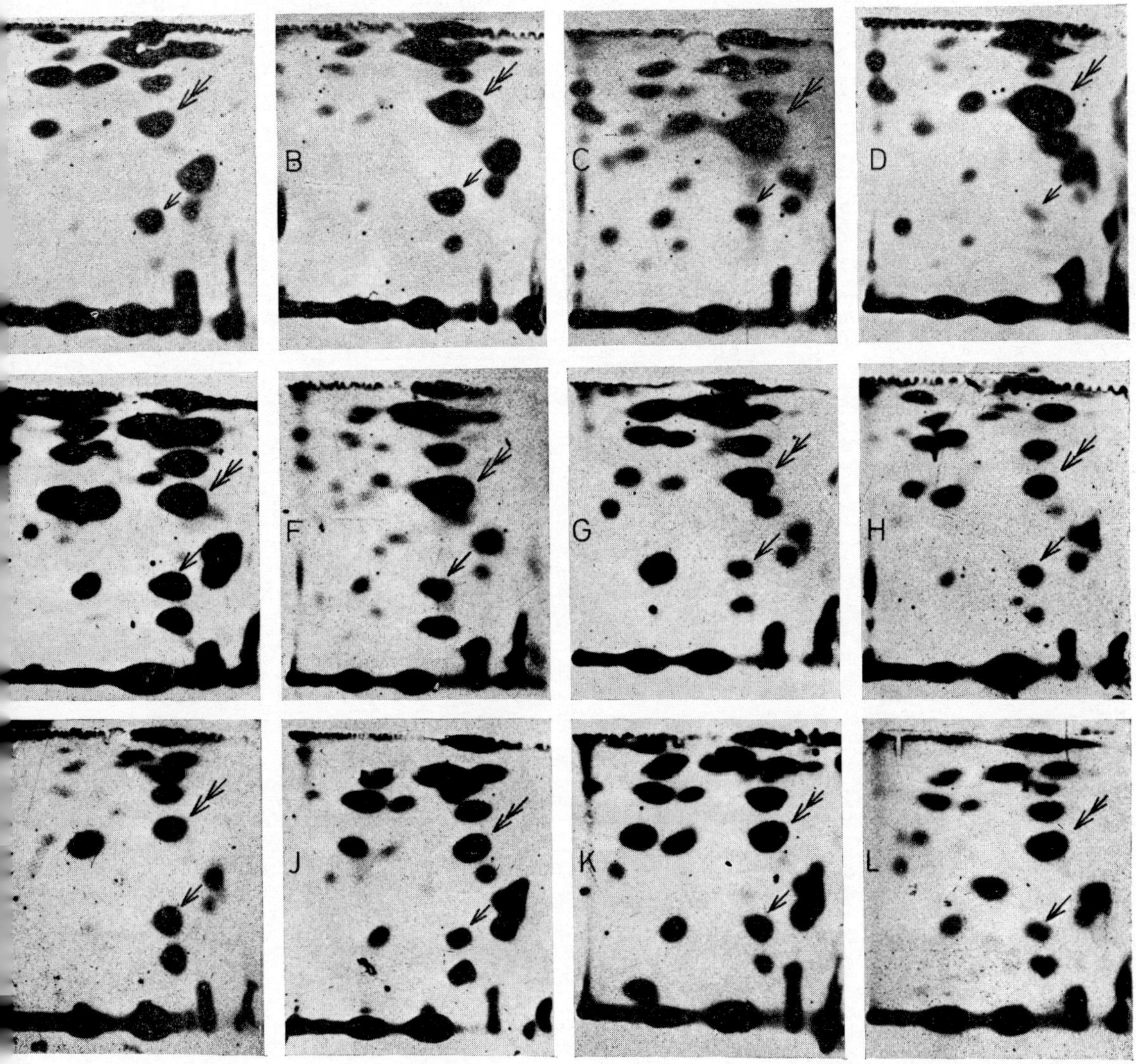

FIG. 55A. Autoradiograms of microchromatograms from nerve extracts of (A) *Pleuronectes platessa*, (B) *Scyllium canicula*, (C) *Echinus esculentus*, (D) *Asterias rubens*, (E) *Locusta migratoria*, (F) *Nephrops norvegicus*, (G) *Carcinus maenus*, (H) *Limax maximus*, (I) *Eledone cirrhosa*, (J) *Buccinum undatum*, (K) *Lumbricus terrestris*, and (L) *Aphrodite aculeata* after having reacted with ^{14}C–dansyl-chloride. The original size of a microchromatogram measured 3 × 3 cm, and was developed in the 1st dimension (horizontal direction) with water/formic acid (100:3 by vol), in the 2nd dimension (vertical direction) with benzene/acetic acid (9:1 by vol) and once more in the 2nd dimension with ethylacetate/methanol and acetic acid (20:1:1 by vol). The positions of glycine (double arrow) and glutamic acid (single arrow) are marked on each chromatogram. (From Osborne, 1972d.)

detection of large amounts of GABA in the dogfish, locust, crab and *Nephrops* nervous system supports the view that this amino acid is an inhibitory transmitter in the vertebrates and arthropods (Krnjević, 1970; Hebb, 1970; Curtis and Johnston, 1970). The functional significance of low amounts of GABA in the nervous systems of annelids, echinoderms and the tunicate (*Ciona*) is not clear because of a lack of experimental data. The presence

TABLE 24. THE CONCENTRATION OF GLYCINE AND GLUTAMIC ACID IN THE NERVOUS SYSTEMS OF THE DOGFISH, PLAICE AND SOME INVERTEBRATES (from Osborne, 1972d)

Animal	Habitat	Group	Tissue	No. of exp.	Glycine content (in μmoles/g wet wt. of tissue)	Glutamic acid content (in μmoles/g wet wt. of tissue)
Pleuronectes platessa (plaice)	Marine	Teleost (boney fish)	Medulla	6	6.1	8.4
Scyllium canicula (lesser spotted dogfish)	Marine	Elasmobranchii (cartilaginous fish)	Medulla	6	10.1	9.3
Echinus esculentus (sea urchin)	Marine	Echinodermata	Radial nerve cord (ectoneural tissue)	6	62.7	2.4
Asterias rubens (starfish)	Marine		Radial nerve cord (ectoneural and hyponeural tissue)	6	73.2	1.5
Locusta migratoria (locust)	Terrestrial	Arthropoda	Brain and ventral nerve cord	6	28.7	25.8
Nephrops norvegicus (Norway lobster)	Marine		Brain and ventral nerve cord	6	25.1	11.2
Carcinus maenas (crab)	Marine		Brain and ventral nerve cord	6	20.1	9.2
Limax maximus (slug)	Terrestrial	Mollusca	Central ganglia	6	4.9	3.6
Eledone cirrhosa (octopus)	Marine		Optic ganglia	6	6.2	6.0
Buccinum undatum (common whelk)	Marine		Central ganglia	6	5.8	4.8
Lumbricus terrestris (earthworm)	Terrestrial	Annelida	Cerebral and suboesophageal ganglia	6	6.6	5.1
Aphrodite aculeata (sea mouse)	Marine		Ventral nerve cord	6	3.9	3.9

TABLE 25. THE CONCENTRATION OF GABA AND TAURINE IN THE NERVOUS SYSTEMS OF THE DOGFISH AND SOME INVERTEBRATES (from Osborne, 1971)

Animal	Habitat	Group	Tissue	No. of exp.	GABA content (in μmoles/g wet wt. of tissue)	Taurine content (in μmoles/g wet wt. of tissue)
Scyllium canicula (lesser spotted dogfish)	Marine	Elasmobranchii (cartilaginous fish)	Cerebellum	4	25	30
Ciona intestinalis (tunicate)	Marine	Urochordata	Cerebral ganglion	4	3	34
Echinus esculentus (sea urchin)	Marine	Echinodermata	Radial nerve cord (ectoneural tissue)	3	0.9	2.9
Asterias rubens (starfish)	Marine		Radial nerve cord ectoneural and hyponeural tissue	4	1.1	3.0
Locusta migratoria (locust)	Terrestrial	Arthropoda	Brain and ventral nerve cord	6	16.8	22.4
Carcinus maenas (crab)	Marine		Brain and ventral nerve cord	5	21.3	49
Nephrops norvegicus (Norway lobster)	Marine		Brain and ventral nerve cord	4	13.6	35
Helix aspersa (garden snail)	Terrestrial	Mollusca	Circumoesophageal ganglia mass	7	0.7	1
Buccinum uudatum (common whelk)	Marine		Circumoesophageal ganglia mass	6	0.9	43.3
Eledone cirrhosa (octopus)	Marine		Optic ganglia	3	1.3	85
Lumbricus terrestris (earthworm)	Terrestrial	Annelida	Cerebral ganglia circumoesophageal commissures, and subpharyngeal ganglia	6	1.2	0.8
Aphrodite aculeata (sea mouse)	Marine		Ventral nerve cord	4	2.6	77

of GABA in the molluscs confirms the findings of Roberts (1964), who showed this amino acid to exist in barely detectable amounts in the brain of an unnamed octopus species. Although it is known that certain cells in the gastropod brain are specifically sensitive to GABA (Gerschenfeld and Lasansky, 1964; Kerkut and Walker, 1961) and that enzymes exist in the brain which are capable of forming GABA from glutamic acid (Osborne *et al.*, 1971), it would seem that not enough evidence is available to suggest that this amino acid serves as a neurotransmitter substance in the molluscs.

Extremely large variations of taurine are known to occur in the vertebrates (Himwich and Agrawal, 1969). It is found in high concentrations in young animals and decreases during the period of maturation (Agrawal *et al.*, 1968). But the functional role of taurine in the vertebrate brain is not clear, although there is some evidence that the level may be affected by a thyroidal function (Ramirez *et al.*, 1966). It has also been shown that the release and uptake mechanisms between GABA and taurine are similar (Davison and Kacmarek, 1971), which supports the idea that taurine can be considered a potential neurotransmitter substance in the vertebrates (Curtis and Johnston, 1970). Large amounts of taurine have previously been discovered in the leg nerves of crabs and lobsters (Koechlin, 1955) and giant nerve fibres of the squid (Lewis, 1952). Koechlin (1955) and Lewis (1952) consider the high concentration of taurine necessary to help balance the high cation concentrations of marine animals. The detection of very large amounts of taurine in *Aphrodite*, *Ciona*, *Buccinum*, *Eledone*, *Carcinus*, *Nephrops* and the dogfish, all of which are marine, compared with the low amounts of the amino acid in the terrestrial earthworm and snail, would seem to be consistent with this view. However, the high taurine concentration in the desert locust is difficult to explain in this context. As far as the marine echinoderms are concerned, it seems probable that their high glycine levels (Osborne, 1972c) fulfill the functional role of taurine in other marine groups.

(iii) *Tryptamine*

Until recently there has been controversy as to whether tryptamine occurs in nervous tissue. The first good evidence for the presence of tryptamine in nervous tissue comes from the studies of Hess *et al.* (1959), though their values were quite high, which was probably due to the fact that tryptophan was carried over in small amounts by their solvent extraction procedure. Since then a number of workers have shown nervous tissue to contain tryptamine, and recently Boulton and Majer (1971), employing high resolution mass spectrometry procedures, obtained unequivocal evidence of tryptamine in rat brain. Because some tryptamine derivatives are known to be hallucinogenic in man (Martin and Sloan, 1970), and since an enzyme capable of N-methylating indolamines has been demonstrated in normal rat brain (Saavendra and Axelrod, 1972), the presence and distribution of tryptamine in the nervous system is of great interest. While analysing the amine and amino acids in snail nervous tissue with the sensitive microdansyl procedure, Osborne (1972c) observed a substance which had the same chromatographic position as authentic dansyl-tryptamine (Osborne and Neuhoff, 1973b). The position of dansyl-tryptamine in relation to some other dansyl metabolites of serotonin is shown in Fig. 55B. Analysis of extracts of nervous tissue from other animals, both vertebrates and invertebrates, revealed a substance to be present which also had the same chromatographic mobility as dansyl-tryptamine. Further proof of the occurrence of tryptamine in the tissue was obtained in the following way. Whole areas on microchromatograms of the various

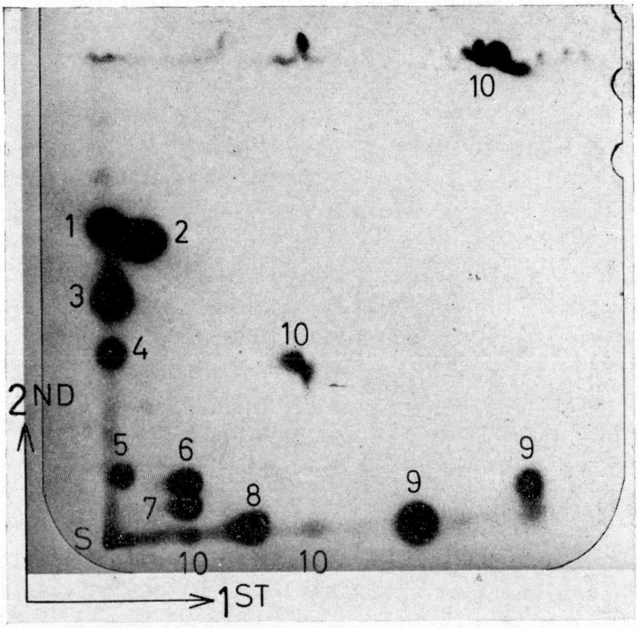

FIG. 55B. An autoradiogram of a microchromatogram (original size 3 × 3 cm) to show the position of ^{14}C–dansyl-tryptamine in relation to some ^{14}C–dansyl metabolites of serotonin. The direction of chromatography is indicated by arrows. 1st dimension, water/formic acid (100:3 by vol) and 2nd dimension, benzene/acetic acid (9:1 by vol). S, starting point; 1, dansyl-bis-serotonin; 2, dansyl-tryptamine; 3, dansyl 5-hydroxyindoleacetic acid; 4, dansyl-5-hydroxytryptophol; 5, dansyl-bis-5-hydroxytryptophan; 6, dansyl-tryptophan; 7, dansyl-N-serotonin; 8, dansyl-N-5-hydroxytryptophan; 9, dansyl-OH; 10, dansyl impurities.

nervous tissues corresponding to the position of dansyl-tryptamine were scraped off and eluted with ethanol. The eluate was then evaporated and applied to new micropolyamide layers and the chromatographic mobility was compared with that of pure dansyl-tryptamine, using five different solvent systems. In all cases they were shown to be the same. Moreover, in some instances the eluate was analysed by fluorometry, and the results from these studies showed the substance to be dansyl-tryptamine.

To measure the content of tryptamine in the various nervous tissues, the spot corresponding to ^{14}C–dansyl-tryptamine on chromatographic plates was isolated and counted in a liquid scintillation spectrometer. The average recovery of added internal standards of tryptamine was found to be 80%; this was taken into account when analysing the results. The following amount of tryptamine was found to be present in the nerve tissues (wet weight) investigated: rat spinal cord 3 nmol/g; mouse brain 1 nmol/g; snail (*Helix pomatia*) brain 1.5 nmol/g; pineal organ of rat 1.8 nmol/g; crab (*Carcinus meanus*) nervous system 0.9 nmol/g. It is of interest to note that the snail brain incubated in ^{14}C–tryptophan or ^{14}C–serotonin showed no incorporation of radioactivity into tryptamine (Osborne and Neuhoff, 1974), and that pretreatment of snails with *p*-chlorophenylalanine or reserpine had no influence upon the tryptamine content. Similar data concerning the tryptamine content in nervous tissue have been found independently by Snodgrass and Horn (1973), who also used labelled dansyl-chloride, together with an initial extraction procedure.

(c) DETECTION OF BIOCHEMICAL CHANGE IN NERVOUS TISSUE OF MINUTE DIMENSIONS

(i) *Effect of electrical stimulation*

As pointed out earlier, the amines and amino acids in nervous tissues may be important not only in general metabolic processes concerned with energy transfer and protein synthesis, etc., but also in the transmission of information across certain synaptic junctions. One approach which might provide clues about the roles of these substances is to study the effects of electrical stimulation on their levels in the brain. For the reasons discussed on page 3, the snail *Helix pomatia* was chosen as the subject for study. An incision was made along the medial dorsal line of the snail to expose the circumoesophageal ganglia. The anterior aorta was canulated and perfused with snail saline (see Osborne, 1972a); the right cerebro-pedal and cerebro-parietal commissures were then cut and those ends which were connected to the suboesophageal ganglia were stimulated electically for 1 hr. Square wave pulses, 15 V in amplitude and of 10 msec duration, were applied at a rate of 3 per sec to the nerves via a radiofrequency unit (Donaldson, 1958) to the Ag–AgCl wires of a plastic electrode.

A comparison of microchromatograms of ^{14}C–dansyl substances in stimulated and unstimulated circumoesophageal ganglia showed that changes occurred in the stimulated animal. The results of several experiments are shown in Table 26. The level of each substance was measured in at least five unstimulated and five stimulated circumoesophageal ganglia, and in some instances ten separate analyses were made. The activities of the separate experiments were summed in order to compare the total counts of the stimulated and unstimulated circumoesophageal ganglia and so establish possible changes in the levels of the different substances. A change of less than 5% was not considered to be of significance. The major changes caused by stimulation were as follows. There was a decrease in the concentrations of tyrosine, tryptophan, ornithine, lysine, phenylalanine, leucine, isoleucine, proline, valine, glycine, asparagine and serine, and perhaps also a slight decrease in serotonin and glutamic acid. In no instance was there a dramatic drop in the level of an amino acid, the greatest reduction being 16% which was recorded for tryptophan. Only three substances showed an increase in level after stimulation. They were alanine, 5-hydroxyindole and an unknown substance (spot 15). The levels of taurine, methionine, GABA, glutamine, threonine, aspartic acid and spot 24 (dansyl-arginine, dansyl-lysine, dansyl-α-amino-histidine and dansyl-cystine) were not influenced by the applied stimulation.

It is of interest to note that Mussini and Morcucci (1962) studied the effect of electro shock on certain amino acids in rat brains and found that three amino acids were profoundly altered; alanine was increased while the serine and threonine concentrations decreased. Alanine is one of the few substances which increases in the snail brain following stimulation. No definite statement can be made about the other two substances because serine has the same chromatographic mobility as asparagine, and threonine as glutamine.

(ii) *LSD-25* (*D-Lysergic acid diethylamide*)

LSD-25, best known for its psychotomimetic properties, is one of the most powerful hallucinogenic compounds known to man (Hoffman, 1961) and is known to influence the serotonin content of the vertebrate brain (Freedman, 1961; Giarman and Freedman,

TABLE 26. PERCENTAGE CHANGE OF DANSYLATED SUBSTANCES IN *Helix pomatia* AFTER ELECTRICAL STIMULATION (from Osborne *et al.*, 1972a)

Spot number	Substance	Number of experiments	Combined counts of single chromatograms from total number of experiments: Brain	Stimulated brain	Percentage change of substance after stimulation
1	Starting point	—	—	—	—
2	Unknown substance	—	—	—	—
3	Taurine	10	9826	9959	1.3 (± 2) increase
4	Dansyl-OH	—	—	—	—
5	N-Tyrosine	10	6996	6563	6.5 (± 3.3) decrease
6	Tryptophan	5	1207	993	16.1 (± 4.1) decrease
7	N-serotonin	5	970	918	5.4 (± 4.2) decrease
8	bis-Ornithine	10	5941	5394	9.2 (± 4.7) decrease
9	bis-Lysine	10	2226	1983	10.9 (± 3.6) decrease
10	Phenylalanine	10	2594	2326	10.3 (± 3.9) decrease
11	Leucine	10	1985	1697	14.5 (± 4) decrease
12	Isoleucine	10	1743	1551	11 (± 3.9) decrease
13	Unknown substance	5	3601	3699	2.7 (± 3) increase
14	bis-Tyrosine	5	1049	1002	4.4 (± 3) decrease
15	Unknown substance	5	1001	1112	11 (± 4.2) increase
16	Proline	10	2688	2397	10.8 (± 4.2) decrease
17	Valine	10	966	876	9.6 (± 2) decrease
18	Methionine	5	854	826	3.3 (± 4) decrease
19	GABA	5	957	991	3.6 (± 3.4) increase
20	Glycine	10	12062	11193	7.9 (± 4.1) decrease
21	Alanine	10	14811	15909	7.4 (± 2.9) increase
22	Glutamine and threonine	5	1681	1629	3.1 (± 2.7) decrease
23	Asparagine and serine	5	1875	1740	7.1 (± 3) decrease
24	Arginine, ε-lysine, α-amino-histidine and cystine	5	8827	8791	0.5 (± 2) decrease
25	Aspartic acid	5	4968	4858	2.2 (± 2) decrease
26	Glutamic acid	5	2568	2426	5.5 (± 1.8) decrease
27	Unknown substance	5	7202	7320	1.6 (± 2) increase
28	5-Hydroxyindole	5	809	1002	23.8 (± 4) increase
29–35	Unknown substances	—	—	—	—

1965). The effect of this drug upon the amines and amino acids from *Helix pomatia* was studied by Osborne and Neuhoff (1972) and Osborne *et al.* (1973d). The anterior aorta of the snail was cannulated just before it enters the central ganglia and perfused either with pure snail saline (Meng, 1960) or with saline containing LSD-25 (10^{-6} g/ml). The perfusion pressure was so arranged that 1 ml of fluid was perfused over a period of 3 hr. In other experiments circumoesophageal ganglion masses were preincubated for 1 hr in snail saline which contained ^{14}C–glucose before exposing the nervous tissue to saline containing LSD-25. This sort of experiment was specifically designed to see whether the drug had any effect on the metabolism of ^{14}C–glutamic acid, ^{14}C–aspartic acid, ^{14}C–alanine or ^{14}C–glutamine, all of which are formed in snail tissue by the metabolism of ^{14}C–glucose (Bradford *et al.*, 1969; Osborne *et al.*, 1971). The main effect of LSD-25 was to deplete completely a substance in the snail brain (see Fig. 56). Only snails obtained just after winter hibernation, i.e. in April and May, have this unknown substance present.

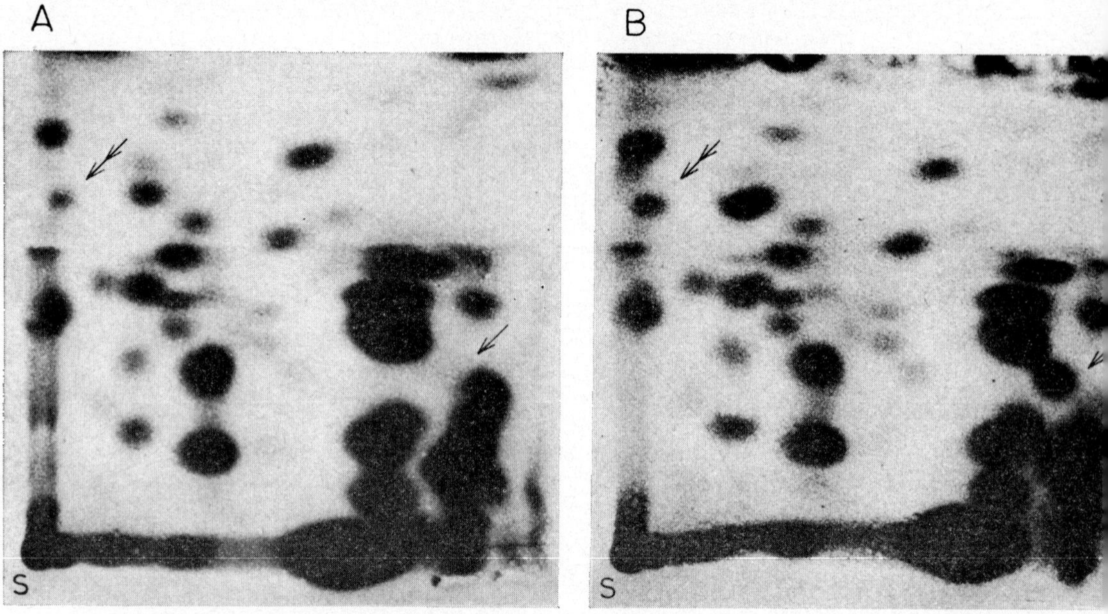

FIG. 56. Autoradiograms of microchromatograms (original size 3 × 3 cm) from extracts of suboesophageal ganglia of LSD-treated (A) and control (B) snails after reaction with ^{14}C–dansyl-chloride. The microchromatogram was developed by ascending chromatography in the 1st dimension (horizontal direction) with water/formic acid (100:3 by vol), in the second dimension (vertical direction) with benzene/acetic acid (9:1 by vol) and again in the 2nd dimension with ethylacetate/methanol/acetic acid (20:1:1 by vol), terminating when the solvent front had migrated a distance of 1.5 cm. The positions of dansyl-bis-serotonin (single arrow) and the substance which is readily depleted by LSD-25 (double arrow) are clearly marked. (From Osborne and Neuhoff, 1972c.)

Other drugs, viz. *p*-chlorophenylalanine, reserpine, and α-methyl-*m*-tyrosine at concentrations of 10^{-6} g/ml had no effect on this substance. The nature of this unknown substance is not clear, though its chromatographic mobility in five different solvent systems is the same as hydroxyproline. However, although examination of the spectra from the dansyl substance showed the unknown to be an amino acid, it could not be confirmed that it was hydroxproline.

The other main effect of LSD-25, even when in high concentrations, was to increase the serotonin level slightly. This is especially interesting in relation to the finding that LSD-25 is known to mimic the excitatory action of serotonin on mollusc heart (Welsh and McCoy, 1957) and muscle (Twarog, 1959) preparations. Also Cottrell (1970b) analysed the effect of LSD-25 on a serotonergic synapse in the snail brain and observed that the drug, at a concentration of 10^{-4} g/ml, blocked the excitatory action of serotonin, whilst at a concentration of 10^{-6} g/ml it potentiated transmission. No alteration in either the content or the turnover rate of glutamic acid, aspartic acid, alanine or glutamine was recorded.

(iii) *Lithium*

Lithium salts, which are effective in the treatment of manic-depressive psychoses (Schou, 1968; Davis and Fann, 1971), have been shown to influence the rate of metabolism of brain neuromediators (Schildkraut *et al.*, 1967; Berl and Clarke, 1972). It is for this reason that the effects of lithium ions on the amine and amino acid content of the snail suboesophageal ganglia were studied (Osborne *et al.*, 1973c). Circumoesophageal ganglia from a number of animals were dissected, allowed to equilibrate in snail saline for 10 min and were then placed individually for 30 min in either 1 ml snail saline (control) or saline where equivalent amounts of lithium ions (0.02–0.05 M) had replaced sodium ions (substituting sodium chloride for lithium chloride). In some experiments circumoesophageal ganglia were pre-incubated in ^{14}C–glucose for 1 hr in order to label aspartic acid, glutamic acid, alanine and glutamine (Osborne *et al.*, 1971) before exposing the ganglia masses to lithium ions (Osborne *et al.*, 1973c).

A slight increase was recorded in the serotonin and tryptophan levels, and a decrease in the glutamic acid, aspartic acid and glutamine (this last substance migrates together with asparagine and threonine on chromatograms). As the concentration of lithium was increased, so these differences became more pronounced. In experiments where the brain was pre-incubated in ^{14}C–glucose, the effect of lithium ions was to stimulate the leakage of radioactive substances into the incubation medium. Examination of these tissues showed that lithium ions released a proportion of ^{14}C–amino acids formed from ^{14}C–glucose depending on the concentration of lithium ions used. The major amino acid to be released was glutamic acid, followed by glutamine and aspartic acid, whilst the concentration of alanine remained stable. The fact that lithium ions primarily affect the serotonin and glutamic acid levels and also their precursors, i.e. tryptophan and glutamine, would seem to support available evidence that these two substances have transmitter roles in the snail nervous system (Kerkut *et al.*, 1965; Cottrell and Laverack, 1968; Gerschenfeld and Stefani, 1968; Gerschenfeld, 1973; Osborne, 1973a).

(iv) *Ontogeny*

It is known that the composition of the free amino acid pool in the brain changes considerably during development (Agrawal *et al.*, 1966). The dansyl microprocedure is convenient for the study of this problem, particularly when changes in discrete brain areas are being followed. The problem essentially is in the definition of defined areas and their removal from animals in the different stages of development. The forebrain and anterior brain stem from 2, 5, 10, 15, 20, 25, 38 and 98-day-old rats were removed and divided along their middle anterior–dorsal axes. The individual pieces were subsequently homogenised in 0.05 M $NaHCO_3$ pH 10, dansylated with ^{14}C–dansyl-chloride and subjected to microchromatography. The composition of the individual dansylated substances on the right- and left-hand sides of the forebrain and anterior brain stem are shown in Table 27, where it can be seen that there is not a great deal of difference between the individual substances on either side of the brain part analysed. However, comparison of the composition of the substances at various stages of development shows a number of differences. The obvious differences can be seen on the autoradiograms of the chromatogram as shown in Fig. 57A. The most evident and dramatic changes are the decrease in hydroxyproline and increase in GABA with age. Many other changes occur, for example, the proportion of lysine to ornithine increases with age. It is tempting to speculate about the increase or decrease in certain substances at the different stages of development. It is almost certain that the progressive restriction of the passage of amino acids from blood to brain at different times of development has an effect on the various compositions of substances in the brain. It is of interest to note the high level of hydroxyproline in the new-born rat and its position as a

TABLE 27. THE PERCENTAGE COMPOSITION OF SUBSTANCES WHICH REACT WITH ^{14}C–DANSYL-CHLORIDE AND OCCUR IN THE RIGHT (R) AND LEFT (L) SIDES OF THE COMBINED FOREBRAIN AND ANTERIOR BRAIN STEM FROM RATS OF DIFFERENT AGES

Spot no.	Dansyl substance	2 Days		5 Days		10 Days		15 Days		20 Days		25 Days		38 Days		98 Days	
		R	L	R	L	R	L	R	L	R	L	R	L	R	L	R	L
1	Taurine	43.04	44.02	45.1	46.56	43.4	44.4	36.4	35.1	33.4	33.6	28.7	27.7	19.9	19.7	17.1	18.0
2	Unknown	0.32	0.25	0.71	0.41	0.42	0.60	0.11	0.21	0.10	0.11	0.18	0.11	0.1	0.15	0.13	0.17
3	Tryptophan	0.26	0.25	0.24	0.21	0.33	0.31	0.19	0.31	0.32	0.29	0.31	0.29	0.35	0.36	0.33	0.34
4	Lysine	1.22	1.13	1.55	1.45	1.90	1.89	1.87	1.78	1.53	1.49	1.71	1.49	1.72	1.87	1.9	1.92
5	Ornithine	0.44	0.43	0.49	0.48	0.51	0.43	0.53	0.49	0.49	0.60	0.61	0.60	0.51	0.46	0.55	0.48
6	Bis-tyrosine	0.77	0.71	0.67	0.77	0.87	0.81	0.91	0.76	0.71	0.65	0.60	0.65	0.51	0.6	0.65	0.6
7	Phenylalanine	0.29	0.25	0.27	0.21	0.39	0.36	0.39	0.38	0.41	0.47	0.41	0.47	0.39	0.38	0.41	0.55
8	Unknown	0.05	0.10	0.08	0.11	0.13	0.09	0.09	0.12	0.11	0.11	0.16	0.11	0.11	0.11	0.19	0.11
9	Methionine	0.32	0.28	0.29	0.31	0.21	0.29	0.21	0.31	0.3	0.29	0.31	0.29	0.32	0.3	0.33	0.29
10	Histidine	0.65	0.61	0.44	0.55	0.47	0.51	0.55	0.61	0.49	0.55	0.51	0.55	0.55	0.59	0.49	0.51
11	Leucine	0.42	0.49	0.38	0.41	0.46	0.44	0.42	0.45	0.53	0.51	0.49	0.51	0.51	0.51	0.61	0.61
12	Isoleucine	0.23	0.27	0.22	0.26	0.21	0.29	0.31	0.31	0.32	0.39	0.31	0.39	0.4	0.41	0.29	0.39
13	Unknown	0.48	0.47	0.84	0.66	0.29	0.31	0.31	0.31	0.28	0.29	0.22	0.29	0.32	0.3	0.4	0.44
14	Unknown	0.56	0.61	0.57	0.49	0.66	0.71	0.99	0.99	1.41	0.31	1.51	0.31	1.9	1.9	1.59	1.6
15	Unknown	0.14	0.15	0.17	0.09	0.24	0.14	0.42	0.47	0.41	0.44	0.48	0.44	0.6	0.7	0.51	0.5
16	Unknown	0.17	0.19	0.27	0.31	0.43	0.33	0.27	0.31	0.10	0.11	0.35	0.11	0.22	0.3	0.24	0.3
17	Unknown	0.68	0.58	0.82	0.81	0.84	0.81	1.05	0.99	0.96	0.91	0.89	0.91	0.61	0.5	0.78	0.81
18	Unknown	1.94	1.77	2.19	2.00	1.79	1.81	1.69	1.8	1.51	1.55	1.34	1.55	1.4	1.5	1.4	1.31
19	Proline	1.41	1.39	1.37	1.37	0.9	1.1	0.8	0.9	0.94	0.88	0.59	0.88	0.61	0.55	0.7	0.66
20	Valine	0.48	0.51	0.33	0.32	0.37	0.36	0.37	0.29	0.48	0.31	0.46	0.31	0.46	0.31	0.54	0.52
21	GABA	2.71	2.81	2.13	2.23	3.6	3.41	6.51	6.51	8.11	8.21	9.99	8.21	12.3	13.3	12.9	13.4
22	Alanine	5.5	5.1	4.9	5.2	4.8	4.9	5.52	5.5	5.32	5.7	6.12	5.7	5.9	6.6	4.9	5.9
23	Ethanolamine	3.6	3.3	3.72	3.63	2.91	2.9	2.91	2.9	2.21	2.22	1.81	2.22	2.0	2.1	2.2	2.1
24	Unknown	0.19	0.17	0.41	0.33	0.69	0.51	0.51	0.61	0.49	0.52	0.71	0.52	0.68	0.66	0.61	0.61
25	Lysine	3.41	3.09	3.77	3.61	3.23	3.5	3.00	3.1	3.61	3.1	3.00	3.4	3.2	3.9	3.5	3.5
26	Glutamic acid and aspartic acid	10.42	11.2	12.41	11.1	12.87	12.9	17.85	16.8	18.69	17.99	21.0	17.99	24.0	25.1	24.3	24.4
27	Unknown	6.81	6.55	5.9	6.1	7.1	6.9	5.9	4.7	3.9	4.4	4.2	4.4	5.1	4.3	4.7	4.9
28	Hydroxyproline	0.58	0.61	0.71	0.72	0.69	0.71	0.42	0.61	0.77	0.51	0	0.51	0	0	0	0
29	Arginine, ϵ-lysine, cystine, glutamine, serine, threonine and α-amino-histidine	12.01	12.71	9.03	9.3	9.29	8.28	9.50	12.38	12.10	13.19	13.03	14.61	15.33	12.54	75	15.08
30	Dansyl-OH	—	—	—	—	—	—	—	—	—	—	—	—	—	—	—	—
31	Dansyl-NH_2	—	—	—	—	—	—	—	—	—	—	—	—	—	—	—	—

dansyl compound on microchromatograms. It may be recalled from the studies on the effect of LSD-25 in snail brain (see page 102) that the substance depleted by the drug is thought to be hydroxyproline, and that furthermore, the substance was only present in

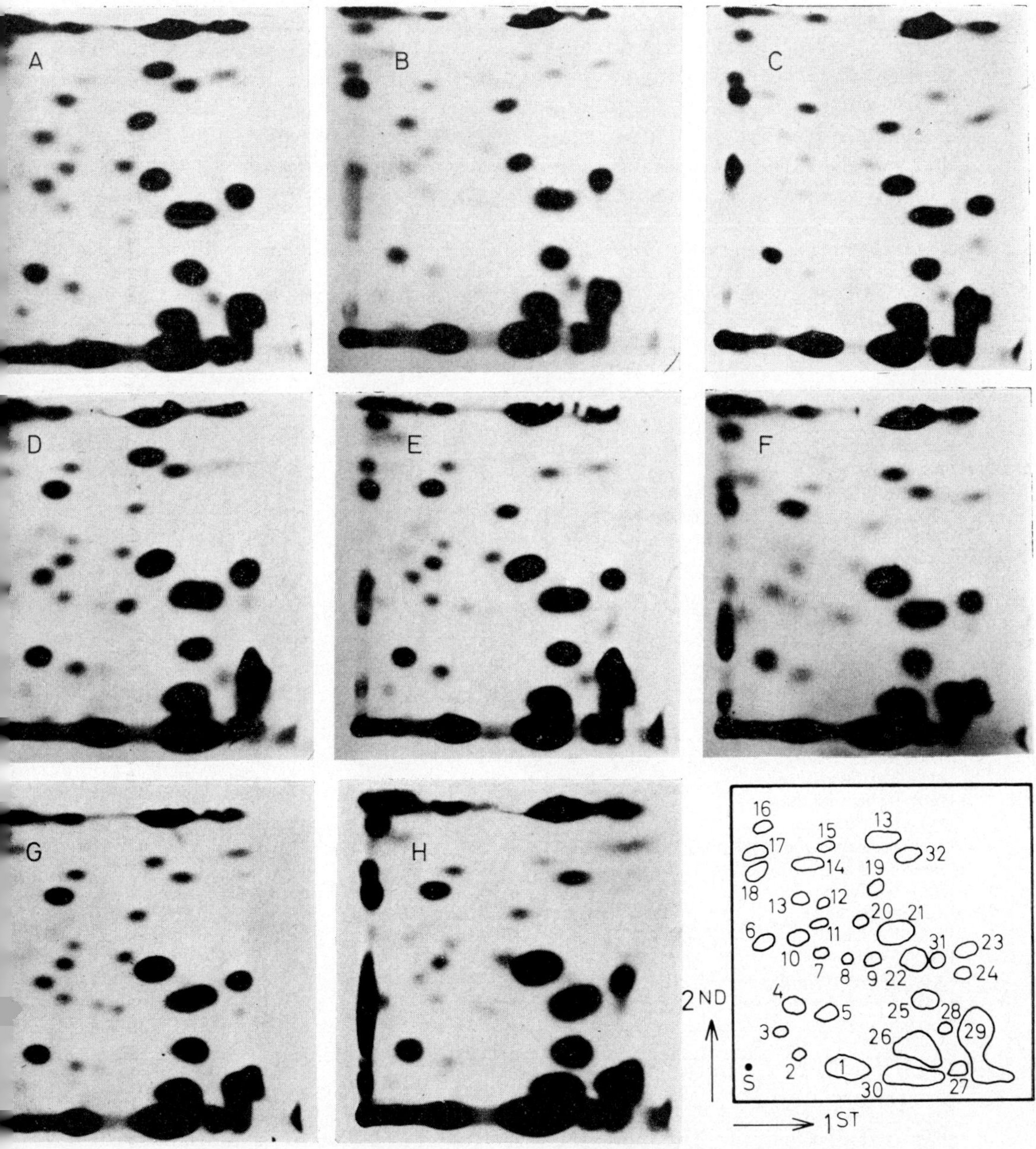

FIG. 57A. Autoradiograms of microchromatograms from extracts of fore-brain and anterior-mid brain stem from 2 (A), 5 (B), 10 (C), 15 (D), 20 (E), 25 (F), 38 (G) and 98 (H) day-old rats. The content of individual substances can be seen by comparing the numbers of the corresponding map with those given in Table 27. Unmarked spots represent either impurities in the purchased dansyl-chloride or in the dansyl-peptides. The direction of chromatography is indicated by arrows and is as described for Fig. 47.

the snail nervous system just after winter hibernation. Could it be that hydroxyproline plays an important role in the development of the nervous system?

(d) OTHER STUDIES INVOLVING THE USE OF MICRODANSYL PROCEDURES

Osborne *et al.* (1971) analysed the content of the free amino acids, GABA, serotonin, 5-hydroxyindole and 5-hydroxytryptophan in the brain (circumoesophageal ganglia), optic tentacle, pharyngeal retractor muscle and heart of *Helix pomatia.* The composition of each of these substances in percentage of total radioactivity in the different tissues is shown in Table 28. A comparison of the amino acids in the four tissues of *Helix pomatia* displays great similarity, though the brain contains significantly less ornithine. In addition

TABLE 28. THE PERCENTAGE COMPOSITION OF SUBSTANCES WHICH REACT WITH ^{14}C–DANSYL-CHLORIDE AND OCCUR IN THE BRAIN, HEART, OPTIC TENTACLE AND PHARYNGEAL RETRACTOR MUSCLE OF *Helix pomatia* (from Osborne *et al.*, 1971)

Substance	Brain	Heart	Optic tentacle	Pharyngeal retractor muscle
Dansyl-taurine	0.67	0.55	1.40	0.35
Dansyl-N-tyrosine	0.31	0.47	0.30	0.49
Dansyl-glutamine + dansyl-threonine + dansyl-asparagine + dansyl-serine	7.11	1.92	5.37	4.31
Dansyl-glutamic acid	5.03	2.27	3.47	2.39
Dansyl-aspartic acid	1.67	0.85	1.57	0.86
Dansyl-cystine	1.12	0.49	2.08	1.22
Dansyl-arginine + dansyl-α-amino-histidine + dansyl-ε-lysine	41.6	18.56	21.36	24.44
Dansyl-N-serotonin	0.42	0.23	0.36	0.3
Dansyl-N-5-hydroxytryptophan	0.47	0.39	0.51	0.22
Dansyl-bis-lysine	0.42	1.11	0.82	1.23
Dansyl-bis-ornithine	1.09	20.26	20.56	29.36
Dansyl-glycine	3.61	1.07	1.63	1.44
Dansyl-bis-tyrosine	0.31	0.30	0.59	0.42
Dansyl-bis-histidine	1.64	0.89	1.00	0.66
Dansyl-phenylalanine	1.79	0.64	0.6	0.50
Dansyl-methionine	0.73	0.41	0.33	0.10
Dansyl-alanine + dansyl-NH_2	17.97	25.62	23.29	17.93
Dansyl-GABA	0.38	0.32	0.36	0.19
Dansyl-valine	1.01	0.91	0.47	0.62
Dansyl-proline	0.72	0.50	0.41	0.66
Dansyl-leucine	1.31	0.91	0.54	0.72
Dansyl-isoleucine	0.88	0.64	0.56	0.39
Dansyl-bis-serotonin	0.34	absent	absent	absent
Dansyl-5-hydroxyindole	0.72	absent	absent	absent
Dansyl-unknown	7.28 (11)	14.96 (14)	12.02 (17)	8.65 (12)

the glycine content of the pharyngeal retractor muscle, heart and optic tentacle is lower than that of the brain. The more abundant amino acids, or not completely separated groups of amino acids, in each of the tissues are alanine; ornithine (except in the brain); glycine; glutamine, threonine and asparagine, histidine (as dansyl-α-amino histidine), lysine (as dansyl-ε-lysine) and cystine; phenylalanine; aspartic acid, and histidine (as dansyl-bis-histidine). A number of other amino acids (leucine, taurine, proline, valine, isoleucine, methionine and tyrosine) are present in all tissues, but to a lesser extent. Of special interest is the presence of low amounts of GABA in all tissues studied, since the substance could not be detected in the snails by other procedures (Kerkut and Cottrell, 1962; Jullien *et al.*, 1961; Cottrell and Osborne, 1969). Other substances present in all tissue samples are serotonin, 5-hydroxytryptophan and 5-hydroxyindole. The brain contains the greatest amount of serotonin, followed by the optic tentacle, heart and pharyngeal retractor muscle. This is consistent with the idea that serotonin has special functions in the nervous system (Cottrell, 1970; Osborne, 1973a; Gerschenfeld, 1973). It is not clear why 5-hydroxyindole occurs in snail tissues though its similarity in structure to 5-hydroxyindoleacetic acid (a major metabolite of serotonin) seems to indicate that the substance is a metabolite in the synthesis of serotonin. The occurrence of ethanolamine is noteworthy since this substance is involved in the metabolism of certain phospholipids, which are thought to participate metabolically in the transport of ions across cellular membranes and also perhaps in the process of learning at the synapse (Hokin and Hokin, 1958; Kandel and Spencer, 1968). The snail brain is known to have a greater phospholipid content than the heart (see page 150).

The amine and amino acid composition of the brain, buccal ganglia and blood of *Helix aspersa* has also been studied (Osborne, 1972c). Autoradiograms of microchromatograms showing the occurrence of ^{14}C–dansylated substances are given in Fig. 57B. A comparison

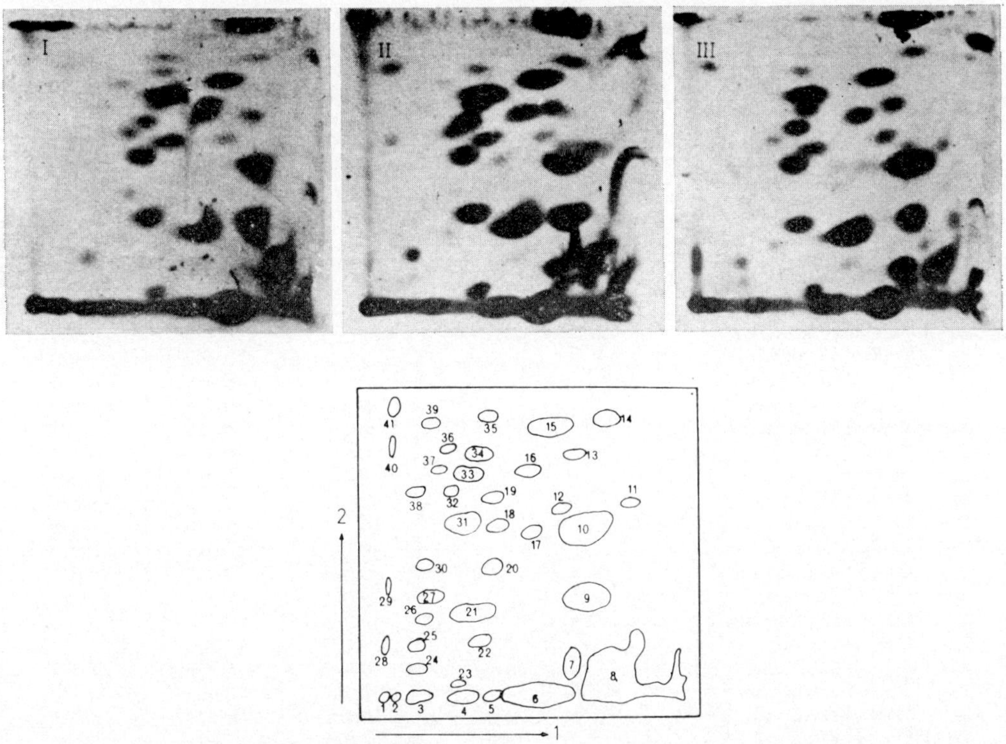

FIG. 57B. Autoradiograms of microchromatograms from extracts of (I) blood, (II) buccal ganglia and (III) circumoesophageal ganglia of *Helix aspersa* after having been reacted with ^{14}C-dansyl chloride. A map is also present to assist in the identification and can be compared with Table 29. Unmarked spots belong to impurities of ^{14}C-dansyl chloride. The direction of chromatography is indicated by the arrows. 1st, water/formic acid (100:3), 2nd, benzene/acetic acid (9:1). (From Osborne, 1972c.)

of amino acids in the blood, brain and buccal ganglia of *Helix aspersa* (see Table 29) reveals great similarities; nervous tissues, however, have higher amounts of ornithine, glutamic acid and aspartic acid. Although glutamic and aspartic acids were not separated in these experiments, it is known from other studies (Osborne *et al.*, 1971) that the glutamic acid content is greater. It is not clear why the ornithine concentrations in the nervous systems are higher than those in the blood, though low in comparison to the heart, optic tentacle

TABLE 29 THE PERCENTAGE COMPOSITION OF SUBSTANCES WHICH REACT WITH ^{14}C–DANSYL-CHLORIDE AND OCCUR IN THE BLOOD, BUCCAL GANGLIA AND BRAIN OF *Helix aspersa* (from Osborne, 1972c)

Spot no.	Substance	Blood	Buccal ganglia	Circumoeso-phageal ganglia
1	Starting point	—	—	—
2	Dansyl-unknown 1	10.245	absent	absent
3	Dansyl-unknown 2	4.338	3.480	2.074
4	Dansyl-taurine + dansyl cystine	absent	1.921	1.510
5	Dansyl-unknown 3	4.906	absent	absent
6	Dansyl-OH	—	—	—
7	Dansyl-glutamic acid + dansyl-aspartic acid	1.631	4.161	4.964
8	Dansyl-glutamine + dansyl-serine + dansyl-threonine + dansyl-argenine + dansyl-ϵ- lysine + dansyl-α-amino-histidine + dansyl-cystine + dansyl-asparagine	31.325	38.590	32.031
9	Dansyl-glycine	5.881	8.085	7.031
10	Dansyl-alanine + dansyl NH_2	21.101	19.035	27.407
11	Dansyl-ethanolamine	absent	0.734	0.215
12	Dansyl-GABA	0.272	0.424	0.407
13	Dansyl-unknown 4	0.590	0.359	0.431
14	Dansyl-unknown 5	0.363	0.342	0.263
15	Dansyl-proline	2.407	3.182	3.153
16	Dansyl-valine	2.007	0.522	0.911
17	Dansyl-unknown 6	0.045	0.136	0.135
18	Dansyl-methionine	0.522	0.195	0.035
19	Dansyl-unknown 7	1.840	0.088	0.743
20	Dansyl-unknown 8	absent	0.220	0.035
21	Dansyl-bis-ornithine	3.974	6.369	5.973
22	Dansyl-unknown 9	0.101	0.097	0.215
23	Dansyl-N-tyrosine	0.796	0.815	1.187
24	Dansyl-N-serotonin	absent	0.179	0.603
25	Dansyl-tryptophan	0.113	0.081	0.323
26	Dansyl-unknown 10	absent	0.052	0.021
27	Dansyl-bis-lysine	1.612	2.904	1.834
28	Dansyl-bis-5-hydroxy-tryptophan	0.031	0.101	1.030
29	Dansyl-unknown 11	0.102	absent	absent
30	Dansyl-unknown 12	0.045	0.092	0.071
31	Dansyl-phenylalanine	3.066	1.152	2.613
32	Dansyl-bis-histidine	0.068	0.212	0.731
33	Dansyl-leucine	1.822	2.240	2.124
34	Dansyl-isoleucine	0.837	1.347	1.003
35	Dansyl-unknown 13	0.522	0.065	0.323
36	Dansyl-unknown 14	0.102	0.121	0.091
37	Dansyl-unknown 15	0.019	0.131	0.131
38	Dansyl-bis-tyrosine	absent	0.016	0.011
39	Dansyl-tryptamine	0.031	0.277	0.278
40	Dansyl-bis-serotonin	absent	0.055	0.079
41	Dansyl-5-hydroxyindole	absent	0.611	0.811

and pharyngeal retractor muscle (see Table 28). Other studies (Jullien *et al.*, 1961; Kerkut and Cottrell, 1962) have shown the glycine levels of nervous tissues from *Helix aspersa* to be lower than reported. Moreover, lysine had not previously been discovered in the blood. The presence of 5-hydroxytryptophan in both blood and nervous tissues seems to demonstrate that gastropods, like the vertebrates (Page and Carlsson, 1970), do not only convert tryptophan to 5-hydroxytryptophan in their nervous systems. All three substances, ethanolamine, 5-hydroxyindole and tryptamine are present in nerve tissues, but absent, with the exception of tryptamine, from the blood. Tryptamine, which can be formed by the decarboxylation of tryptophan and is not clearly evident in *Helix pomatia*, probably represents a metabolite of serotonin synthesis (see above), although no information is available about its formation in nervous tissue. The higher level of serotonin in the brain in relation to that observed in the buccal ganglia is consistent with findings from amine-histochemistry (Osborne and Cottrell, 1971). Since serotonin is thought to be a transmitter substance in gastropods (Cottrell and Laverack, 1968; Cottrell, 1970a; Gerschenfeld, 1973; Osborne, 1973a), it is not entirely surprising to find that this substance is absent from the blood.

The *in vitro* metabolism of ^{14}C–glucose and ^{14}C–glutamic acid by snail nervous tissue has been investigated by Osborne *et al.* (1971). The results showed that the snail brain and heart have the enzymes present to form ^{14}C–GABA from ^{14}C–glutamic acid. Since the mammalian nervous system is characterised by the rapid formation of glutamic acid, aspartic acid, alanine, glutamine and GABA from glucose (e.g. see Gaitonde *et al.*, 1965), only the metabolism of these amino acids was followed. It can be seen from Table 30 that part of the labelled glucose is incorporated into glutamic acid, glutamine, alanine and

TABLE 30. POOL SIZES OF AMINO ACIDS (EXPRESSED IN TERMS OF CPM) IN BRAIN AND HEART OF *Helix pomatia*, AFTER INCUBATION WITH U-^{14}C–GLUCOSE

A blank of 30 cpm is already subtracted. Corresponding spots of six chromatograms were combined and counted (from Osborne *et al.*, 1971)

Amino acid	Counts in brain after 2 hr incubation	Counts in brain after 4 hr incubation	Counts in heart after 2 hr incubation	Counts in heart after 4 hr incubation
GABA	2.6	3.6	2.1	4.1
Glutamine	25	66.2	5.3	15.6
Glutamic acid	50.9	75.4	47.3	67.4
Alanine	340	397	156	233
Aspartic acid	14.5	30.2	10.2	21.4

aspartic acid in both heart and brain tissues after 2 or 4 hr incubation in U-^{14}C-glucose. The counts recorded from GABA were too low when compared with the background activity (30 cpm) to be of any significance. Furthermore, in other experiments two-dimensional chromatograms of the same extracts were exposed to sensitive film (Dupont Dosimeter film 046/545) for 41 days. Even after this long exposure, autoradiograms still recorded nothing in the position of GABA. The metabolism of glucose into glutamic acid, glutamine, alanine and aspartic acid is similar in both heart and brain tissue in quantity.

The major metabolite is alanine, carrying four to five times more radioactivity than glutamic acid which is also present in fairly high amounts. When either of the two snail tissues is incubated with U-^{14}C–glutamic acid as substrate, GABA is formed (see Table 31). The amount of radioactivity associated with GABA increases only slightly when the incubation period is extended from 2 to 4 hr. Again, quantitatively, the radioactive metabolites of each sample are similar. Compared with glucose as substrate, the effect of glutamic acid results in the formation of glutamine as the major metabolite and in increased amounts of aspartic acid in comparison to the alanine content.

TABLE 31. POOL SIZES OF AMINO ACIDS (EXPRESSED IN TERMS OF CPM) IN BRAIN AND HEART OF *Helix pomatia* AFTER INCUBATION IN U-^{14}C–GLUTAMIC ACID

A blank of 30 cpm is already substracted. Corresponding spots of six chromatograms were combined and counted (from Osborne *et al.*, 1971)

Amino acid	Counts in brain after 2 hr incubation	Counts in brain after 4 hr incubation	Counts in heart after 2 hr incubation	Counts in heart after 4 hr incubation
GABA	27.1	29.4	17.7	22
Glutamine	161	200	95.5	200
Alanine	84	93.3	42.4	75.2
Aspartic acid	49.3	53	25.9	40

Dolezalova *et al.* (1973) employed the microprocedure in an attempt to detect putative transmitter substances in the individual ganglia of the snail *Helix pomatia*. The very important physiological amines and amino acids such as serotonin, tyramine, adrenaline, noradrenaline, dopamine and dopa were found in all the ganglia investigated. Furthermore, GABA was proved to be present by the use of mass spectrometry. Several other indole-derivatives, such as tryptophan, 5-HTP, 6-HTP and 5,6-HTP were also found. The actual composition of these and other substances varied in the different ganglia studied, though no variations in the left and right ganglia were recorded. These data confirmed previous findings (see Osborne, 1973a, 1972c) where it had been shown that serotonin, tyramine, tryptamine, noradrenaline, dopamine, GABA, tryptophan and 5-HTP occur in the central nervous tissues of the snail. The presence of adrenaline, 6-HTP, 6-HT and 5,6-HTP, however, is so far a unique discovery which probably needs confirmation through other techniques, such as mass spectrometry.

The dansyl procedure has also been applied to the study of the distribution and axonal transport of free amino acids and related compounds in the dorsal sensory neuron of the rat (Roberts *et al.*, 1973). Major quantitative differences in the amino acids were not recorded in the dorsal and ventral roots, though the dorsal roots contained very much more glutamine, threonine and arginine. After an 8-hr period of dorsal root ligation *in vivo*, an apparent selective accumulation of alanine, glutamic acid and alanine occurred. These findings were interpreted as being compatible with the idea that glutamic acid is a transmitter substance at the terminals of the primary efferent fibres.

The concept of molecular coding of neuronal information has been progressing steadily during the last decade. Perhaps one of the most significant findings in this field is that of Ungar and his collaborators (e.g. Ungar, 1972; Ungar *et al.*, 1972), who isolated a special substance termed scotophobin from brains of rats trained to avoid the dark. The microdansyl procedure is an important method in the isolation process of scotophobin, which is a pentadecapeptide. Because of the sensitivity of the dansyl procedure, no more than 125 mg of brain from animals trained to avoid the dark was sufficient to produce a chromatogram clearly showing the presence of a substance which did not occur in nervous tissues of untrained rats. This substance was eventually identified as scotophobin, and when injected into donor rats, induced them to avoid the dark.

(e) DETERMINATION OF AMINO ACIDS IN PEPTIDES AND PROTEINS

The microdansyl procedure is particularly useful for the analysis of amino acids in small quantities of proteins or peptides. A single disc from a fractionated polyacrylamide gel, for example, is first eluted by treatment with glycerol, followed by freezing with liquid nitrogen (Neuhoff and Schill, 1968). The eluted small sample of protein is then dialysed in a microchamber (see page 55) against 0.05 M $NaHCO_3$ (pH 7–8) for 4 to 6 hr to remove the glycine from the electrophoresis buffer. A proportion of the sample can than be dansylated and checked by microchromatography for its purity. The remainder of the purified protein is freeze-dried and transferred to a micro-Pyrex glass tube. For 30 μg of sample about 30 μl of 6 N HCl is added, the tube is sealed under a nitrogen atmosphere and then hydrolysed for 24 hr at 100°C. During this period some amino acids (methionine, cysteine, cystine, tryptophan) are partially destroyed and should be accounted for in the final analysis. Subsequent to the hydrolysis, the sample is freeze-dried, resuspended in a small quantity of 0.05 M $NaHCO_3$ pH 10 and subjected to chromatographic analysis as dansyl substances (Neuhoff *et al.*, 1969).

Another use of the dansyl procedure is that for the analysis of N-terminal and C-terminal amino acids in proteins and peptides, which can be combined with the Edman degradation procedure for sequence determination of peptides and proteins. Gray and Hartley (1963) were the first to use dansyl-chloride to identify N-terminal amino acids of proteins. Later on, the method developed by Gray (1967), which combined dansylation with the Edman degradation procedure, allowed determination of protein sequence. The advantage of this method over others lies not only in its high sensitivity, but also in its execution. By scaling down the procedure, Burton and Hartley (1970) determined the sequence of as little as 1 to 2 mg of methionyl-t-RNA synthetase from *Escherichia coli*, and also showed lysine to be associated with the enzyme's catalytic site. The publication by Neuhoff *et al.* (1970) provides a very good example of the use of the microdansylation procedure as a method in the determination of C- and N-terminal amino acids in less than 0.2 mg starting material. In their analysis, purified RNA polymerase from *Escherichia coli* was treated with sodium-dodecyl sulphate and oxidised with hydrogen peroxide for the analysis of the N-terminal amino acid. By examination of the dansylated amino acids they were able to demonstrate that methionine is at the N-terminal group, which is in agreement with the findings of Burgess (1969). The C-terminal amino acid of RNA polymerase was determined by using carboxypeptidase A, which splits the C-terminal amino acid off. Subsequent analysis of this amino acid by the microdansyl procedure showed it to be glycine, which agreed with the independent findings of Möckel and Barnard (1969).

7. The Possibility of Using Dansyl-chloride to Detect Noradrenaline, Adrenaline, Octopamine and Related Compounds in the Microscale

Our knowledge of the catecholamines (dopamine, noradrenaline, adrenaline) has expanded enormously in the last 20 years, and the data available show clearly that dopamine, noradrenaline, and the related substance octopamine have very specific and important functions within nervous systems (see *British Medical Bulletin*, Vol. 29, No. 2, 1973). The great advances made in our understanding of the biochemistry of catecholamines can be attributed partly to the development of two very sensitive assay procedures. The first is the conversion of small amounts of catecholamines by oxidisation to fluorescent trihydroxyindole derivatives (Ehrlen, 1948; Lund, 1949) and the second the ethylenediamine condensation method of Weil-Malherbe and Bone (1952). Though these two procedures have been improved continuously (see e.g. Anton and Sayre, 1962) in order to increase their sensitivity even further, they suffer from one serious drawback, namely the difficulty of measuring several catecholamines with large variations in concentration in the same sample without an extensive primary separation procedure. Moreover, any separation procedure involves a considerable risk of destruction of these easily oxidisable substances (Iversen, 1967).

A number of attempts have therefore been made to produce better procedures for the determination of catecholamines. These include mainly enzymatic radioactive techniques (Saelens *et al.*, 1967; Aizawa and Yamada, 1969; Engelman *et al.*, 1968) and gas chromatographic procedures where the methods involve the separation of non-derivative catecholamines (Fales and Pinsano, 1962), Schiff base derivatives of catecholamines (Brochmann-Hanssen and Baerheim-Svendsen, 1962), trimethyl silyl ethers of catecholamines (Sen and McGeer, 1963; Horning *et al.*, 1967), acetylated catecholamines (Brooks and Horning, 1964), cyclic boronated catecholamines (Antony *et al.*, 1970) and silyated catecholamines at their hydroxy groups in combination with the formation of either a Schiff base (Capella and Horning, 1966) or with acetylations (Brooks and Horning, 1964). However, the problem of instability of the catecholamines and several of their derivatives at a slightly raised temperature has meant that it has not yet been possible to use the gas chromatography methods with a great deal of certainty, whilst enzymatic radioactive techniques suffer from the disadvantage that careful isolation of the catecholamine derivatives is necessary to ensure reliable results.

Theoretically the very sensitive dansyl procedure can be applied to study the catecholamines, as dansyl derivatives of catecholamines can be separated in the normal scale (Seiler, 1970; Seiler and Wiechmann, 1970). However, analysis of catecholamines in the microscale using dansyl-chloride has proved difficult, mainly because at least five different dansyl derivatives from each single catecholamine are formed (see Fig. 58B), thus in the first place decreasing the sensitivity of the procedure for catecholamines as compared with glycine for example, which forms a single dansyl derivative, and in the second place producing problems in the identification of the individual dansyl-catecholamine derivatives. Moreover, since the amino acid content in biological tissue is so high in comparison to the catecholamines, it is often impossible to distinguish the dansyl derivatives of the latter on microchromatograms, where the dansyl-amino acids predominate. It is, of course, possible to extract first the catecholamines from a sample before dansylation, but this process decreases the sensitivity of the method, which is not compatible with the idea of a microprocedure. Clearly it would be a great step forward if a method could be found to

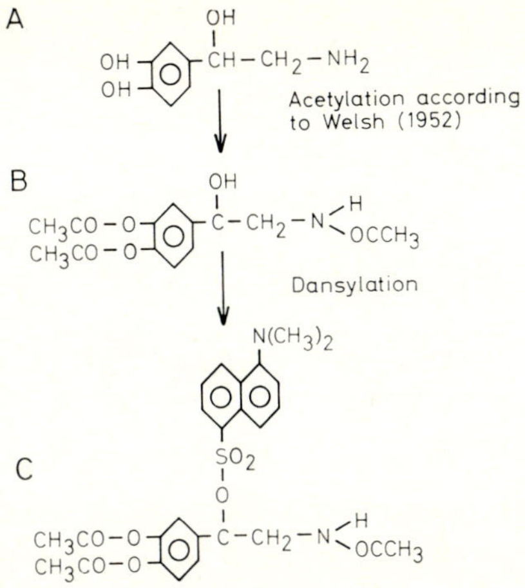

FIG. 58A. Theoretical reaction possibility for obtaining a single highly fluorescent product of noradrenaline. By acetylating noradrenaline (A) according to the method of Welsh (1952), the 3,4,N-triacetyl derivative of noradrenaline is formed (B) and this product would produce a single highly fluorescent derivative (C) when reacted with dansyl-chloride.

produce a single dansyl derivative from each catecholamine, as then all the advantages in using dansyl-chloride could be exploited. In theory this would be feasible if the catecholamines were reacted with a substance to produce a derivative which has only one reactive group available to react with dansyl-chloride. One possibility is to produce triacetyl derivatives of adrenaline and noradrenaline according to the procedure of Welsh (1952, 1955), and then to react the derivatives with dansyl-chloride, thus yielding a single derivative for each catecholamine (see Fig. 58A). This procedure would have the following advantages: (a) the acetylation process stabilises the labile catecholamines, (b) during the acetylation procedure, any amino acids present will also be acetylated, which will prevent them from reacting with dansyl-chloride, and (c) the individual dansyl products of catecholamines can be easily separated. In such a procedure, however, care would have to be taken not to acetylate the catecholamines to too great an extent, for tetra-acetyl derivatives of noradrenaline and adrenaline are also easily formed (Bretschneider, 1946). In order to ascertain whether it is indeed possible to adapt the microdansyl procedure to detect catecholamines by the procedure outlined in Fig. 58A, the following experiment was carried out: 10 mg of catecholamine or octopamine was dissolved in 0.5 ml double distilled water. 10 μl acetic anhydride was added to the solution and immediately afterwards 30 mg $NaHCO_3$ in small portions. When the evolution of carbon dioxide had ceased the mixture was extracted with 3 $\times$ 1 ml chloroform and the extract washed with 3 $\times$ 1 ml water. Thereafter the chloroform extract was concentrated in a current of air and the wet residue freeze-dried. The freeze-dried catecholamine triacetates were then dissolved in 5 ml methanol. 0.1 ml of this solution was mixed with 0.4 ml 0.05 M $NaHCO_3$ pH 13, and 0.5 ml dansyl-chloride (1 mg/ml in acetone) and incubated for 30 min at 20°C. A small amount of the dansylated solution was applied to a 3 $\times$ 3 cm polyamide layer and chromatographed as already described

The chromatographic separation of the dansylated–acetylated derivatives of octopamine, noradrenaline, and adrenaline is shown in Fig. 58B. It can be clearly seen that a single dansylated–acetylated octopamine derivative is formed, whilst in the instance of

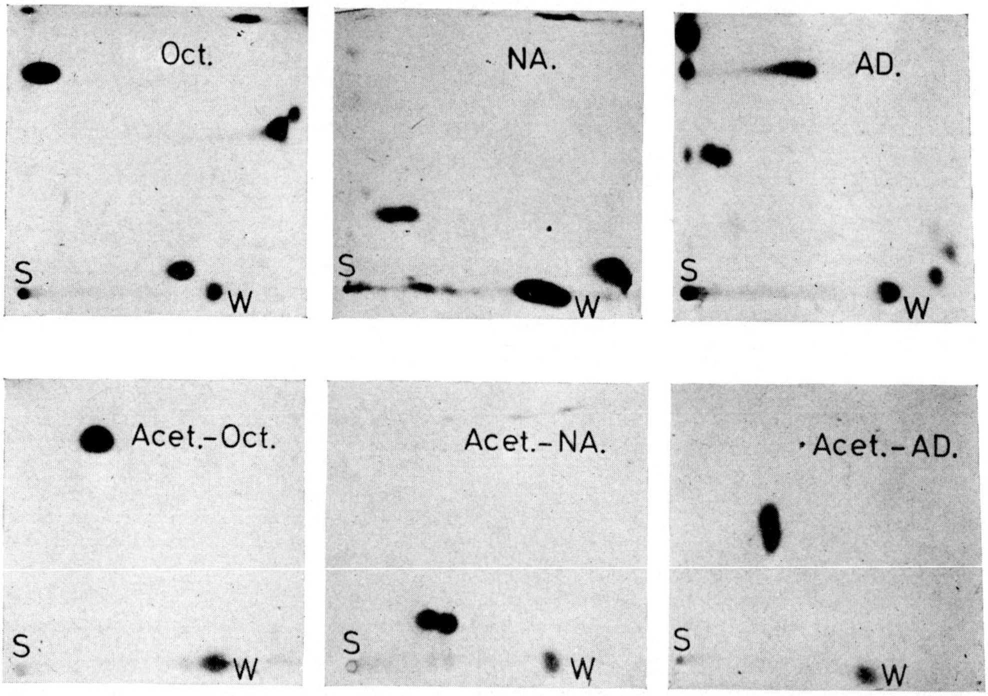

Fig. 58B. Autoradiograms of microchromatograms (original size 3 × 3 cm) to show the positions of octopamine (Oct.), noradrenaline (NA), adrenaline (AD), 4,N-diacetyl octopamine (Acet.-Oct.), 3,4,N-triacetyl noradrenaline (Acet.-NA) and 3,4,N-triacetyl adrenaline (Acet.-AD), which had been reacted with ^{14}C–dansyl-chloride. Each microchromatogram was developed in an ascending way using two solvent systems; 1st development in the horizontal direction was with water/formic acid (100:3 v/v) and the 2nd development was in the vertical direction with benzene/acetic acid (9:1 v/v). The starting points (S) and the positions of dansyl-OH (W) are indicated. It can be clearly seen that in each instance fewer dansylated-acetylated products of the individual substances are produced than the amount of products formed by dansylation alone.

both noradrenaline and adrenaline two derivatives in each case with approximately the same chromatographic mobility are present. Since analysis of the acetylated compounds of the catecholamines by the method of proton spectrometry showed only triacetyl derivatives of noradrenaline and adrenaline to be formed, it would appear that the dansylation procedure causes a partial breakdown of the triacetyl derivatives, resulting in the formation of two derivatives from a single catecholamine. An explanation for this may be that the triacetyl derivatives become unstable at alkaline pH, a pH necessary for the dansylation process. Work is now proceeding in this laboratory to resolve this problem.

Even at this stage, it is obvious that this approach is exciting, for the results so far achieved show it indeed possible to produce a highly sensitive microprocedure for specific catecholamines, octopamine and other amines. In trial experiments where tissue extract, together with standard amounts of noradrenaline was acetylated and dansylated according to the procedure illustrated in Fig. 58A, the results showed only noradrenaline to be identified on microchromatograms as a dansylated substance. However, these experiments also demonstrated that the actual conditions of dansylation will have to be very carefully studied, since on some microchromatograms an additional dansyl derivative spot, also belonging to noradrenaline, was formed.

References

AGRAWAL, H. C., DAVIS, J. M. and HIMWICH, W. A. (1966) Post-natal changes in free amino acid pool of rat brain. *J. Neurochem.* **13**, 607–615.

AGRAWAL, H. C., DAVIS, J. M. and HIMWICH, W. A. (1968) Maturational changes in amino acids in CNS of different mammalian species. In: *Recent Advances in Biological Psychiatry* (Ed. J. Wortis), Vol. 10, pp. 258–265. Plenum Press, New York.

AIZAWA, Y. and YAMADA, K. (1969) Double isotope derivative method for determination of noradrenaline in rat brain. *Jap. J. Pharmacol.* **19**, 474–476.

ANTON, A. H. and SAYRE, D. F. (1962) A study of the factors affecting the aluminium oxide trihydroxyindole procedure for the analysis of catecholamines. *J. Pharmacol. Exp. Ther.* **138**, 360–375.

ANTONY, G. M., BROOKS, C. J. W. and MIDDLEDITCH, B. S. (1970) The use of boronate derivatives in the characterisation of catecholamines and related β-hydroxyamines by gas–liquid chromatography mass spectrometry. *J. Pharm. Pharmacol.* **22**, 205–213.

APRISON, M. H., DAVIDOFF, R. A. and WERMAN, R. (1970) Glycine: its metabolic and possible transmitter roles in nervous tissue. In: *Handbook of Neurochemistry* (Ed. A. Lajtha), Vol. III, pp. 381–393. Plenum Press, New York.

BERL, S. and CLARKE, D. D. (1972) Effects of Li^+ on the metabolism in brain of glutamate, glutamine, aspartate, and GABA from $L-{}^{14}C$ acetate *in vitro*. *Brain Research*, **36**, 203–213.

BOULTON, A. A. and MAJER, P. J. (1971) Identification and distribution of some primary aromatic non-catecholic amines in brain. Abst. Common. 3rd Int. Meeting Int. Soc. Neurochem. Budapest.

BRADFORD, H. F., CHAIN, E. B., CORY, H. T. and ROSE, S. P. R. (1969) Glucose and amino acid metabolism in some invertebrate nervous systems. *J. Neurochem.* **16**, 969–978.

BRETSCHNEIDER, H. (1946) Über die Acetylierung von Adrenalin. *Mh. Chem.* **66**, 355–367.

BRIEL, G. (1972) Mikroanalyse von Ameisensäuren als ^{14}C-Dansyl-Verbindungen und ihre Anwendung zur Bestimmung von freien Aminosäuren des Zentralnervensystems. Inaugural-Dissertation, Göttingen.

BRIEL, G., NEUHOFF, V. and OSBORNE, N. N. (1971) Determination of amino acids in single identifiable nerve cells of *Helix pomatia*. *Intern. J. Neuroscience*, **2**, 129–136.

BRIEL, G., NEUHOFF, V. and MAIER, M. (1972) Microanalysis of amino acids and their determination in biological material using dansyl-chloride. *Hoppe-Seyler's Z. Physiol. Chem.* **353**, 540–553.

BROCHMANN-HANSSEN, E. and BAERHEIM-SVENDSON, A. (1962) Separation and identification of sympatomimetic amines by gas–liquid chromatography. *J. Pharm. Sci.* **51**, 938–941.

BROOKS, C. J. W. and HORNING, E. C. (1964) Gas chromatographic studies of catecholamines, tryptamines and other biological amines. *Analyt. Chem.* **36**, 1540–1549.

BURGESS, R. R. (1969) Separation and characterisation of the sub-units of ribonucleic acid polymerase. *J. Biol. Chem.* **244**, 6178–6188.

BURTON, C. J. and HARTLEY, B. S. (1970) Chemical studies on methionyl-t-RNA synthetase from *Escherichia coli*. *J. Mol. Biol.* **52**, 165–178.

CAPELLA, P. and HORNING, E. C. (1966) Separation and identification of derivatives of biological amines by gas–liquid chromatography. *Analyt. Chem.* **38**, 316–321.

CASOLA, L. and di METTEO, G. (1972) Studies on the dansylation reaction by use of ^{14}C-dansyl chloride: application to analysis of free amino acids in the rat optic nerve. *Analyt. Biochem.* **49**, 416–429.

COTTRELL, G. A. (1970a) Direct postsynaptic responses to stimulation of serotonin-containing neurons. *Nature*, **225**, 1060–1062.

COTTRELL, G. A. (1970b) Actions of LSD-25 and reserpine on a serotonergic synapse. *J. Physiol.* **208**, 28–29.

COTTRELL, G. A. (1971) Synaptic connections made by two serotonin-containing neurons in the snail (*Helix pomatia*) brain. *Experientia*, **27**, 813–815.

COTTRELL, G. A. and LAVERACK, M. S. (1968) Invertebrate pharmacology. *Ann. Rev. Pharmacol.* **8**, 273–298.

COTTRELL, G. A. and POWELL, B. (1971) Formation of serotonin by isolated serotonin-containing neurons and by isolated non-amine-containing neurons. *J. Neurochem.* **18**, 1695–1697.

COTTRELL, G. A. and OSBORNE, N. N. (1969) Localisation and mode of action of cardioexcitatory agents in molluscan hearts. *Experimentia* (Suppl. 15) (Ed. F. V. McCann), pp. 220–231.

COTTRELL, G. A. and OSBORNE, N. N. (1970) Subcellular localisation of serotonin in an identified serotonin-containing neuron. *Nature*, **225**, 470–472.

COTTRELL, G. A., MACON, J. and SZCZEPANIAK, A. C. (1972) Glutamic acid mimicking of synaptic inhibition on the giant serotonin neuron of the snail. *Brit. J. Pharmacol.* **45**, 684–688.

CROWSHAW, K., JESSUP, J. and RAMWELL, P. W. (1967) Thin-layer chromatography of 1-dimethylamino-naphthalene-5-sulphonyl derivates of amino acids present in superfusates of cat cerebral cortex. *Biochem. J.* **103**, 79–85.

CURTIS, D. R. and JOHNSTON, G. A. (1970) Amino acid transmitters. In: *Handbook of Neurochemistry* (Ed. A. Lajtha), Vol. IV, pp. 115–131. Plenum Press, New York.

DAVIS, J. M. and FANN, W. E. (1971) Lithium. *Ann. Rev. Pharmacol.* **11**, 285–302.

DAVISON, A. N. and KACMAREK, L. K. (1971) Taurine—a possible neurotransmitter? *Nature*, **234**, 107–108.

DOLEZALOVA, H., GIACOBINI, E. and STEPITA-KLAUCO, M. (1973) An attempt to identify putative neurotransmitter molecules in the central nervous system of the snail. *Int. J. Neuroscience*, **5**, 53–60.

DONALDSON, P. E. R. (1958) *Electronic Apparatus for Biological Research*. Butterworths, London.

EHRLEN, I. (1948) Fluorometric determination of adrenaline. *Farm Rev.* (*Stockh.*), **47**, 242–252.

ENGELMAN, K., PORTNOY, B. and LOVENBERG, W. (1968) A sensitive and specific double isotope derivative method for the determination of catecholamines in biological specimens. *Amer. J. Med. Sci.* **255**, 259–268.

FALES, H. M. and PINSANE, J. J. (1962) Gas chromatography of biologically important amines. *Analyt. Biochem.* **3**, 337–342.

FERGUSON, J. C. (1970) An autoradiographic study of the translocation and utilisation of amino acids by starfish. *Biol. Bull.* **138**, 14–25.

FREEDMAN, D. X. (1961) Effects of LSD-25 on brain serotonin. *J. Pharmacol.* **134**, 160–166.

GAINER, H. (1972) Patterns of protein synthesis in individual, identified molluscan neurons. *Brain Research*, **39**, 369–385.

GAITONDE, M. K., DAHL, D. and ELLIOTT, K. A. C. (1965) Entry of glucose carbon into amino acids of rat brain and liver *in vivo* after injecting uniformly ^{14}C-labelled glucose. *Biochem. J.* **94**, 345–352.

GERSCHENFELD, H. M. (1973) Chemical transmission in invertebrate central nervous systems and neuromuscular junctions. *Physiol. Rev.* **53**, 1–119.

GERSCHENFELD, H. M. and LASANSKY, A. (1964) Action of glutamic acid and other naturally occurring amino acids on snail central neurons. *Int. J. Neuropharmacol.* **3**, 301–314.

GERSCHENFELD, H. M. and STEFANI, E. (1968) Evidence for an excitatory transmitter role of serotonin in molluscan central synapses. *Adv. in Pharmacol.* **6A**, 470–472.

GIARMAN, N. J. and FREEDMAN, D. X. (1965) Biochemical aspects of the actions of psychotomimetic drugs. *Pharmacol. Rev.* **17**, 1–25.

GLOWINSKI, J. and AXELROD, J. (1964) Inhibition of uptake of tritiated norepinephrine in the intact rat brain by imipramine and structurally related compounds. *Nature*, **204**, 1339–1340.

GRAY, W. R. (1964) Ultra-micro methods for investigation of protein structure. Ph.D. Thesis, St. John's College, University of Cambridge.

GRAY, W. R. (1967) Dansyl-chloride procedure. In: *Methods in Enzymology* (Ed. C. H. W. Hirs), pp. 139–151. Academic Press, New York.

GRAY, W. R. and HARTLEY, B. S. (1963) A fluorescent end-group reagent for proteins and peptides. *Biochem. J.* **89**, 59p.

GROS, C. and LABOUESSE, B. (1969) Study of the dansylation reaction of amino acids, peptides and proteins. *European J. Biochem.* **7**, 463–470.

HARTLEY, B. S. (1970) Strategy and tactics of protein chemistry. *Biochem. J.* **119**, 805–822.

HEBB, C. (1970) CNS at the cellular level: identity of transmitter agents. *Ann. Rev. Physiol.* **32**, 165–192.

HESS, S. M., REDFIELD, B. G. and UDENFRIEND, S. (1959) The effect of monoamine oxidase inhibitors and tryptophan on the tryptamine content of animal tissues and urine. *J. Pharmac. exp. Ther.* **127**, 178–181.

HIMWICH, W. A. and AGRAWAL, H. C. (1969) Amino acids. In: *Handbook of Neurochemistry* (Ed. A. Lajtha), Vol. I, pp. 33–48. Plenum Press, New York.

HOFFMANN, A. (1961) Chemical, pharmacological and medieval aspects of psychotomimetics. *J. Exp. Med. Sci.* **5**, 31–51.

HOKIN, L. E. and HOKIN, M. R. (1958) Acetylcholine and the exchange of inositol and phosphate in brain phosphoinositide. *J. Biol. Chem.* **223**, 818–821.

HORNING, M. G., MOSS, A. M. and HORNING, E. C. (1967) A new method for the separation of the catecholamines by gas–liquid chromatography. *Biochim. Biophys. Acta*, **148**, 597–600.

IVERSEN, L. L. (1967) *The Uptake and Storage of Noradrenaline*, pp. 11–13. Cambridge University Press, London.

JEQUIER, E., LOVENBERG, W. and SJOERDSMA, A. (1967) Tryptophan-hydroxylase inhibition: the mechanism by which *p*-chlorophenylalanine depletes rat brain serotonin. *Molec. Pharmacol.* **3**, 274–278.

JULLIEN, A., CARDOT, J. and RIPPLINGER, J. (1961) La composante aminoacide libre du coeur d'*Helix pomatia*. *C. R. Soc. Biol. Paris*, **155**, 819–820.

KANDEL, E. R. and SPENCER, W. A. (1968) Cellular neurophysiological approaches in the study of learning. *Physiol. Rev.* **48**, 65–124.

KANDEL, E. R. and TAUC, L. (1966) Anomalous rectification in metacerebral giant cells and its consequences for synaptic transmission. *J. Physiol.* **183**, 287–304.

KERKUT, G. A. (1969) Neurochemistry of invertebrates. In: *Handbook of Neurochemistry* (Ed. A. Lajtha), Vol. II, pp. 539–562. Plenum Press, New York.

KERKUT, G. A. and COTTRELL, G. A. (1962) Amino acids in blood and nervous system of *Helix aspersa*. *Comp. Biochem. Physiol.* **5**, 227–230.

KERKUT, G. A. and WALKER, R. J. (1961) The effect of drugs on neurons of the snail *Helix aspersa*. *Comp. Biochem. Physiol.* **3**, 143–160.

KERKUT, G. A., LEAKE, L. D., SHAPIRA, A., COWAN, S. and WALKER, R. T. (1965) The presence of glutamate in nerve-muscle perfusates of *Helix*, *Carcinus* and *Periplaneta*. *Comp. Biochem. Physiol.* **15**, 485–502.

KERKUT, G. A., SHAPIRA, A. and WALKER, R. J. (1967) The transport of ^{14}C-labelled material from CNS $\rightleftharpoons$ muscle along a nerve trunk. *Comp. Biochem. Physiol.* **23**, 729–748.

KERKUT, G. A., FRENCH, M. C. and WALKER, R. J. (1970) The location of axonal pathways of identifiable neurons of *Helix aspersa* using the dye procion yellow M-4R. *Comp. Biochem. Physiol.* **32**, 601–690.

KOE, B. K. and WEISSMAN, A. (1966) *p*-Chlorophenylalanine: a specific depletor of brain serotonin. *J. Pharmacol. Exp. Therap.* **154**, 499–516.

KOECHLIN, B. A. (1955) On the chemical composition of the axoplasm of squid giant nerve fibres with particular reference to its ion pattern. *J. Biophys. Biochem. Cytol.* **1**, 511–529.

KRAVITZ, E. A., SLATER, C. R., TAKAHASHI, K., BOWNDS, M. D. and CROSSFELD, R. M. (1970) Excitatory transmission in invertebrates—glutamate as a potential neuromuscular transmitter compound. In: *Excitatory Synaptic Mechanisms* (Eds. P. Anderson and J. K. S. Jansen), pp. 85–93. Universitetsforlaget, Oslo.

KRNJEVIĆ, K. (1970) Glutamate and γ-aminobutyric acid in brain. *Nature*, **228**, 119–124.

LEWIS, P. R. (1952) The free amino acids of invertebrate nerve. *Biochem. J.* **52**, 330–338.

LOVENBERG, W. (1972) Serotonin synthesis in the brain. 5th Congress on Pharmacol., San Francisco, USA, pp. 250–251 (Invited presentations).

LUND, A. (1949) Fluorometric determination of adrenaline in blood. A new sensitive and specific method. *Acta Pharmacol.* (*Kbh*), **5**, 231–247.

MALYUK, V. I. and MAISKII, V. A. (1970) Incorporation of radioactive methionine into giant neurons of the mollusc, *Planorbis corneus*. *Zh. Evol. Biokim. Fiziol.* **6**, 12–18.

MARTIN, W. R. and SLOAN, J. W. (1970) Effects of infused tryptamine in man. *Psychopharmacologia*, **18**, 231–237.

MENG, K. (1960) Untersuchungen zur Störung der Herztätigkeit beim *Helix pomatia*. *Zool. Fahr*, **68**, 193–204.

MÖCKEL, W. and BARNARD, E. A. (1969) Amino acids around an active center histidine in a turtle chymotrypsin. *Biochim. Biophys. Acta*, **194**, 622–624.

MUSSINI, E. and MORCUCCI, F. (1962) Free amino acids in brain after treatment with psychotropic drugs. In: *Amino Acid Pools* (Ed. J. T. Holden), pp. 486–492. Elsevier, Amsterdam.

NEADLE, D. J. and POLLITT, R. J. (1965) The formation of 1-dimethyl-aminonaphthalene-5-sulphonamide during the preparation of 1-dimethylaminonaphthalene-5-sulphonyl-amino acids. *Biochem. J.* **97**, 607–608.

NEUHOFF, V. (1973) Microdetermination of amino acids and related compounds. In: *Micromethods in Molecular Biology* (Ed. V. Neuhoff), Vol. 12, pp. 85–147. Springer Berlin, Heidelberg, New York.

NEUHOFF, V. and WEISE, M. (1969) Unpublished observations.

NEUHOFF, V. and WEISE, M. (1970) Determination of picomole quantities of γ-aminobutyric acid (GABA) and serotonin. *Arzneim. Forsch.* (*Drug Research*), **20**, 368–372.

NEUHOFF, V., van der HAAR, F., SCHLIMME, E. and WEISE, M. (1969) Zweidimensionale Chromatographie von Dansyl-Aminosäuren im pico-Mol-Bereich, angewandt zur direkten Charakterisierung von Transfer-Ribonucleinsäuren. *Hoppe-Seyler's Z. Physiol. Chem.* **350**, 121–128.

NEUHOFF, V. and SCHILL, W. B. (1968) Kombinierte Mikro-Disk-Elektrophorese und Mikro-Immunpräzipitation von Proteinen. *Hoppe-Seyler's Z. Physiol. Chem.* **349**, 795–800.

NEUHOFF, V., WEISE, M. and STERNBACH, H. (1970) Micro-analysis of pure deoxyribonucleic acid dependent ribonucleic acid polymerase from *Escherichia coli*. VI. *Hoppe-Seyler's Z. Physiol. Chem.* **351**, 1395–1401.

NEUHOFF, V., BRIEL, G. and MAELICKE, A. (1971) Characterisation and microdetermination of histidine as its dansyl compounds. *Arzneim. Forsch.* (*Drug Research*), **21**, 104–107.

OSBORNE, N. N. (1971) Occurrence of GABA and taurine in the nervous systems of the dogfish and some invertebrates. *Comp. Gen. Pharmacol.* **2**, 433–438.

OSBORNE, N. N. (1972a) The *in vivo* synthesis of serotonin in an identified serotonin-containing neuron of *Helix pomatia. Intern. J. Neuroscience*, **3**, 215–219.
OSBORNE, N. N. (1972b) Effect of electrical stimulation on the *in vivo* metabolism of glucose and glutamic acid in an identified neuron. *Brain Research*, **41**, 237–241.
OSBORNE, N. N. (1972c) Serotonin, free amino acids and related substances occurring in the blood and nervous tissue of *Helix aspersa* (Grstropod Mollusca). *Comp. Gen. Pharmacol.* **3**, 171–177.
OSBORNE, N. N. (1972d) Occurrence of glycine and glutamic acid in the nervous system of two fish species and some invertebrates. *Comp. Biochem. Physiol.* **43B,** 579–585.
OSBORNE, N. N. (1973a) The analysis of amines and amino acids in micro-quantities of tissue. In: *Progress in Neurobiology* (Eds. G. A. Kerkut and J. W. Phillis), Vol. 1, part 4, pp. 301–332. Pergamon Press, Oxford.
OSBORNE, N. N. (1973b) Tryptophan metabolism in characterised neurons of *Helix. Brit. J. Pharmacol.* **49**, 546–549.
OSBORNE, N. N. (1973c) Microbiological and physiological studies on an identified serotonergic neuron in the snail *Helix pomatia. Malacologia*, **14,** 97–106.
OSBORNE, N. N. (1973d) Application of the dansyl procedure to study the metabolism and accumulation of 5-hydroxytryptamine in characterised neurons of *Helix pomatia. Brit. J. Pharmacol.* **49**, 170P–171P.
OSBORNE, N. N. and COTTRELL, G. A. (1970) Transport of amines along the visceral nerve of *Helix pomatia. Z. Zellforsch.* **109**, 171–179.
OSBORNE, N. N. and COTTRELL, G. A. (1971) Distribution of biogenic amines in the slug *Limax maximus. Z. Zellforsch.* **112**, 15–30.
OSBORNE, N. N. and COTTRELL, G. A. (1972a) Amine and amino acid microanalysis in two identified snail neurons with known characteristics. *Experientia*, **28**, 656–658.
OSBORNE, N. N. and COTTRELL, G. A. (1972b) The effect of optic tentacle removal on the transmitter content of the giant serotonin cell of *Helix aspersa. J. Neurochem.* **19**, 2363–2368.
OSBORNE, N. N. and NEUHOFF, V. (1972c) Complete depletion of a substance in the snail brain after treatment with LSD-25. *Naturwissenschaften*, **59**, 371.
OSBORNE, N. N. and NEUHOFF, V. (1973a) Neurochemical studies on characterised neurons. *Naturwissenschaften*, **60**, 78–87.
OSBORNE, N. N. and NEUHOFF, V. (1973b) Microchromatographic detection of tryptamine in nervous tissue. *Naturwissenschaften*, **60**, 392.
OSBORNE, N. N. and NEUHOFF, V. (1974) *In vitro* experiments on the metabolism, accumulation and release of serotonin in the nervous system of the snail *Helix pomatia. J. Neurochem.* **22**, 363–371.
OSBORNE, N. N., BRIEL, G. and NEUHOFF, V. (1971) Distribution of GABA and other amino acids in different tissues of the gastropod mollusc *Helix pomatia*, including *in vitro* experiments with ^{14}C-glucose and ^{14}C-glutamic acid. *Intern. J. Neuroscience*, **1**, 265–272.
OSBORNE, N. N., POWELL, B. and COTTRELL, G. A. (1972a) The effect of electrical stimulation on the levels of free amino acids and related compounds in the snail brain. *Brain Research*, **41**, 379–386.
OSBORNE, N. N., BRIEL, G. and NEUHOFF, V. (1972b) The amine and amino acid composition in the Retzius cells of the leech *Hirudo medicinalis. Experientia*, **28**, 1015–1018.
OSBORNE, N. N., PRIGGEMEIER, E. and NEUHOFF, V. (1973a) The effect of imipramine and nialamide on the accumulation of serotonin in snail (*Helix pomatia*) nervous tissue. *Experientia*, **29,** 1351–1352.
OSBORNE, N. N., SZCZEPANIAK, A. C. and NEUHOFF, V. (1973b) Amines and amino acids in identified neurons of *Helix pomatia. Int. J. Neuroscience*, **5**, 125–131.
OSBORNE, N. N., MAIER, M. and NEUHOFF, V. (1973c) A preliminary investigation into the effect of LSD-25 and lithium on the amines and amino acids in the snail central ganglia. *Biochem. Soc. Trans.* **1**, 118–120.
OSBORNE, N. N., WAEHNELDT, T. V. and NEUHOFF, V. (1973d) Effect of LSD-25 on the contents of amines, amino acids and proteins of the snail (*Helix pomatia*) brain. *Comp. Gen. Pharmacol.* **4,** 259–266).
PAGE, I. H. and CARLSSON, A. (1970) Serotonin. In: *Handbook of Neurochemistry* (Ed. A. Lajtha), Vol. IV, pp. 251–261. Plenum Press, New York.
RAMIREZ, A. E., de GUGHELMONE, A. E. R. and GOMEZ, C. J. (1966) Influence of neonatal hypothyroidism on amino acids in developing rat brain. *J. Neurochem.* **13**, 1017–1025.
ROBERTS, E. (1964) Comparative aspects of the distribution of ninhydrin-reactive constituents in nervous tissue. In: *Comparative Neurochemistry* (Ed. D. Eichter), pp. 167–178. Pergamon, Oxford.
ROBERTS, P. J., KEEN, P. and MITCHELL, J. F. (1973) The distribution and axonal transport of free amino acids and related compounds in the dorsal sensory neuron of the rat, as determined by the dansyl reaction. *J. Neurochem.* **21**, 199–210.
RUDE, S., COGGESHALL, R. E. and van ORDEN III, L. S. (1969) Chemical and ultrastructural identification of 5-hydroxytryptamine in an identified neuron. *J. Cell. Biol.* **41**, 832–854.

SAAVENDRA, J. M. and AXELROD, J. (1972) Psychotomimetic N-methylated tryptamines. Formation in brain *in vivo* and *in vitro*. *Science*, **175**, 1365–1366.

SAELENS, J. K., SCHOEN, M. S. and KOVACSICS, G. B. (1967) An enzyme assay for norepinephrine in brain tissue. *Biochem. Pharmacol.* **16**, 1043–1049.

SANGER, F. (1945) The free amino groups of insulin. *Biochem. J.* **39**, 507–515.

SEGAWA, T. and KURUMA, I. (1968) The influence of drugs on the uptake of 5-hydroxytryptamine by nerve ending particles of rabbit brain stem. *J. Pharm. Pharmacol.* **20**, 320–322.

SEILER, N. (1970) Use of the dansyl reaction in biochemical analysis. In: *Methods of Biochemical Analysis* (Ed. D. Glick), Vol. 18, pp. 259–337.

SEILER, N. and WIECHMANN, M. (1970) TLC analysis of amines as their dans-derivatives. In: *Progress in Thin-layer Chromatography and Related Methods*, Vol. 1, pp. 94–144. Ann Arbor–Humphrey, Ann Arbor, Mich.

SEN, N. P. and McGEER, P. L. (1963) Gas chromatography of phenolic and catecholic amines as the trimethyl silyl esters. *Biochem. Biophys. Res. Commun.* **13**, 390–393.

SCHILDKRAUT, J. J., SCHANBERG, S. M., BREESE, G. R. and KOPIN, I. J. (1967) Norepinephrine metabolism and drugs used in the affective disorders: a possible mechanism of action. *Amer. J. Psychiat.* **124**, 600–608.

SCHOU, M. (1968) Lithium in psychiatric therapy and prophylaxis. *J. Psychiat. Res.* **6**, 67–95.

SHIELDS, P. J. and ECCLESTON, D. (1972) Effect of electrical stimulation of rat mid-brain on 5-hydroxytryptamine synthesis as determined by a sensitive radioisotope method. *J. Neurochem.* **19**, 265–272.

SNODGRASS, S. R. and HORN, A. S. (1973) An assay procedure for tryptamine in brain and spinal cord using its [^{3}H]-dansyl derivative. *J. Neurochem.* **21**, 687–696.

SPIVAK, V. A., SCHERBUKHIN, V. V., ORLOV, V. M. and VARSHAVSKY, La. M. (1971) Quantitative ultramicroanalysis of amino acids in the form of their DNS-derivatives, II. On the use of the dansylation reaction for quantitative estimation of amino acids. *Analyt. Biochem.* **39**, 271–281.

STEPHENS, G. C. and SCHINSKE, R. A. (1961) Uptake of amino acids by marine invertebrates. *Limnol. and Oceanog.* **6**, 175–181.

TANGE, J. and HARTLEY, B. S. (1970) Amino acid sequences around the disulphide bridges and methionine residues of porcine pepsin. *Biochem. J.* **118**, 611–623.

TWAROG, B. M. (1959) The pharmacology of a molluscan smooth muscle. *Brit. J. Pharmacol.* **14**, 404–407.

UNGAR, G. (1972) Molecular coding of information in the nervous system. *Naturwissenschaften*, **59**, 85–91.

UNGAR, G., DESIDERIO, D. M. and PARR, W. (1972) Isolation, identification and synthesis of a specific-behaviour inducing brain peptide. *Nature*, **238**, 198–202.

USHERWOOD, P. N. R. (1971) Evidence for release of glutamate from arthropod motor nerve endings. *Proc. Int. Union Physiol. Science*, **8**, 251–252.

WEBER, G. (1952) Polarisation of the fluorescence of macromolecules. 2. Fluorescent conjugates of ovalbumin and bovine serum albumin. *Biochem. J.* **51**, 155–167.

WEIL-MALHERBE, H. and BONE, A. D. (1952) The chemical estimation of adrenaline-like substance in blood. *Biochem. J.* **51**, 311–318.

WEISE, M. and EISENBACH, G. M. (1972) Quantitative determination of amino acids in the 10^{-14} molar range by scanning photometry. *Experientia*, **28**, 245–247.

WELSH, L. H. (1952) O^3,O^4,N-triacetyl-(−)-epinephrine. *J. Amer. Chem. Soc.* **74**, 4967–4968.

WELSH, L. H. (1955) The analysis of solutions of epinephrine and norepinephrine. *J. Amer. Pharm. Ass.* **44**, 507–514.

WELSH, J. H. and McCOY, A. G. (1957) Actions of d-lysergic acid diethylamine and its 2-bromo derivative on the heart of *Venus mercenaria*. *Science*, **125**, 348.

WERMAN, R. (1971) Electrophysiological aspects of the pharmacology of CNS inhibition. *Proc. Int. Union. Physiol. Science*, **8**, 188–189.

WOODS, K. R. and WANG, K. T. (1967) Separation of dansyl-amino acids by polyamide chromatography. *Biochim. Biophys. Acta*, **133**, 369–370.

ZANETTA, J. P., VINCENDON, G., MANDEL, P. and GOMBOS, G. (1970) The utilisation of 2-dimethylaminonaphthalene-S-sulphonyl-chloride for quantitative determination of free amino acids and partial analysis of primary structure of proteins. *J. Chromat.* **51**, 441–458.

CHAPTER 8

Microdetermination of Phospholipids

1. Introduction

As discussed on page 1, the biochemical characterisation of the neuron is necessary in order to elucidate its function in the dynamics of the nervous system. Among the many biochemical components that constitute the nerve cell, to know the precise lipid composition may be a prerequisite for the understanding of the structure and function of the neuron, since much evidence has accumulated to implicate lipids in the molecular organization of membranes and in the actual role of neuronal excitation.

The lipids in vertebrate nervous tissue can be divided roughly into polar lipids, e.g. phospholipids, and apolar lipids, e.g. cholesterol. Of these, the phospholipids, which are the prime concern of this chapter, make up about 25% of the dry weight of mammalian brain. This is probably the reason why most students of phospholipid biochemistry use the brain as their source of material. The well-characterised phospholipids in the brain are phosphatidylserine (1,2-diacyl-sn-glycero-3-phosphoryl-L-serine), phosphoinositides, sphingomyelin (N-acyl-4-sphingenine-1-phosphorylcholine), phosphatidylcholine (1,2-diacyl-sn-glycero-3-phosphorylcholine), ethanolamine plasmalogen (1-*cis*-alk-1′-enyl-2-acyl-sn-glycero-3-phosphorylethanolamine), phosphatidylethanolamine (1,2-diacyl-sn-glycero-3-phosphoryl-ethanolamine), lysophosphatidylethanolamine and cardiolipin (diphosphatidyl-glycerol). Phosphatidylserine was first isolated by Folch (1942) and occurs in nervous tissue in the form of a salt with potassium as the predominant cation. It can be isolated in a pure form by following the procedure of Sanders (1967). Though Folch and Wooley (1942) isolated phosphoinositides it was Hörhammer *et al.* (1959) who demonstrated that phosphoinositides consists of a mixture of three components, viz. mono-, di- and triphosphoinositides. These observations have since been confirmed by a number of workers and it has moreover been shown that triphosphoinositides is particularly labile and is hydrolysed in the brain very quickly after death. The actual structure of sphingomyelin was only determined in 1960 (Zabin, 1960), though its existence was known of before the turn of the twentieth century (Thudichum, 1884). Pure samples of sphingomyelin can be prepared by applying the method of Carter *et al.* (1947). Phosphatidylcholine, known sometimes as lecithin, was in fact the first phospholipid to be isolated from brain tissue (see Ansell and Hawthorne, 1964), mainly because of its high levels, which account for 25% of total vertebrate brain tissue phospholipid. This contrasts with phosphatidylethanolamine, which was shown to be a brain constituent only in 1913 and which can be isolated by the technique of Ansell and Spanner (1971). The observation that the brain contains ethanolamine plasmalogen was even later (Feulgen and Bersin, 1939), and it was not until 1957 that Rapport *et al.* demonstrated it to be a lysoplasmalogen, i.e. deacylated pure plasmalogen. Ethanolamine plasmalogen can be isolated in a fairly pure form according to the procedure of Ansell and Spanner (1963). Finally, cardiolipin and lysophosphatidylethanolamine both occur in the brain, but only in minute amounts. Indeed there is controversy as to whether the latter is not in fact an artefact brought about by the isolation procedure, though recently enzymes which can form this lipid have been detected in brain tissue (Gallai-Hatchard *et al.*, 1962). Though the precise chemical configuration of cardiolipin is not known, it can be isolated in a relatively pure form after the method of Courtada *et al.* (1967).

The normal procedure for analysing phospholipids in nervous tissues can be divided into the following methodological steps: Extraction; separation of lipids from non-lipid contaminants; chromatographic separation and identification and quantification of each phospholipid. For the extraction procedure, chloroform/methanol (2:1 by vol) is normally used (Folch *et al.*, 1957). The extraction can be intensified by multiple rather than by a single extraction, or by the addition of an antioxidant (Rouser *et al.*, 1967), or by extraction in a nitrogen atmosphere. The separation of lipid from non-lipid contaminants is usually accomplished by solvent partition (Folch *et al.*, 1957) or partition column chromatography (Rouser *et al.*, 1967). Fractionation of the individual lipids is mostly by chromatography, either column or thin layer or a combination of the two. As thin-layer chromatography is more rapid than column chromatography, it is the more common method. Most of the lipid classes in nervous tissues can be easily separated by thin layer chromatography, following the procedure of Rouser *et al.* (1967). However, there are several advantages in using column to thin layer chromatography. Not only can larger amounts of the individual lipids be isolated in a shorter time, but the columns can also be used to concentrate minor components which may not be detectable by thin-layer chromatography alone. The advantage and limitations of the various chromatographic procedures used for the separation of lipids have been fully described by Rouser *et al.* (1967).

The final step in phospholipid analysis is that of identification and quantitative determination. Various substances react with phospholipids and the reaction product can be used to identify the individual phospholipids (see page 126), and often the same reactions serve for quantification, by either weighing or measuring the intensity of colour produced.

The purpose of this chapter is to describe procedures recently developed in this laboratory for the analysis of phospholipids at the cellular level (Schiefer and Neuhoff, 1971; Althaus and Neuhoff, 1973). It will become evident that although these methods are similar to the macroprocedures, certain steps are alone peculiar to the former. The degree of sensitivity is greater than that of the micromethods lately evolved by Kleinig and Lempert (1970) and Svetashev and Vaskovsky (1972), who used 6 × 6 cm silica gel thin layer plates to detect nanomole quantities of individual phospholipids.

2. Theoretical and Practical Considerations

The quantitative fluorometric analysis of individual phospholipids in the microscale can be summarised in the following way:

Homogenise nerve tissue in chloroform/methanol (2:1 by vol), leave 1 hr then centrifuge

↓

Remove solvent system and repeat homogenisation process with new chloroform/methanol (2:1 by vol) and centrifuge, Repeat once more

↓

Combine supernatants and reduce volume to minute amount by evaporating with nitrogen gas

↓

Apply sample to a microplate of silica gel

↓

Chromatograph sample using one, two or more solvent systems

↓

Visualise phospholipids by exposing chromatography plate to iodine vapour and mark their positions on the plate

↓

Allow iodine to evaporate from chromatogram, then remove individual spots corresponding to phospholipids and place in microtubes

↓

Elute individual phospholipids from silica gel in tubes with chloroform/methanol (2:1 by vol)—repeat each extraction five times

↓

Pool chloroform/methanol solutions from a single phospholipid and transfer to a quartz cuvette

↓

Dry samples in a cuvette with a stream of nitrogen and analyse the content fluorometrically by measuring the fluorescence after the addition of cyclohexane/rhodamine/acetic acid solution.

(a) ADVANTAGES OF MICROCHROMATOGRAPHY

Thin-layer chromatography of phospholipids on the microscale has a number of advantages when compared to macromethods. Not only can very small amounts of phospholipids be fractionated but the whole chromatographic procedure can be completed in only a fraction of the time (about 8 min) required for macroprocedures. In addition to these points, the quality of separation of the individual spots is much better. This is due to the small distance required for development, so minimising diffusion of the phospholipid to surrounding areas and resulting in small definite spots. Moreover, solvent systems which migrate slowly can conveniently be used on the microscale, while the same solvent can prove impracticable for normal procedures because of the long developing times.

(b) PREPARATION OF MICROCHROMATOGRAPHY PLATES

Cover glasses (24 × 48 mm or 24 × 32 mm) or plates of thin glass cut to measure 45 × 26 mm are carefully cleaned in chromic and sulphuric acid and thereafter rinsed in running water and finally a chloroform/methanol mixture. The microplates are then held at one end with cross-action tweezers (see Fig. 59) and dipped into a slurry suspension of silica gel in chloroform and placed horizontally so that they can air dry. The gel suspension is made by mixing 30 g of silica gel (Camag D-O. Muttenz, Switzerland) with 130 ml chloroform and shaking it vigorously in a screw-cap bottle. Alternatively, the suspension may be obtained by homogenisation of the silica gel in chloroform using an Ultra-turrax. The thickness of the silica gel/chloroform suspension on the microplates should be about 50 μm, otherwise the layer can shed off on drying. A slight increase in the amount of

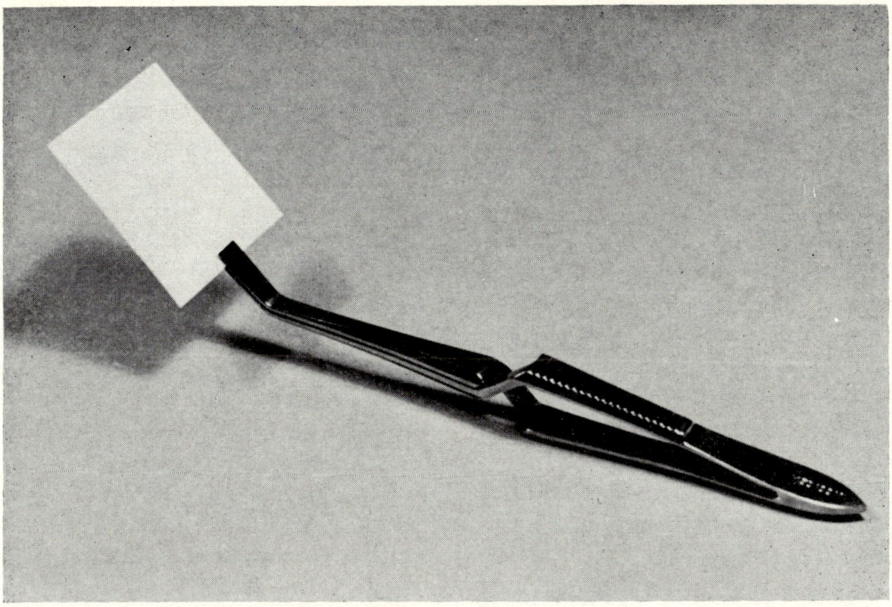

FIG. 59. This photograph shows a pair of cross-action forceps holding a 3 × 3 cm polyamide layer.

chloroform used results in thinner layers, though the actual thickness of the layer will also depend on how long the plate is dipped into the suspension. It is advisable to use the prepared plates within 30 min or else the gel layer often absorbs too much water and the separation is impaired.

Of importance is the nature of the silica gel, which should be of finest quality. In practice, the D-O quality obtainable from Camag (Muttenz, Switzerland) is suitable, although often different batches can vary. Should the particle size be too large or inhomogeneous, it is best to purify the silica gel by placing it in a measuring cylinder and an homogeneous suspension achieved by shaking it vigorously in distilled water. After allowing the suspension to stand until about half of the silica has settled, the remaining suspension is decanted and the water removed in a rotary evaporator. The last traces of water can be removed by treating the silica with methanol and evaporating off the alcohol (see Neuhoff, 1973). One difficulty that arises with silica gel plates is that often the fragile silica layer is damaged in the chromatographic procedure. It is particularly susceptible during the spraying of the plates for the detection of certain phospholipids. This can be partly overcome by adding Plaster of Paris to the silica gel when preparing the plates. In these instances the gel suspension is made up of 30 g silica gel, 3 g $CaSO_4.2H_2O$ (Plaster of Paris) and 130 ml chloroform/methanol/water (30:20:4 by vol). The plates are coated as described above, then held for about 4 sec over steam and finally air-dried. Thus a fairly stable plate is produced, allowing the use of spray reagents.

Another way of stabilising the fragile silica layer is to use waterglass. This procedure, developed by Althaus and Neuhoff (1973) has the advantage of fixing the silica gel very tightly to the glass sheet and increasing the separating quality of the silica gel. This is carried out by simply chromatographing the silica gel covered plates with waterglass/water (1:20 by vol) in an ascending way and thereafter drying the plate. The impregnated plates are then precleaned by ascending chromatography, but in the opposite direction to that previously, using chloroform/methanol/acetic acid (6:3:3 by vol). The top 4 mm of the layer is then scraped off to remove any impurities which migrate with the chloroform/methanol/acetic acid front. After a final drying of the plates they can be stored until required. The advantage of this technique is not only the complete stabilisation of the silica layer, but also the possibility of making autoradiograms of separated labelled phospholipids.

(c) CHROMATOGRAPHY

Cover glasses measuring 32 × 24 mm and spread with thin silica layers are best developed in the first dimension with the solvent system chloroform/methanol/water (65:25:4 by vol) and in the second dimension n-butanol/acetic acid/water (3:1:1 by vol). Chromatograms are developed in covered beakers which are partially lined with fitted paper soaked in the developing solvent in order to saturate the atmosphere within the beaker. After each chromatography the plate is thoroughly dried by a stream of nitrogen or air produced from a hair-dryer. Using these solvent systems, neutral lipids, cerebroside ,sulfatide, phosphatidylethanolamine, phosphatidylcholine, sphingomyelin and cardiolipin can be clearly fractionated, though phosphoinositides and phosphatidylserine have similar chromatographic positions (see Fig. 60). A similar separation of phospholipids on silica

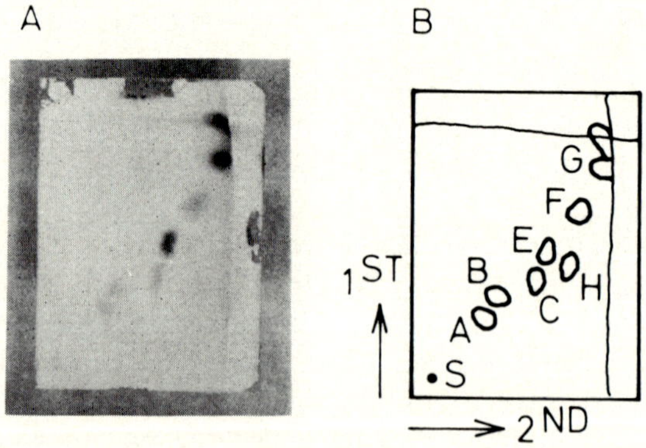

FIG. 60. A photograph of a two-dimensional microchromatogram (A) and a corresponding map (B) of a total lipid extract from a punched-out portion of cat's spinal cord stained with iodine vapour. The original size of the thin-layer chromatogram measured 32 × 24 mm, and it was developed in the 1st dimension with chloroform/methanol/water (65:25:4 by vol) and in the 2nd dimension with n-butanol/acetic acid/water (60:20:20 by vol). Substances are identified as (S) starting point, (A) sphingomyelin, (B) phosphatidylcholine, (C) phosphoinositide and phosphatidylserine, (E) phosphatidylethanolamine, (F) cerebroside, (G) neutral lipids and (H) sulfatide.

gel layers partially stabilised with Plaster of Paris can be achieved by use of chloroform/methanol/cyclohexane/water (65:30:10:4 by vol) for the first dimension and n-butanol/chloroform/acetic acid/water (60:10:20:20 by vol) for the second dimension. In order to resolve phosphoinositides and phosphatidylserine in particular, layers partly stabilised with Plaster of Paris can be developed in chloroform/methanol/25% NH_3 (60:35:5 by vol) for the first chromatography and chloroform/methanol/25% NH_3 (35:60:5 by vol) for the second chromatography in the second dimension (see Fig. 61).

Waterglass impregnated silica gel plates measuring 24 × 48 mm can conveniently be used to separate both phospholipids and neutral lipids by one dimensional chromatography,

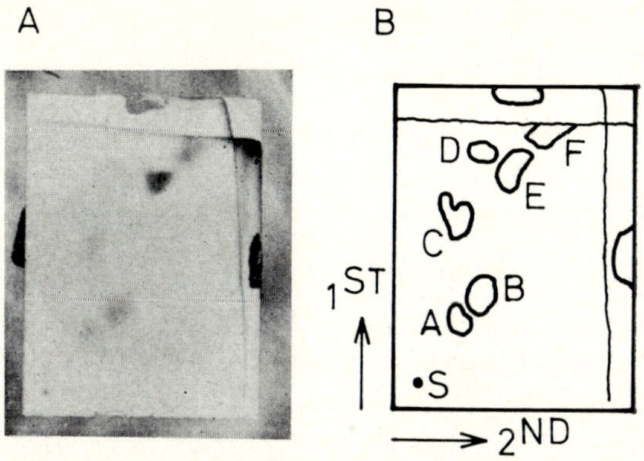

FIG. 61. A photograph of a two-dimensional microchromatogram (A) and a corresponding map (B) of a total lipid extract from a punched out portion of cat's spinal cord stained with iodine vapour. The original size of the thin-layer chromatogram measured 32 × 24 mm, and it was developed in the 1st dimension with chloroform/methanol/7 N NH_4OH (60:35:5 by vol) and in the 2nd dimension with chloroform/methanol 7 N_4OH (35:60:5 by vol). Substances are identified as (S) starting point, (A) sphingomyelin, (B) phosphatidylcholine, (C) phosphatidylserine, (D) phosphoinisitide, (E) phosphatidylethanolamine and (F) cerebrosides.

using a step-wise technique of development (Althaus and Neuhoff, 1973). The first chromatography (ascending) is allowed to proceed until the solvent front has migrated a distance of 20 mm. The solvent system used is chloroform/methanol/water/acetic acid (65:35:3.6:2 by vol). After careful drying of the plate in a stream of nitrogen, a second development in the same direction is carried out. This time the chromatography mixture is chloroform/methanol/water/acetic acid/acetone (60:25:5.3:10:5 by vol) and the front is allowed to migrate a distance of 30 mm before termination. The plate is once again dried and a third chromatography in the same direction carried out by the use of n-hexane/diethylether/acetic acid (70:3:1 by vol), and the development is allowed to proceed to the top edge of the plate. By employing this procedure (see Fig. 62) a good separation of sphingomyelin, phosphatidylcholine, phosphoinositides, phosphatidylserine, phosphatidylethanolamine and cardiolipin can be achieved. It is worthwhile noting that phosphoinositides and phosphatidylserine are also resolved.

Separation of neutral fats, free fatty acids, cholesterol, cholesterol esters and phospholipids can be accomplished on a single microplate (Hubmann, 1973). In this procedure a microscale slide cut down to 45 × 26 mm and covered with a slurry of silica gel proves sufficient. The silica gel on the back of the plate is wiped off, the thin-layer microplate precleaned by chromatography (ascending) with n-butanol/acetic acid/water/chloroform/methanol/hexane (6:2:2:6:10:8 by vol) and then air-dried. After applying the sample to one corner of the plate, a distance of 1.5 cm below the upper margin is clearly marked. The chromatogram is then developed with the solvent system chloroform/methanol/water (65:25:4 by vol), terminating the development when the solvent front has reached the marking. The plate is dried and a second chromatography in the same direction but with the solvent system n-hexane/diethylether/acetic acid (82:20:2 by vol) is carried out, this time allowing the solvent front to reach the upper edge of the plate. When dried the chromatogram is then turned through 90° and a third chromatography using the solvent system n-hexane/diethylether/acetic acid (85:20:2 by vol) is effected. When the

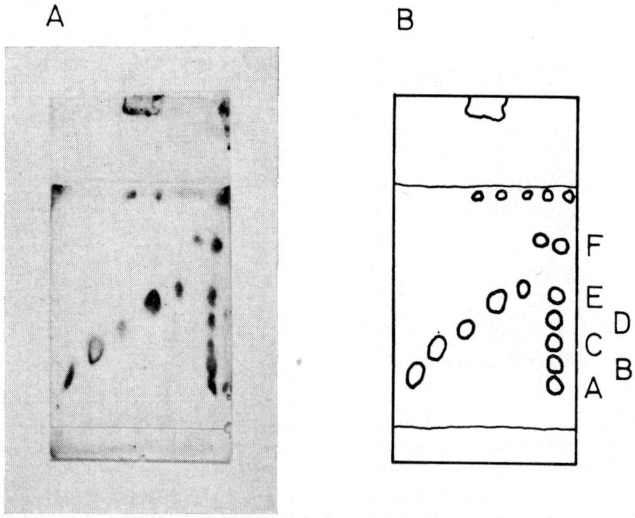

FIG. 62. A photograph of a one-dimensional microchromatogram (A) and a corresponding map (B) of a standard solution of phospholipids and individual phospholipids, stained with iodine vapour. The original size of the thin-layer chromatogram measured 48 × 24 mm, and was developed in the solvent systems as described in the text. Substances are identified as (A) sphingomyelin, (B) phosphatidylcholine, (C) phosphoinositide, (D) phosphatidylserine, (E) phosphatidylethanolamine and (F) cardiolipin. (Photograph by courtesy of H. H. Althaus.)

solvent front reaches the upper edge the chromatography is terminated and the chromatogram dried. One part of the plate is then isolated so that the other can still be used for chromatography by removing the silica gel with a spatula in the manner shown by the scheme in Fig. 63. The solvent system n-butanol/acetic acid/water (3:1:1 by vol) is used to resolve the part of the plate designated 'X' by development in the second dimension and allowing the chromatography to proceed to the upper edge. The scheme shown in Fig. 63 clearly demonstrates the various processes in this procedure, while the chromatograms in Fig. 64 illustrate its use for the separation of lipids from human serum.

(d) IDENTIFICATION OF LIPIDS

Most of the lipids on microchromatograms can be visualised as yellow spots in an iodine atmosphere and identified by comparison with chromatograms of authentic lipids. This is normally carried out by placing the microchromatogram in a Petri dish together with a few crystals of iodine and covering the dish. The lipid spots are then identifiable within 5 min. The advantage of using iodine is that the phospholipid spots can be marked around their perimeters with a very fine needle, and the iodine then allowed to evaporate away from the spots simply by removing the chromatogram from the iodine vapour. The evaporation process can take as little as 10 min. This then makes it possible to remove the spots from the chromatogram, elute the phospholipids and measure the content of each by fluorometry.

Phospholipids on microchromatograms can also be visualised by use of appropriate sprays. Though microplates partially stabilised with Plaster of Paris or permanently fixed by waterglass impregnation to 'hold' the silica gel on the plates are most suited for the use

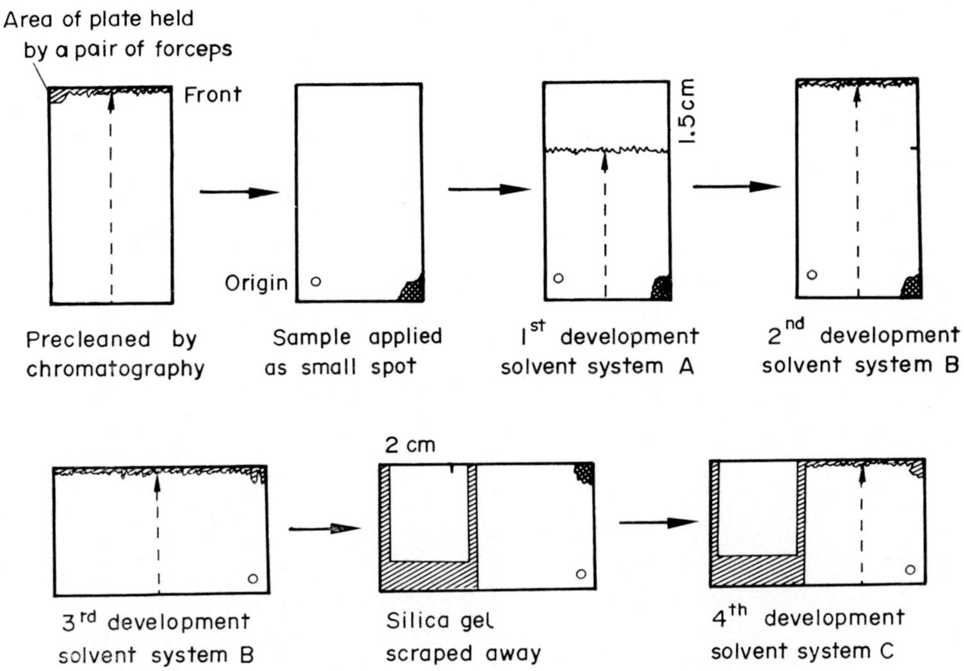

FIG. 63. The schematic procedure for the two-dimensional two-step chromatography of lipids in the microscale (see Fig. 64). (Photograph by courtesy of F. H. Hubman.)

of the sprays, normal silica gel covered microplates may also be sprayed if great care is taken. A particularly useful spray reagent is the phosphatic solution described by Hanes and Isherwood (1949) which detects as little as 50 ng phospholipid. In practice the plate is sprayed with a spray gun from a distance of about 30 cm, dried in warm air and visualised under ultraviolet light. The phospholipids can then be identified as blue spots on a white background. The following spray reagents are especially suitable for detecting phospholipids:

(i) *Phosphate spray, after Hanes and Isherwood* (*1949*)

This is a mixture containing 5 ml of 60% perchloric acid, 10 ml N HCl and 20 ml of 5% (w/v) ammonium molybdate solution made up to 100 ml with water. This may be stored in a polythene bottle at 4°C for very long periods without deterioration.

(ii) *Phosphate spray, after Dittmer and Lester* (*1964*)

(a) 40.1 g of MoO_3 is added to 1 litre of 25 N H_2SO_4 and the mixture is gently boiled until the solid has dissolved.

(b) 1.78 g of powdered molybdenum is added to 500 ml of solution (a), the mixture is boiled gently for 15 min, cooled and decanted.

The spray reagent is then made up with 1 vol solution (a) and 1 vol of solution (b) to 4 vol of water. This yellow-greenish solution is used to visualise the phospholipids. Exposure to ultraviolet light is not required.

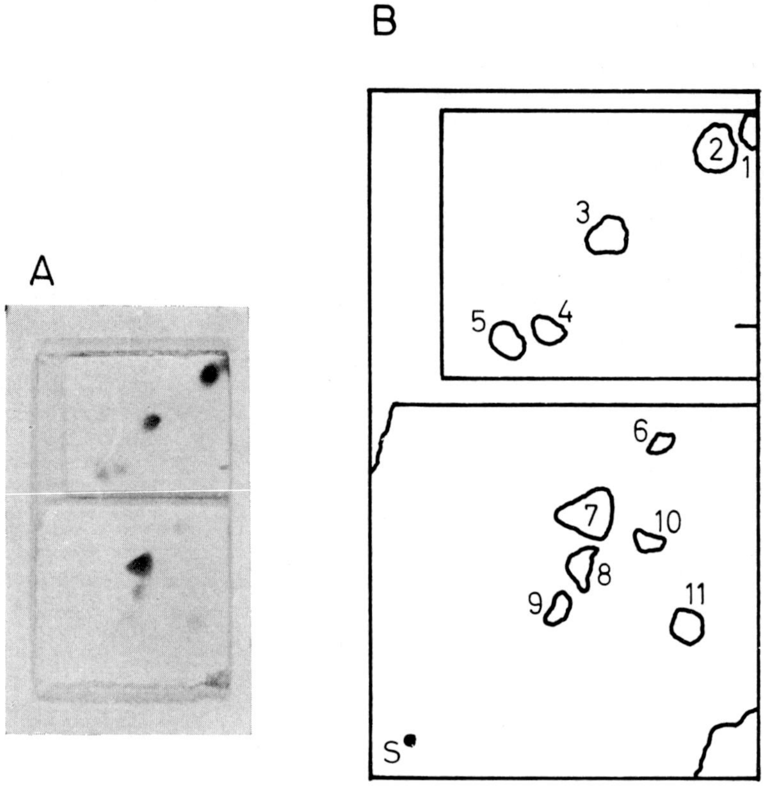

FIG. 64. Two-dimensional, two-step microchromatography (original size of plate 45 × 26 mm) of the total lipid extract from human serum (A) and a map (B) visualised by iodine vapour. Solvent systems used are as described in the text and the direction of chromatography with each solvent system is as described in Fig. 63. S, starting point; 1, front; 2, cholesterol esters; 3, triglycerides; 4, free fatty acids; 5, cholesterol; 6, cephalin; 7, lecithin; 8, sphingomyelin; 9, lysolecithin; 10 and 11, unidentified substances. (Photograph by courtesy of F. H. Hubman.)

(iii) *Phosphate spray, after Vaskovsky and Kostetsky (1968)*

(a) 16 g of ammonium molybdate are dissolved in 120 ml water.

(b) To 40 ml of concentrated HCl are added 10 ml mercury and 80 ml solution (a), the mixture is shaken gently for 30 min and then filtered.

The spray reagent is made up with 40 ml solution (a), 200 ml concentrated H_2SO_4 and the whole solution (b). This is then diluted to 1 litre with water. Ultraviolet light is necessary to observe the phospholipid spots clearly.

(iv) *Dragendorff reagent, after Skidmore and Entenman (1962)*

(a) 1.7 g of $Bi(NO_3)_3 . 5\,H_2O$ are dissolved in 100 ml of 20% (v/v) acetic acid.

(b) 40 g of KI are dissolved in 100 ml of water.

The spray reagent consists of 4 ml of solution (a) and 1 ml solution (b) diluted to 20 ml with water. The spots are observed under ultraviolet light.

(v) *Rhodamine spray, after Jones et al. (1966)*

The mixture consists of 100 mg Rhodamine B, 35 mg 2-7-dichlorofluoresceine, 150 ml diethylether, 70 ml 95% ethanol and 16 ml water. Chromatograms are sprayed, allowed to dry and then viewed under ultraviolet light. Lipids fluoresce orange or orange-purple on a green background.

(e) QUANTIFICATION

Fluorometry is used to quantify the micro amounts of individual phospholipids. The procedure is based on the fluorometric analysis of the fluorescing phospholipid–rhodamine complex, formed by the interaction of an acid phosphatide with the basic dye. It is known that cholesterol and triglycerides neither form a fluorescent–rhodamine complex, nor change the apparent shape of the fluorescent spectrum of a phosphatide–rhodamine complex. Since each phosphatide reacts slightly differently with rhodamine 6GO, the conditions of reaction between the various phospholipids and rhodamine to produce maximum fluorescence are also slightly different.

The actual fluorescence spectrum of rhodamine 6GO dissolved in chloroform and diluted in cyclohexane is shown in Fig. 65. The excitation maximum of rhodamine is at 523 and

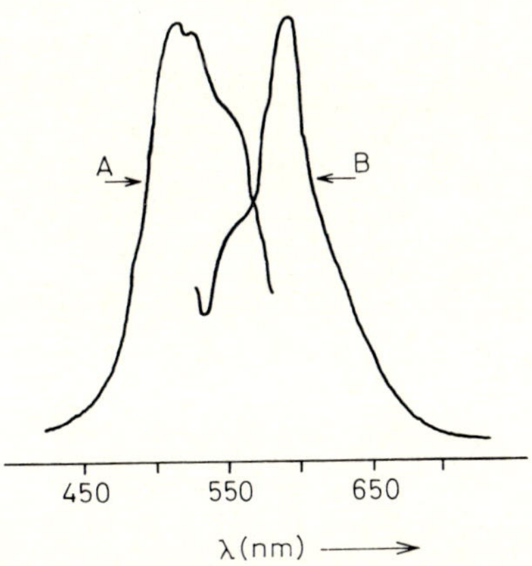

FIG. 65. Excitation (A) and emission (B) spectra of rhodamine 6GO (after Schiefer and Neuhoff, 1971) The cuvette contained 2 ml cyclohexane and 20 μl rhodamine solution (10 mg/25 ml chloroform). In order to obtain the excitation spectrum, the emission wavelength was set at 527 nm. The bandwidth for both monochromators was 100 Å.

537 nm and the emission maximum at 608 nm. The addition of acetic acid causes a shift in the excitation maximum to 527 and 555 nm. When the phosphatides are added to the cuvette, similar types of fluorescence are recorded as shown in Fig. 66, where phosphatidylethanolamine was added, though detailed analysis (Schiefer and Neuhoff, 1971) revealed that the maxima of the individual phospholipid-rhodamine complexes differ slightly (see Table 32). This variation is probably due to differences in the composition of

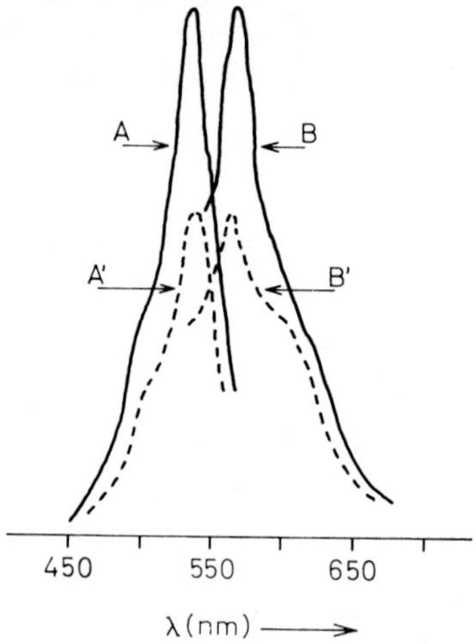

FIG. 66. Excitation (A and A′) and emission (B and B′) spectra of phosphatidylethanolamine (11.6 μg) before (A′ and B′) and after (A and B) the addition of 2 μl acetic acid (after Schiefer and Neuhoff, 1971). The cuvette contained 2 ml cyclohexane, 20 μl rhodamine solution, and in the case of the upper curve, 2 μl acetic acid. The spectra were measured using a 25 Å bandwidth on both monochromators. For the emission spectra the excitation wavelength was set at 532 nm; for the excitation spectra the emission wavelength was set at 558 nm.

TABLE 32. OPTIMAL REACTION MIXTURES AND CONDITIONS FOR THE DETERMINATION OF PHOSPHATIDES (after Schiefer and Neuhoff, 1971)

Lipid	Wave-length of excitation max. nm	Wave-length of emission max. nm	Concentr. range μg	Cyclo-hexane ml	Rhodamine (10 mg/ 25 ml $CHCl_3$) μl	Acetic acid μl
Phosphatidylcholine	535	563	0.01– 5 1 – 50	2 2	5 20	1 1
Phosphatidyl-ethanolamine*	532	558	0.01– 0.5 0.1 – 1 0.5 – 50	2 2 2	2 5 20	0.2 0.2 0.2
Phosphatidylserine	532	560	0.01– 1 0.1 – 5 1 – 30	2 2 2	2 5 20	— — —
Phosphatidyl-inositol	528	557	0.01– 2 0.1 – 30	2 2	5 20	1 2
Sphingomyelin	533	565	0.01– 0.09 0.09– 2 1 – 30	2 2 2	2 10 20	0.2 1 2
Cardiolipin	530	557	0.01– 20	2	20	2
Total lipid extract from guinea-pig brain	534	558	0.03– 0.2 nM P 0.15–130 nM P	2 2	10 20	1 2

* The procedure used did not separate any plasmalogen from phosphatidylethanolamine.

the fatty acids responsible for the hydrophobic interactions between the dye and the phosphatide. The fluorescence maximum of the individual phospholipid–rhodamine complex is also increased by the addition of acetic acid as shown in Fig. 66. The acetic acid probably influences the ionic and steric arrangements of the complex.

The optimal reaction conditions between the individual phospholipids and rhodamine 6GO at the maximum excitation and emission wavelengths can be determined in the following way. The mixture in the cuvette besides the phospholipid consists of cyclohexane, rhodamine 6GO mixture (10 mg/25 ml chloroform) and acetic acid. The cyclohexane, which is the dispersing solution, increases the sensitivity of the analysis by about twenty-fold compared to iso-octane, which is normally used. Acetic acid is particularly important for the analysis of sphingomyelin and cardiolipin (sodium salts) and also increases the sensitivity of the determination (see Fig. 66), while the rhodamine solution is required to form the fluorescent complex with the phospholipid. In practice, the conditions for maximum sensitivity are simply achieved by varying the rhodamine solution and acetic acid for definite amounts of phospholipid and cyclohexane and observing the fluorescence at the maximum excitation and emission wavelengths. As can be seen from Table 32, the optimal conditions required depend not only on the type of lipid but also on its concentration.

Before the analysis of tissue samples, standard curves of the individual phospholipids using the reaction conditions in Table 32 should be determined. Each set of standard curves should be defined for a particular fluorometer as there are often discrepancies between the exact wavelength settings on different instruments and these should be rechecked from time to time. It is possible to obtain calibration data from consecutive measurements in one cuvette by a stepwise addition of phospholipid or total phospholipid mixture dissolved in chloroform until the rhodamine 6GO is exhausted. Single identifications give identical results. The actual fluorescence (F) is measured in arbitrary units at the emission maximum and must be corrected by considering the 'blank' fluorescence, the volume of the cuvette and the amplification of the instrument,

i.e.
$$F = \frac{(\text{test value} - \text{'blank' value}) \times \text{volume}}{\text{amplification}}.$$

The 'blank' value is obtained by measuring the fluorescence of a cuvette's contents without lipids, in the conditions of the lipid measurement.

The fluorescence data can then be plotted on a log–log chart, and the standard curves calculated according to the method of linear regression: $\log y = \log b + a \times \log x$. The calculated values of the regression line, the standard error of coefficients, the standard error of observation and the corrected coefficient for each standard curve can also be calculated. Figures 67–73 show standard curves of phosphatidylcholine (Fluka), phosphatidylethanolamine (Koch-light), phosphatidylserine (Koch-light), phosphoinositides (Koch-light), sphingomyelin (Koch-light), cardiolipin (Koch-light) and total phospholipid extract from guinea-pig brain obtained by use of a Turner 210 absolute spectrofluorometer. Although it is possible to use either the luminescence or fluorescence modes on a Turner spectrofluorometer, the fluorescence mode at the excitation band width (25 nm) was used for this analysis (see Schiefer and Neuhoff, 1971).

In the analysis of phospholipids in tissue samples fractionated by microchromatography, each phospholipid is eluted from the silica gel and the content determined by fluorometry. This procedure allows as little as 0.01 μg phospholipid to be quantitatively determined.

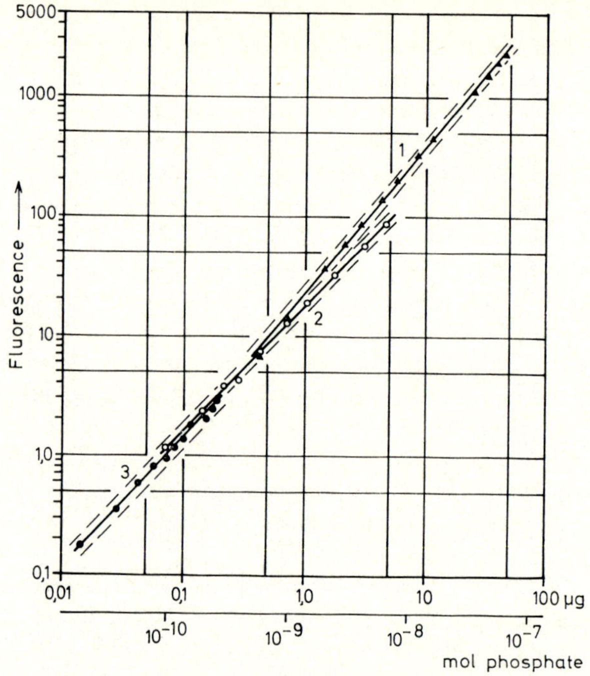

FIG. 67. Standard curves for the fluorometric determination of phosphatidylcholine (after Schiefer and Neuhoff, 1971). The excitation wavelength was set at 535 nm. The fluorescence was measured at the emission maximum (563 nm).

Curve 1: 2 ml cyclohexane, 20 μl rhodamine solution, 1 μl acetic acid.
Calculations: $\lg y = [(1.3518 \pm 0.0149) + (1.2199 \pm 0.0148)] \lg x$
$= \pm 0.0343, r = 0.9992.$

Curve 2: 2 ml cyclohexane, 5 μl rhodamine solution, 1 μl acetic acid.
Calculations: $\lg y = [(1.2586 \pm 0.0098) + (1.0466 \pm 0.0162)] \lg x$
$= \pm 0.0286, r = 0.9990.$

Curve 3: 2 ml cyclohexane, 5 μl rhodamine solution, 1 μl acetic acid.
Calculations: $\lg y = [(1.2501 \pm 0.4928) + (1.0960 \pm 0.0438)] \lg x$
$\sigma = \pm 0.0526, r = 0.9913.$

It is possible to increase this sensitivity further by *in situ* measurements (see Hezel, 1973). *In situ* measurements are based on the idea of subjecting the thin-layer microchromatography plate directly to quantitative photometry and measuring the individual phospholipids on the plate by absorption photometry in remission or alternatively by fluorometry. Using a Zeiss chromatogram spectrophotometer, Hezel (1973) was able to demonstrate the possibility of analysing 10 pg of individual substances. *In situ* measurements, besides being more sensitive, are also more accurate (less than $\pm 1\%$ inaccuracy) and quicker to carry out, and the individual substances are not required to be eluted from the microplate.

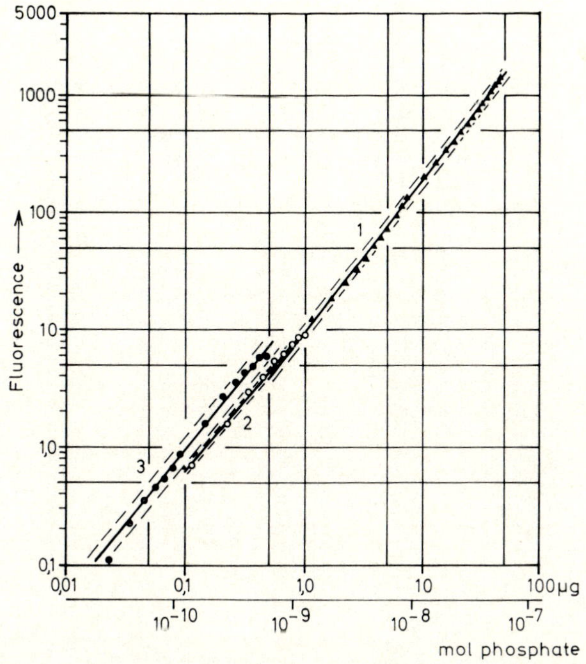

FIG. 68. Standard curves for the fluorometric determination of phosphatidylethanolamine (after Schiefer and Neuhoff, 1971). The excitation wavelength was set at 532 nm. The fluorescence was measured at the emission maximum (558 nm).

Curve 1: 2 ml cyclohexane, 20 μl rhodamine solution, 0.2 μl acetic acid.
Calculations: $\lg y = [(0.9666 \pm 0.0136) + (1.3017 \pm 0.0121)] \lg x$
$= \pm 0.0329, r = 0.9990.$

Curve 2: 2 ml cyclohexane, 5 μl rhodamine solution, 0.2 μl acetic acid.
Calculations: $\lg y = [(0.9961 \pm 0.0147) + (1.1925 \pm 0.0334)] \lg x$
$= \pm 0.0295, r = 0.9973.$

Curve 3: 2 ml cyclohexane, 2 μl rhodamine solution, 0.2 μl acetic acid.
Calculations: $\lg y = [(1.2367 \pm 0.03978) + (1.2573 \pm 0.0366)] \lg x$
$\sigma = \pm 0.0683, r = 0.9946.$

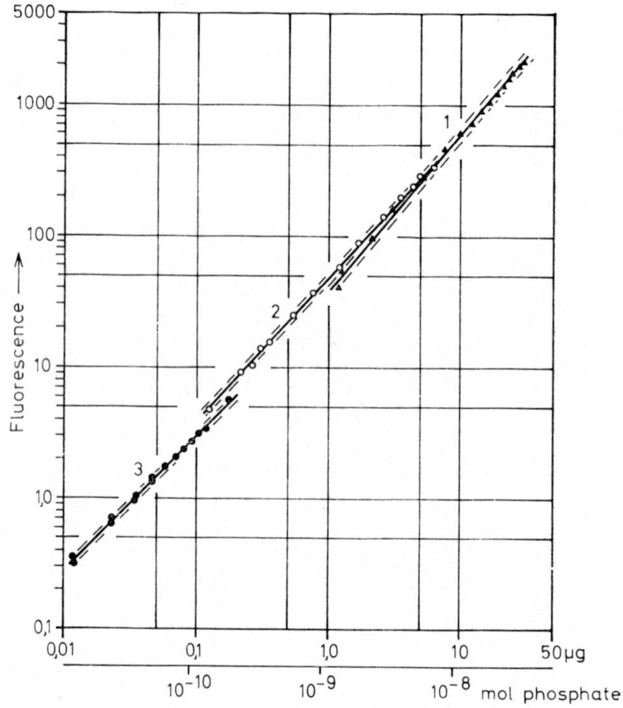

FIG. 69. Standard curves for the fluorometric determination of phosphatidylserine (after Schiefer and Neuhoff, 1971). The excitation wavelength was set at 532 nm. The fluorescence was measured at the emission maximum (560 nm).

Curve 1: 2 ml cyclohexane, 20 μl rhodamine solution.
Calculations: $\lg y = [(1.5818 \pm 0.0191) + (1.1950 \pm 0.0173)] \lg x$
$= \pm 0.0283, r = 0.9986.$

Curve 2: 2 ml cyclohexane, 5 μl rhodamine solution.
Calculations: $\lg y = [(1.6879 \pm 0.0049) + (1.0900 \pm 0.0091)] \lg x.$
$= \pm 0.0174, r = 0.9996.$

Curve 3: 2 ml cyclohexane, 2 μl rhodamine solution.
Calculations: $\lg y = [(1.4963 \pm 0.0169) + (1.0151 \pm 0.0123)] \lg x.$
$\sigma = \pm 0.0162, r = 0.9991.$

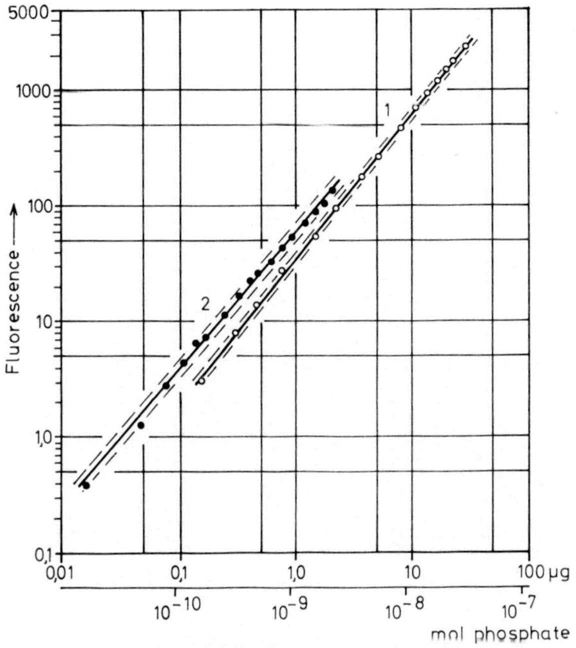

FIG. 70. Standard curves for the fluorometric determination of phosphatidylinositol (after Schiefer and Neuhoff, 1971). The excitation wavelength was set at 528 nm. The fluorescence was measured at the emission maximum (557 nm).

Curve 1: 2 ml cyclohexane, 20 μl rhodamine solution, 2 μl acetic acid.
Calculations: $\lg y = [(1.5223 \pm 0.0071) + (1.2404 \pm 0.0077)] \lg x$
$= \pm 0.0215, r = 0.9998.$

Curve 2: 2 ml cyclohexane, 5 μl rhodamine solution, 1 μl acetic acid.
Calculations: $\lg y = [(1.7458 \pm 0.0140) + (1.1654 \pm 0.0186)] \lg x$
$\sigma = \pm 0.0449, r = 0.9981.$

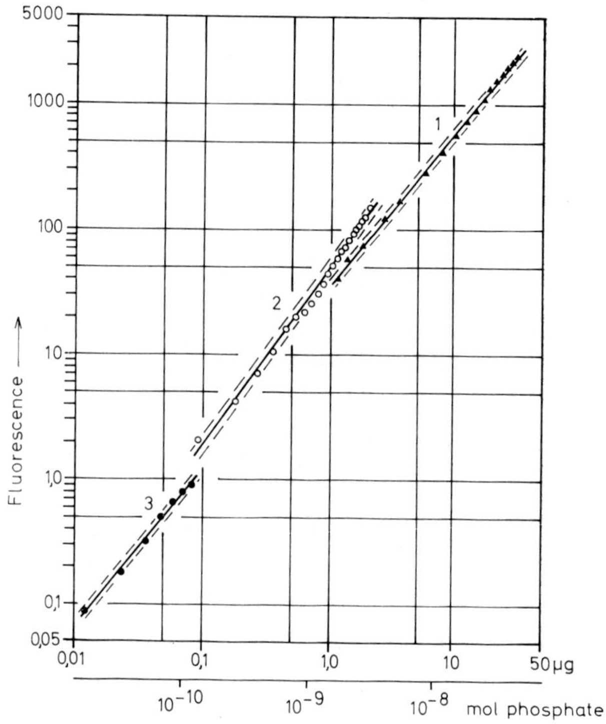

FIG. 71. Standard curves for the fluorometric determination of sphingomyelin (after Schiefer and Neuhoff, 1971). The excitation wavelength was set at 533 nm. The fluorescence was measured at the emission maximum (565 nm).

Curve 1: 2 ml cyclohexane, 20 μl rhodamine solution, 2 μl acetic acid.
Calculations: $\lg y = [(1.4952 \pm 0.0153) + (1.2301 \pm 0.0142)] \lg x$
$= \pm 0.0273, r = 0.9990.$

Curve 2: 2 ml cyclohexane, 10 μl rhodamine solution, 1 μl acetic acid.
Calculations: $\lg y = [(1.6435 \pm 0.0093) + (1.4070 \pm 0.0261)] \lg x$
$= \pm 0.0429, r = 0.9966.$

Curve 3: 2 ml cyclohexane, 2 μl rhodamine solution, 0.2 μl acetic acid.
Calculations: $\lg y = [(1.3067 \pm 0.0611) + (1.2402 \pm 0.0430)] \lg x$
$\sigma = \pm 0.0314, r = 0.9970.$

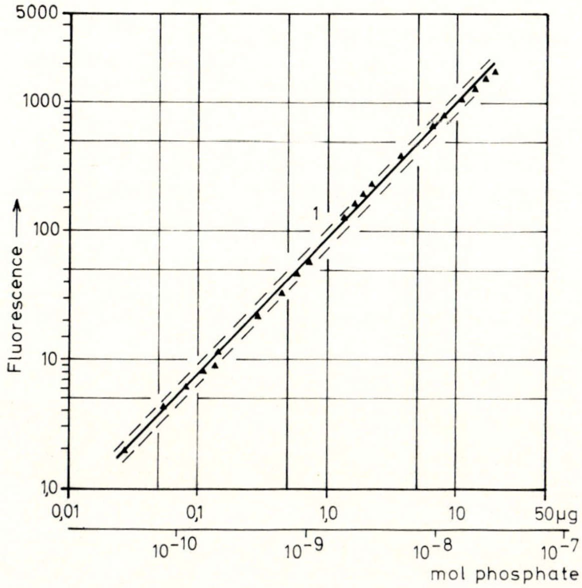

FIG. 72. Standard curve for the fluorometric determination of cardiolipin (after Schiefer and Neuhoff, 1971). The excitation wavelength was set at 530 nm. The fluorescence was measured at the emission maximum (557 nm).

2 ml cyclohexane, 20 µl rhodamine solution, 2 µl acetic acid.

Calculations: $\lg y = [(1.9449 \pm 0.0087) + (1.0613 \pm 0.0104)] \lg x$
$= \pm 0.0417, r = 0.9990.$

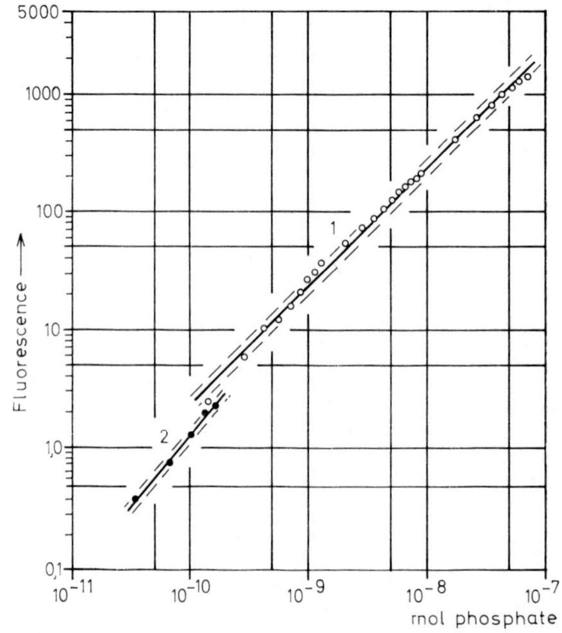

FIG. 73. Standard curves for the fluorometric determination of total phospholipid extract from guinea pig brain (after Schiefer and Neuhoff, 1971). The excitation wavelength was set at 534 nm. The fluorescence was measured at the emission maximum (558 nm).

Curve 1: 2 ml cyclohexane, 20 μl rhodamine solution, 2 μl acetic acid.
Calculations: $\lg y = [(1.3616 \pm 0.0118) + (0.9873 \pm 0.0119)] \lg x$.
$= \pm 0.0458, r = 0.9982$.

Curve 2: 2 ml cyclohexane, 10 μl rhodamine solution, 1 μl acetic acid.
Calculations: $\lg y = [(1.2892 \pm 0.0697) + (1.1652 \pm 0.0644)] \lg x$.
$= \pm 0.0356, r = 0.9954$.

(f) REMARKS ABOUT FLUOROMETRY

Any fluorometer can be used for the fluorometric analysis of phospholipids (for theory of fluorescence measurements, see e.g. Udenfriend, 1969). It is advisable to record the full spectrum rather than just the pen deflection at a known fluorescence maximum. In this way impurities can be detected. The actual origin of impurities can be endogenous, that is, associated with the tissue sample. Alternatively, they may originate from the chemicals or glassware. It is advisable therefore to use the purest grade chemicals possible, i.e. analar. Furthermore, on no account should solutions be pipetted from stock bottles, since this is often a source of error. Stock solution bottles should be tightly corked and kept preferably in the dark, and care must be taken that the stoppers are not contaminated, which can occur if they lie on a bench top, when the solutions are being transferred.

In order to keep microtubes, cuvettes, etc., clean, the protocol below can be followed.

1. Store in 11 N HNO_3 (prepared from fuming HNO_3 diluted with four times glass distilled water in ratio 1:1).
2. Rinse briefly with tap water.
3. Rinse thoroughly with glass distilled water.
4. Rinse with ethanol and repeat two or three times.
5. Rinse with chloroform and repeat two or three times.
6. Rinse with cyclohexane and repeat two or three times.
7. Dry under a stream of nitrogen.

(g) AUTORADIOGRAPHY

Both ^{14}C and ^{32}P radioactive precursors can be used to incorporate radioactivity into individual lipids. These substances can be located by autoradiography of microchromatograms. Microchromatograms stabilised by waterglass are particularly advantageous for use in autoradiography. Autoradiograms are prepared with either Agfa Gevaert Denture Ultra Rapid L film or Dupont Dosimeter film 046/545. Exposure time varies with the isotope and degree of labelling of radioactive components. Radioactive phosphorus rapidly darkens the film and the process need only take a few hours. ^{14}C–radioactivity is weaker and requires a longer time (1–6 days). However, the actual time necessary is proportional to the radioactivity on the plates and very much longer periods may be essential.

One of the greatest advantages of autoradiography is that the position of the phospholipids can be found without the use of dyes, so making it possible for the compounds to be eluted from the plate for further analysis. Moreover, the sensitivity of autoradiography for detecting small amounts of labelled substances far exceeds that of the dyes required for identification.

3. Microanalysis of Individual Phospholipids in Nervous Tissue

(a) EXTRACTION

Phospholipids are extracted from nervous tissue by homogenisation in 15 vol (v/w) chloroform/methanol (2:1 by vol). Homogenisation with a dentist's drill for 5 min at top speed (24,000 rpm, see page 53) is sufficient. Depending on the amount of nervous tissue, homogenisation is carried out either in Drummond capillaries heat-sealed at one end, or in microtubes (see page 24). After homogenisation, the microcap or microtube and contents are allowed to stand for 30 min and then centrifuged at about 40,000 *g* for 5 min. The organic phase containing the phospholipids is then removed with a fine capillary (see page 23). The pellet is extracted twice more with chloroform/methanol (2:1 by vol) by the same procedure. For the fluorometric determination of total phospholipids the organic extracts are combined in a quartz cuvette, dried under a stream of nitrogen and analysed after the addition of appropriate amounts of cyclohexane, rhodamine and acetic acid (see Table 32). For the analysis of individual phospholipids the chloroform/methanol extracts are combined in the smallest practicable size of glass tube and concentrated to a minute volume (less than 2 μl) with a stream of nitrogen.

(b) MICROCHROMATOGRAPHY

The concentrated phospholipid extract is applied to a microchromatographic plate using a finely drawn out capillary (see page 23). To ensure that the sample spot is small (less than 1 mm in diameter) the application is carried out under stereomicroscopic vision. The actual application spot should be about 3 mm from the glass edges in one corner of the microchromatogram for the separation of phospholipids in two-dimensional chromatography, or alternatively, 3 mm above the lower edge for one-dimensional chromatography. Microchromatograms should not be overloaded with phospholipids, the maximum allowable for two-dimensional chromatography being about 10 μg phospholipid. The actual quality of separation of the phospholipids depends a great deal upon the way in which the sample is applied. As pointed out earlier, the application should be, if possible, less than 1 mm in diameter. This is achieved by a capillary of appropriate diameter (see page 18), so that the amount of extract released from the capillary tip (which is controlled by pressure produced from the mouth) is automatically absorbed into a 1 mm diameter silica gel area. The capillary tip should just touch the silica layer in order to assist in the application, but care must be taken that the fragile layer is in no way damaged. It takes about 10 min to apply a 2 μl sample of phospholipids to a single microchromatographic plate. Any traces of phospholipid remaining in either the pipette capillary shaft or the microtube which contained the extract can be rinsed with a little chloroform/methanol and applied to the chromatographic plate.

Microchromatograms are developed in 50 ml beakers containing a 1 mm layer of the solvent to be used. As in all chromatographic procedures, the point of application of the phospholipids must be above the solvent layer. Chromatography is carried out in covered beakers, the insides of which are lined about two-thirds of the way up with filter paper soaked in the appropriate solvent system, in order to produce a saturated atmosphere for chromatography. When developing microchromatograms, care should be taken that the top edge of the plate leans against the glass surface of the beaker and not the saturated filter paper, otherwise the silica layer will tend to absorb solvent from the filter paper, thus disturbing the ascending chromatographic procedure. A number of different solvent systems can be used to develop the microchromatogram (see page 126), depending on the problems investigated. As the solvent systems are very labile, they must be renewed regularly and should not be kept, even in stoppered containers, for more than 6 hr. Small changes in the concentrations of the components in a solvent mixture drastically alter the pattern of separation of the individual phospholipids. Microchromatograms must be carefully dried between each development and this is best done by using a stream of nitrogen or a hair-dryer. The former is better as there is always the possibility of destroying the phospholipids by the air produced from a hair-dryer.

(c) IDENTIFICATION AND ELUTION

After the development of a microchromatogram, the plate is placed in an enclosed petri dish containing a few crystals of iodine. It is advisable to lay the plate on a small disc (e.g. plastic lid) in the petri dish so that it can be easily lifted out after the identification of the phospholipids. When the individual phospholipids are to be clearly located on the microchromatogram as yellow-brown spots (this takes about 5 min), the plate is removed from the petri dish and the perimeters of the spots marked out with a very fine glass needle. The plate is then allowed to stand in an atmosphere free of iodine until the iodine coloration has completely evaporated. It is important to let all traces of iodine disappear, since even minute quantities of the substance interfere with the fluorometric measurements. It is therefore recommended that the plate be exposed to iodine vapour for the shortest time possible, otherwise the evaporation process can take quite a long time.

After the iodine has completely evaporated, the silica gel containing the individual phospholipids is removed from the microchromatogram with a microshovel (see Fig. 74)

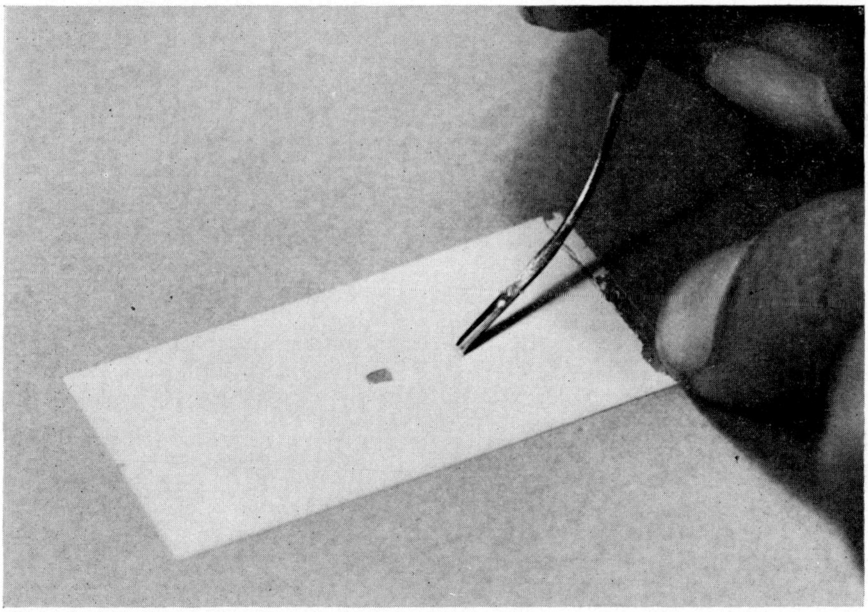

FIG. 74. This photograph shows the type of microshovel used for removing individual areas from a microglass layered with silica gel. Each area removed corresponds to the position of a single phospholipid, the perimeter of which had been marked with a fine glass needle.

and transferred to clean microtubes (inner diameter 2 mm, length *ca.* 40 mm). It is often necessary to use a microshovel together with a fine scalpel in order to remove all grains of silica gel from the plate. The microtubes containing the individual phospholipids are then half-filled with chloroform/methanol (2:1 by vol), agitated on a whirlmixer, and briefly centrifuged. The organic supernatant is transferred to a quartz cuvette by means of a suitable pipette capillary. The elution of the silica gel is repeated another four times; each time the combined supernatant is placed into the same cuvette. According to Schiefer and Neuhoff (1971), 90–105% of the amount of lipids applied to a microchromatogram are recovered from the silica gel by this process. Small traces of silica gel accidentally introduced into the cuvettes do not influence the fluorometric measurements so long as they do not form suspensions

(d) FLUOROMETRIC DETERMINATION

Combined organic extracts in the cuvettes are carefully dried with a stream of nitrogen delivered through a capillary. Care must be taken that too fast a stream of nitrogen does not spray any of the extract out of the cuvette. It is essential to dry the extract completely, as minute traces of water affect the fluorometric determination. 2 ml of cyclohexane is then added to the cuvette together with an appropriate amount of acetic acid and rhodamine 6G solution (see Table 32). The rhodamine 6G solution (10 mg/25 ml chloroform) should be prepared daily and stored in the dark. The solutions in the cuvette are shaken carefully but thoroughly and subjected to fluorometric measurements. It is best to use quartz cuvettes with fitted Teflon stoppers, so that the thorough shaking required before the fluorometric measurement can be done without any loss of solution.

4. Requirements for the Analysis of Tissue Samples

The following is a list of the apparatus, glassware, chemicals and sundries which are necessary for the microanalysis of phospholipids:

(a) APPARATUS AND INSTRUMENTS

Spectrofluorometer
Scintillation counter
Special 'cross-action' forceps (see Fig. 59) for handling microplates
Microshovel
Whirlmixer
Centrifuge
Homogeniser (see page 53)
Hair-dryer
Microforge for making tubes, capillaries, etc.
Micropipettes (see page 18)
Dissecting microscope
Dissecting instruments
Suitable freeze-dryer that can also be used as a vacuum evaporator
Pasteur pipettes
Ultra-turrax
Rotary evaporator

(b) GLASSWARE

Microtubes (see page 34)
Stoppered quartz cuvettes
Petri dishes with covers
Beakers (50 ml), at least six
Petri dish covers for beakers

Drummond microcaps, 1, 2, 5, 10 μl
Pipettes, series from 1 to 10 μl

(c) CHEMICALS AND SOLUTIONS

Silica gel DO (Camag, Muttenz, Switzerland)
Waterglass (sodium silicate solution DAB 6 quality) N-butanol (analar)
Chloroform (analar)
Methanol (analar)
Acetic acid (analar)
Ammonium solution (analar)
N-hexane (analar)
Diethylether (analar)
Double quartz glass distilled water
Rhodamine 6GO (Chroma, Stuttgart, Germany
Iodine crystals
Cyclohexane (analar)
Scintillation liquid (4 g PPO and 0.1 g POPOP per litre)
^{14}C–glucose and $NaH_2{}^{32}PO_4$ for incorporation experiments
Absolute ethanol
11 N HNO_3
Standard purified samples of lipids
Plaster of Paris

(d) SUNDRIES

Film for autoradiography (see page 75)
Coverslips 32 $\times$ 24 mm and 24 $\times$ 48 mm
Pieces of glass 45 $\times$ 26 mm
Scintillation vials
Filter paper
Nitrogen gas
Parafilm
Rubber tubing for pipette capillaries
Polythenc stoppered 250-ml bottles
Developer for autoradiograms
Spirit flame for sealing Drummond capillaries
Glass needle
Ice bucket
Plasticine to hold capillaries or microtubes

5. Application of the Procedure

(a) CHARACTERISED CELLS

Schiefer and Neuhoff (1971) used the microprocedure for the analysis of total phospholipids in single nerve cells (perikarya plus parts of their dendrites and axons) isolated from beef brain stem *Nucleus hypoglossus* by free-hand dissection. In order to avoid bacterial contamination occurring during the dissection, which would interfere with the lipid analysis, all solutions were freshly prepared and care was taken that the instruments, glassware and wires used were fat-free and sterile. Single cells were isolated by free-hand dissection (see page 10) and directly transferred into quartz cuvettes containing 1 ml chloroform/methanol (2:1 by vol). After 2 hr at room temperature the contents were evaporated to dryness by a stream of nitrogen and the fluorescence spectra measured (see Fig. 75). The total

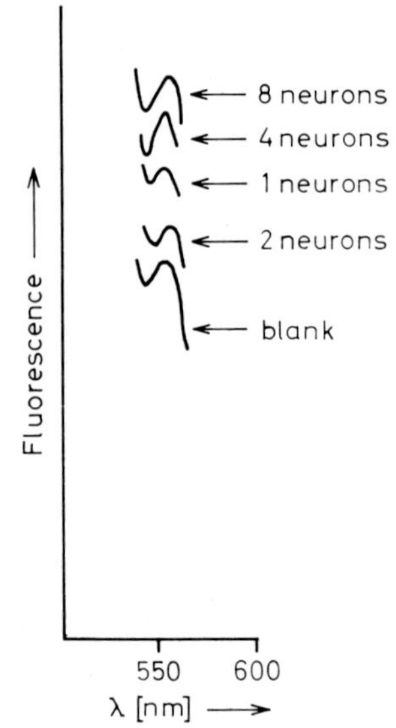

FIG. 75. Emission spectra of total phospholipid content from isolated neurons from beef strain stem *Nucleus hypoglossus* (from Schiefer and Neuhoff, 1971). The excitation wavelength was set at 534 nm, the bandwidth of the emission monochromator at 100 Å, the bandwidth of the excitation monochromator at 25 Å and the apparatus amplification at a magnification of 10.

phospholipid per cell was calculated by comparison with a standard curve of the total phospholipid mixture of guinea-pig brain. Analysis of four different cells showed them to contain 32.2, 7.2, 12.9 and 7.7 ng phospholipid respectively. Since the dry weight of these cells is 2–10 ng as found by X-ray absorption (Hydén and Rosengren, 1962), it can be assumed that the cells should contain 0.5–2.5 ng phospholipid based on the calculation that a quarter of a cell's dry weight is due to phospholipids (Brante, 1949). Schiefer and Neuhoff (1971) argued that the high content of phospholipids found in the cells is due to the fact that the neurons isolated have long axons and dendrites and that they are also very rich in phospholipids. This explains not only the higher content, but also the different values obtained for individual cells. The actual microchromatographic separation of individual phospholipids from a single neuron (GSC) from *Helix pomatia* has also been achieved (Althaus *et al.*, 1973). In these experiments the snail nervous system was perfused with snail's Ringer (Meng, 1960) containing ^{14}C–glucose for a period of 3 hr, and three GSCs (see page 79) dissected. The cells were then placed directly on the starting-point of a 24 × 48 mm waterglass impregnated silica gel plate, where a very minute amount of silica had previously been pin-pricked away to make a little cavity for the cells. After allowing the cells to dry onto the glass plate of the microchromatogram, the phospholipids were extracted from the cells on the chromatogram using chloroform/methanol (2:1 by vol) as described in Fig. 76. The chromatogram was then developed in an ascending fashion using chloroform/methanol/water/acetic acid (65:35:3.6:2 by vol) and terminating the chromatography when the front had migrated a distance of 20 mm. The microchromatogram was then dried and developed in the same direction with chloroform/methanol/water/acetic acid/acetone (60:25:5.3:10:5 by vol), this time allowing the solvent front to reach a distance of 30 mm. Thereafter the plate was dried once more, and a third chromatography in the same direction but with the solvent system n-hexane/diethylether/acetic acid (70:3:1 by vol) was allowed to proceed to the top edge of the plate. On exposure of the microchromatogram to iodine vapour a number of phospholipids were easily identifiable (see Fig. 76). These included phosphatidylcholine, phosphoinositides, phosphatidylserine and

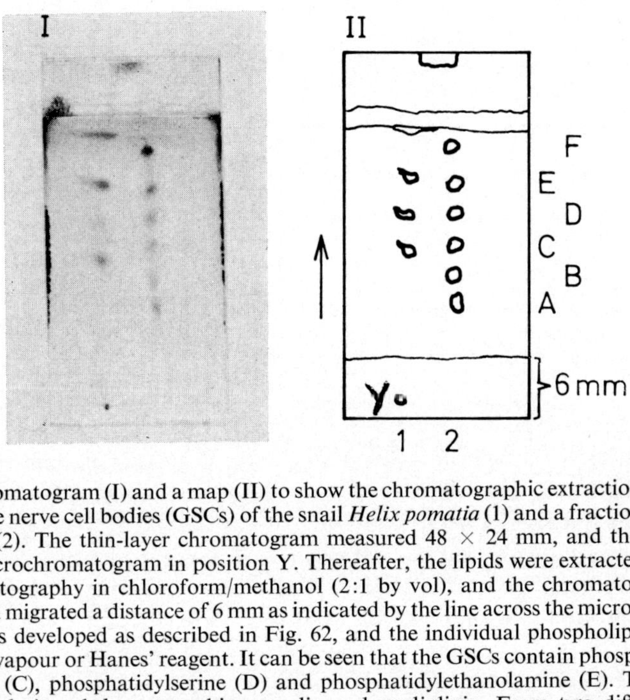

FIG. 76. Microchromatogram (I) and a map (II) to show the chromatographic extraction and fractionation of lipids from three nerve cell bodies (GSCs) of the snail *Helix pomatia* (1) and a fractionation of a mixture of phospholipids (2). The thin-layer chromatogram measured 48 × 24 mm, and the cells were placed directly on the microchromatogram in position Y. Thereafter, the lipids were extracted from the cells by ascending chromatography in chloroform/methanol (2:1 by vol), and the chromatography terminated when the front had migrated a distance of 6 mm as indicated by the line across the microplate. After drying, the microplate was developed as described in Fig. 62, and the individual phospholipids were visualised with either iodine vapour or Hanes' reagent. It can be seen that the GSCs contain phosphatidylcholine (B), phosphoinositides (C), phosphatidylserine (D) and phosphatidylethanolamine (E). The spots A and E in the standard solutions belong to sphingomyelin and cardiolipin. From two different experiments, where the individual phospholipids were removed from the microchromatograms and analysed by spectrophotometry, it was deduced that a single GSC contains 0.68 μg phosphatidylcholine, 0.57 μg phosphatidylethanolamine, 0.07 μg phosphoinositides and 0.12 μg phosphatidylserine. (After Althaus *et al.*, 1973.)

phosphatidylethanolamine. Moreover, autoradiography of the plate showed these phospholipids to incorporate ^{14}C-radioactivity from glucose.

The microbiochemical procedure for analysing phospholipids has also been used to study the incorporation of ^{32}P into individual phospholipids of the GSCs of *Helix pomatia* (Osborne *et al.*, 1972). In these experiments the snail nervous system was perfused with snail saline containing $NaH_2{}^{32}PO_4$ (specific activity 221 mCi/mM); the concentration of perfused NaH_2PO_4 was 80 μCi/ml perfused saline. After perfusion with the labelled substance (1 ml for 90 min) the brain was rinsed in pure saline for 5 min. Giant neurons (GSCs) were then dissected and placed in a 5-μl Drummond microcap which contained chloroform/methanol (2:1 by vol) and their phospholipids extracted. The GSC organic extract together with unlabelled standard phospholipids were then applied to a single microchromatographic plate (32 × 24 mm) layered with silica gel. After chromatography in two solvent systems, first dimension chloroform/methanol/water (65:25:4 by vol), second dimension n-butanol/acetic acid/water (3:1:1 by vol), individual spots were identified with iodine vapour, removed and suspended in 10 ml scintillation liquid for counting. Similar experiments were carried out on whole tissue for comparison. As can be seen in Table 33, the pattern of ^{32}P incorporation into brain and GSC is similar: phosphoinositides and phosphatidylserine > phosphatidylethanolamine > phosphatidylcholine > sphingomyelin (see Table 33). However, the radioactivity associated with the phosphoinositides and phosphatidylserine components in the GSCs is almost twice as high as that occurring

TABLE 33. INCORPORATION OF ^{32}P INTO THE PHOSPHOLIPIDS OF THE BRAIN (AN ALIQUOT SEPARATED AND IDENTIFIED ON A SINGLE MICROCHROMATOGRAM) AND IDENTIFIED NEURONS (A TOTAL OF SIX NEURONS) (after Osborne *et al.*, 1972)

	Brain (counts/min)	Neurons (counts/min)	Ratio of radioactivity in brain : neurons
Sphingomyelin	47.2	10.9	4.33:1
Phosphatidylcholine	99.8	24.6	4.1 :1
Phosphoinositides and phosphatidylserine	739	364.1	4.02:2
Phosphatidylethanolamine	172	36.4	4.7 :1

Results are expressed in counts/min, with background activites of 35 counts/min already substracted.

for the same substances in the brain. The importance of the phosphoinositides in nervous tissue is well established. They appear to be involved not only in post-synaptic events accompanying transmission of nerve impulses, but also in the axonal membrane, possibly because of their affinity with divalent cations (Hawthorne and Kai, 1970). In addition, acetylcholine, noradrenaline and electrical stimulation are known to increase the incorporation of ^{32}P into the phosphoinositides (Hokin, 1969; Friedel and Schanberg, 1971).

(b) NERVOUS TISSUE

Quantitative data on the phospholipid composition of the snail brain obtained by fluorometry after two-dimensional chromatography (see Fig. 77) are shown in Table 34. It can be seen that the sum of the phosphoinositides and phosphatidylserine amounts to 43.4% of the total phospholipids though this figure is probably too high because of the assumption made in the calculations. Phosphatidylcholine and phosphatidylethanolamine represent respectively 25.8% and 19.9% of the phospholipids, while the sphingomyelin composition is only 10.8%. Moreover, analysis of the total phospholipid content in the snail brain and other tissues (see Table 35) showed clearly that the brain contains higher levels. This is also the case in the vertebrates, where analysis of homologous tissues has shown the nervous system to have most phospholipids (Ansell and Hawthorne, 1964; Friedel and Schanberg, 1971). These studies also indicated that phospholipids are particularly associated with nervous tissue, as the concentration of phospholipids in brain, heart, liver and

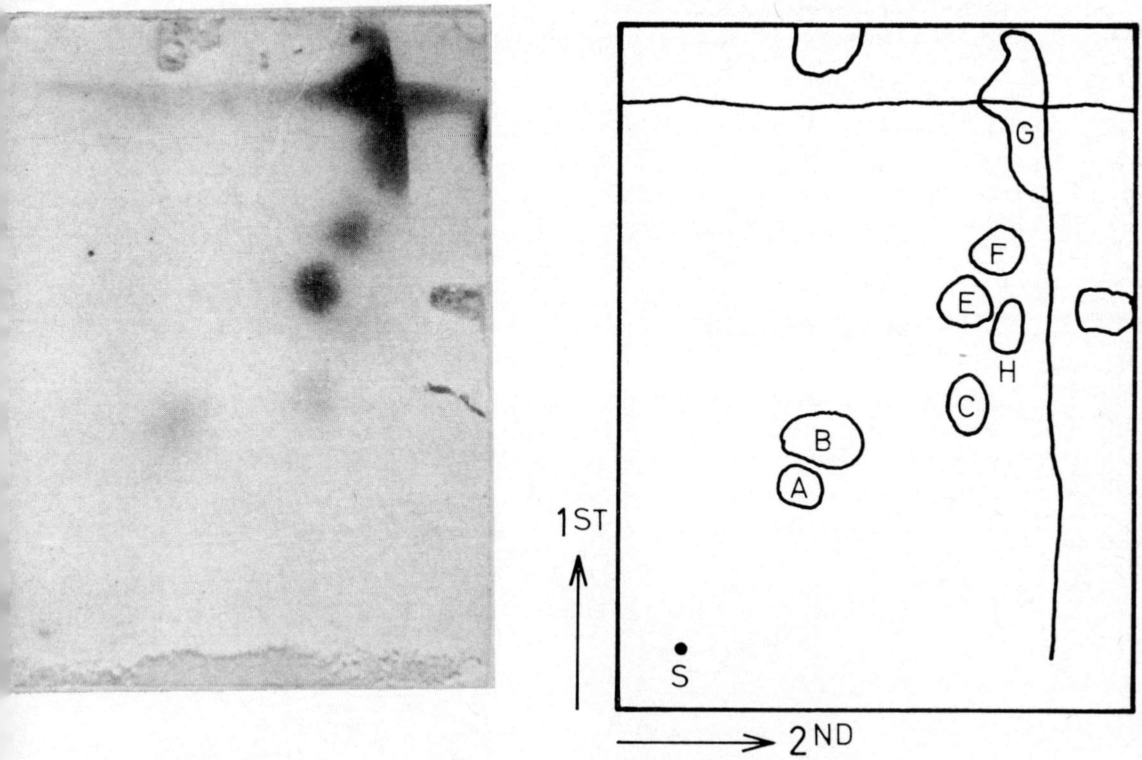

FIG. 77. A photograph of a two dimensional microchromatogram (*left*) and a corresponding map (*right*) of a total lipid extract from circumoesophageal ganglia of the snail *Helix pomatia*. The original size of the thin layer chromatogram measured 32 × 24 mm, and was developed in the 1st dimension with chloroform/methanol/water and in the 2nd dimension with n-butanol/acetic acid/water. Substances are identified as (G) neutral lipids, (F) cardiolipin contaminated with cerebroside, (E) phosphatidylethanolamine, (H) sulphatide, (C) phosphoinositides and phosphatidylserine, (B) phosphatidylcholine and (A) sphingomyelin. (From Osborne *et al.*, 1972.)

TABLE 34. CONTENT OF INDIVIDUAL PHOSPHOLIPIDS IN THE BRAIN OF THE SNAIL EXPRESSED IN μg PHOSPHOLIPID/100 μg WET WEIGHT OF TISSUE (after Osborne *et al.*, 1972)

	Experiment 1	Experiment 2	Experiment 3	Experiment 4	Mean values ± S.E.M. for four experiments	Composition (%)
Sphingomyelin	0.166	0.148	0.172	0.178	0.165 ± 0.013	10.83
Phosphatidylcholine	0.358	0.414	0.419	0.388	0.394 ± 0.028	25.92
Phosphoinositides and phosphatidyl-serine	0.646	0.672	0.689	0.641	0.662 ± 0.023	43.41
Phosphatidyl-ethanolamine	0.292	0.311	0.297	0.317	0.304 ± 0.011	19.95

The molecular weight of a phospholipid molecule was taken to be 800.

kidney of the snail closely resembles the distribution of nervous elements (Bullock and Horridge, 1965) and of a 'nerve-specific' protein (Osborne *et al.*, 1971). In addition, the comparatively low amounts of phospholipids in the snail brain in relation to the greater

TABLE 35. TOTAL PHOSPHOLIPID CONTENT IN FOUR TISSUES OF THE SNAIL EXPRESSED IN μg PHOSPHOLIPID/100 μg WET WEIGHT OF TISSUE (after Osborne *et al.*, 1972)

	Experiment 1	Experiment 2	Experiment 3	Experiment 4	Experiment 5	Mean values ± S.E.M. for five experiments
Brain	1.4	1.36	1.39	1.38	1.47	1.4 ± 0.042
Heart	0.852	0.86	0.852	0.868	0.892	0.865 ± 0.016
Liver	0.688	0.691	0.71	0.728	0.68	0.699 ± 0.019
Kidney	0.418	0.395	0.429	0.46	0.476	0.436 ± 0.032

The molecular weight of a phospholipid molecule was taken to be 800.

quantities found in the nervous systems of higher molluscs, e.g. the cephalopods (McColl and Rossiter, 1951), and to the even larger concentrations in the much more complex vertebrates (Brante, 1949), show that these substances are probably involved in the more complicated behaviour patterns connected with the nervous system. However, the distributions of the individual phospholipids in the various tissues are similar. Figure 78

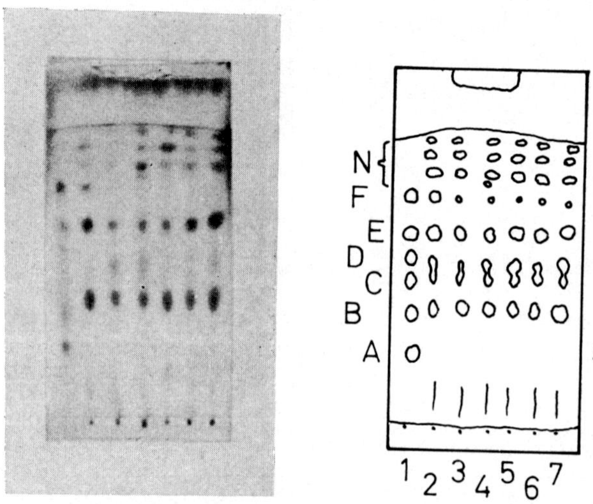

FIG. 78. A photograph of a one-dimensional microchromatogram (*left*) and a corresponding map (*right*) of a number of extracts from the snail *Helix pomatia*. The conditions of chromatography and the chromatographic mobility of each substance are as described in Fig. 62. N, neutral lipids; 1, extract of pure phospholipids; 2, extract of heart; 3, extract of peripheral nerve; 4, extract of brain; 5, extract of kidney; 6, extract of digestive gland; 7, extract of albumen gland. (Photograph by courtesy of H. H. Althaus.)

shows a one-dimensional fractionation of lipids extracted from a number of tissues of the snail, where it can be seen that, as in the brain, the major phospholipids in the other tissues are phosphatidylcholine, phosphoinositides, phosphatidylserine and phosphatidylethanolamine, while sphingomyelin is barely visible on the photograph. It is interesting to note that the neutral lipid content in brain and peripheral nerve is much lower than in other tissues.

Schiefer and Neuhoff (1971) analysed the phospholipid composition of minute tissue pieces 'punched out' from cat's spinal cord anterior column. One piece, of 90.8 μg wet weight, gave a content of 3.8 μg phospholipid per 100 μg wet weight, and a second piece weighing 118.6 μg gave 6.2 μg phospholipid per 100 μg wet weight. According to the data of Brante (1949), grey matter contains 3.3 μg phospholipid and white matter 8.4 μg phospholipid per 100 μg wet weight, so it was not surprising that there were differences in the punched-out spinal cord samples as the pieces contained varying amounts of grey and white matter. Analysis of individual phospholipids from a sample punched from the cat's spinal cord (162 μg wet weight) revealed the phosphatide composition as: phosphatidylcholine 2.0 μg (2.4 μg), phosphoinositides and phosphatidylserine 0.7 μg (0.3 μg), phosphatidylethanolamine 1.7 μg (1.7 μg), sphingomyelin 1.1 μg (1.2 μg). These data agree well with the values quoted by Brante (1949), which are given in the brackets.

Since 'post-tetanic potentiation' is considered to offer a possible paradigm for the process of learning at synapses (Eccles and McIntyre, 1953; Kandel and Spencer, 1968), and because lipids constitute an essential part of cellular membranes (Stoeckenius and Engelman, 1969), Althaus *et al.* (1972) studied the effects of post-tetanic potentiation on the monosynaptic reflexes in the spinal cord of the cat. Only the right α-motor neurons in the L7–S1 region of the spinal cord were affected by the monosynaptic reflex, while the left side was used as a control. Analysis of punched-out samples of grey matter in these two areas of the L7–S1 region showed that in the side stimulated there was an increase in sphingomyelin (14%) and phosphoinositides/phosphatidylserine (5%), in contrast to a decrease in the phosphatidylcholine (18%) while the phosphatidylethanolamine content remained unchanged.

References

ALTHAUS, H. H. and NEUHOFF, V. (1973) One-dimensional microchromatography of phospholipids and neutral lipids on sodium silicate-impregnated silica gel layers. *Hoppe-Seyler's Z. Physiol. Chem.* **354,** 1073–1076.

ALTHAUS, H. H., BRIEL, G., DAMES, W. and NEUHOFF, V. (1972) Neurochemische Mikroanalysen des Rückenmarks der Katze nach posttetanischer Potenzierung monosynaptischer Reflexe. In: *Nervensystem und biologische Information.* SFB 33, Göttingen, Germany.

ALTHAUS, H. H., OSBORNE, N. N. and NEUHOFF, V. (1973) Mikrochromatographische Extraktion und Fraktionierung von Lipiden einzelner Nervenzellen von *Helix pomatia. Naturwissenschaften* **60,** 553–554.

ANSELL, G. B. and HAWTHORNE, J. N. (1964) *Phospholipids.* Elsevier, Amsterdam.

ANSELL, G. B. and SPANNER, S. (1963) The alkaline hydrolysis of the ethanolamine plasmogen in brain tissue. *J. Neurochem.* **10,** 941–950.

ANSELL, G. B. and SPANNER, S. (1971) Purification and assay of phospholipids. In: *Methods of Neurochemistry* (Ed. R. Fried), Vol. 1, pp. 31–81. Marcell Dekker, Inc., New York.

BRANTE, G. (1949) Studies on lipids in the nervous system with special reference to quantitative chemical determination and topical distribution. *Acta Physiol. Scand.* **18,** Suppl. 63.

BULLOCK, T. H. and HORRIDGE, G. A. (1965) The Mollusca. In: *Structure and Function of the Nervous System,* Vol. 11, pp. 1273–1518. Freeman, San Francisco.

CARTER, H. E., HANES, W. J., LEDYARD, W. E. and NORRIS, W. P. (1947) Biochemistry of the sphingolipids. 1. Preparation of sphingolipids from beef strain and spinal cord. *J. Biol. Chem.* **169,** 77–82.

COURTADA, S., MARINETTI, G. V. and STOTZ, E. (1967) The structure and abundance of rat tissue phospholipids. *Biochem. Biophys. Acta,* **137,** 121–134.

DITTMER, J. C. and LESTER, R. L. (1964) A simple, specific spray for the detection of phospholipids on thin-layer chromatograms. *J. Lipid Res.* **5,** 126–127.

ECCLES, J. C. and McINTYRE, A. K. (1953) The effect of disuse and of activity on mammalian spinal reflexes. *J. Physiol.* **121,** 492–516.

FEULGEN, R. and BERSIN, Th. (1939) Zur Kennntnis des Plasmalogens. IV. Mitteilung. Eine neuartige Gruppe von Phosphatiden (Acetalphosphatide). *Hoppe-Seyler's Z. Physiol. Chem.* **260,** 217–245.

FOLCH, J. (1942) Brain cephalin, a mixture of phosphatides. Separation from it of phosphatidylserine, phosphatidylethanolamine and a fraction containing an inositol phospholipid. *J. Biol. Chem.* **146,** 35–44.

FOLCH, J. and WOOLEY, D. W. (1942) Inositol, a constituent of a brain phosphatide. *J. Biol. Chem.* **142**, 963–964.

FOLCH, J., LEES, M. and SLOANE-STANLEY, G. H. (1957) A simple method for the isolation and purification of total lipids from animal tissues. *J. Biol. Chem.* **226**, 497–509.

FRIEDEL, R. O. and SCHANBERG, S. M. (1971) Incorporation *in vivo* of intracisternally injected ^{32}P into phospholipids of rat brain. *J. Neurochem.* **18**, 2191–2200.

GALLAI-HATCHARD, J., MAGEE, W. L., THOMPSON, R. H. S. and WEBSTER, G. R. (1962) The formation of lysophosphatides from diacyl phosphatides by brain preparations. *J. Neurochem.* **9**, 545–551.

HANES, C. S. and ISHERWOOD, F. A. (1949) Separation of the phosphoric esters on the filter paper chromatogram. *Nature*, **164**, 1107–1112.

HAWTHORNE, J. N. and KAI, M. (1970) Metabolism of phosphoinositides. In: *Handbook of Neurochemistry* (Ed. A. Lajtha), Vol. III, pp. 491–505. Plenum Press, New York.

HEZEL, U. (1973) Direkte quantitative Photometrie an Dünnschicht-Chromatogrammen. *Angew. Chem.* **85**, 334–342.

HOKIN, L. E. (1969) Effect of norepinephrine on ^{32}P incorporation into individual phosphatides in slices from different areas of the guinea pig brain. *J. Neurochem.* **16**, 127–134.

HÖRHAMMER, L., WAGNER, H. and RICHTER, G. (1959) Zur papierchromatographischen Auftrennung von Phosphatiden. 1. Mitteilung. *Biochem. Z.* **331**, 155–161.

HUBMANN, F. H. (1973) Unpublished data.

HYDÉN, H. and ROSENGREN, B. (1962) A comparison of two integrating X-ray absorption methods for the determination of total cell mass. *Biochim. Biophys. Acta*, **60**, 638–640.

JONES, D., BOWLER, D. E., GRESHAM, G. A. and HOWARD, A. N. (1966) An improved spray reagent for detecting lipids on thin-layer chromatograms. *J. Chromatog.* **23**, 172–174.

KANDEL, E. R. and SPENCER, W. A. (1968) Cellular neurophysiological approaches in the study of learning. *Physiol. Rev.* **48**, 65–134.

KLEINIG, H. and LEMPERT, U. (1970) Phospholipid analysis on a micro scale. *J. Chromatog.* **53**, 595–597.

McCOLL, J. D. and ROSSITER, R. J. (1951) Lipids in the nervous system of the squid *Loligo pealii*. *J. Exp. Biol.* **28**, 116–124.

MENG, K. (1960) Untersuchungen zur Störung der Herztätigkeit bei *Helix pomatia*. *Zool. Fahr.* **68**, 539–566.

NEUHOFF, V. (1973) Microdetermination of phospholipids. In: *Micromethods in Molecular Biology* (Ed. V. Neuhoff), Vol. 14, pp. 149–178. Springer, Berlin.

OSBORNE, N. N., ANSORG, R. and NEUHOFF, V. (1971) Micro-disc electrophoretic separation of soluble proteins from nervous and other tissues of *Helix* (Pulmonata, Mollusca). *Int. J. Neuroscience*, **1**, 259–264.

OSBORNE, N. N., ALTHAUS, H. H. and NEUHOFF, V. (1972) Phospholipids in the nervous system of the gastropod mollusc *Helix pomatia*, and the *in vivo* incorporation of ^{32}P into the phospholipids of identified neurons. *Comp. Biochem. Physiol.* **43B**, 671–679.

RAPPORT, M. M., LERNER, B., ALONZO, N. and FRANZL, R. E. (1957) The structure of plasmalogens. II. Crystalline lysophosphatidal ethanolamine (acetal phospholipid). *J. Biol. Chem.* **225**, 859–867.

ROUSER, G., KRITCHEVSKY, G. and YAMAMOTO, A. G. (1967) Column chromatographic and associated procedures for separation and determination of phosphatides and glycolipids. In: *Lipid Chromatographic Analysis* (Ed. G. V. Marinetti), Vol. 1, pp. 99–162. Marcel Dekker, New York.

SANDERS, H. (1967) Preparative isolation of phosphatidylserine from brain. *Biochim. Biophys. Acta*, **144**, 485–487.

SCHIEFER, H. G. and NEUHOFF, V. (1971) Fluorometric microdetermination of phospholipids on the cellular level. *Hoppe-Seyler's Z. Physiol. Chem.* **352**, 913–926.

SKIDMORE, W. D. and ENTENMAN, C. (1962) Two-dimensional thin-layer chromatography of rat liver phospholipids. *J. Lipid Res.* **3**, 471–475.

STOECKENIUS, W. and ENGELMAN, D. M. (1969) Current models for the structure of biological membranes. *J. Cell Biol.* **42**, 613–646.

SVETASHEV, V. I. and VASKOVSKY, V. E. (1972) A simplified technique for thin-layer microchromatography of lipids. *J. Chromatog.* **67**, 376–378.

THUDICHUM, J. L. W. (1884) *A Treatise on the Chemical Constitution of the Brain*. Balliére, Tindall & Cox, London.

UDENFRIEND, S. (1969) *Fluorescence Assay in Biology and Medicine*. Academic Press, New York.

VASKOVSKY, V. E. and KOSTETSKY, E. Y. (1968) Modified spray for the detection of phospholipids on thin-layer chromatograms. *J. Lipid Res.* **9**, 396.

ZABIN, I. (1960) In: *The Lipids* (Ed. D. J. Hanahan), Chapter 5. Wiley, New York.

CHAPTER 9

Microelectrophoresis of Proteins

1. Introduction

One of the major problems in neurobiology is that of isolating and identifying specific proteins in the nervous system and determining their function. Nervous tissues are very rich in proteins and their exceedingly high turn-over rate indicates their importance (Roberts *et al.*, 1970; D'Monte *et al.*, 1970). Moreover, special protein species can be characteristically associated with certain diseases, such as encephalitis (Alvord, 1968), or with the developmental structure of the animal (Grossfeld and Shooter, 1971). Specific neuronal components like myelin, synaptosomes, neurofilaments, neurotubules have all been shown to contain certain characteristic proteins. However, it is the two acid proteins S-100 and 14–3–2 which have attracted a great deal of interest, mainly because they appear to be specific to the nervous system, their concentration in the brain being at least a thousand-fold higher than in other organs (Moore and Perez, 1968). The 14–3–2 protein level is higher in grey matter than in white, and all the evidence available suggests the protein is primarily neuronal in origin (Cicero *et al.*, 1970). In contrast, there is less S-100 protein in grey than in white matter. From the studies of Hydén and McEwen (1966) it was concluded on the basis of fluorescent antibody studies on cells from Deiters' nucleus that S-100 was glial in origin, although some interaction with the nuclei of the neurons occurred. In behaviour experiments (transfer of handedness in rats) it was found that the amount of S-100 in the CA3 area of the hippocampus (see Fig. 4) of the rat increases in trained animals, and that S-100 is specifically correlated with learning (Hydén and Lange, 1970). The possibility that proteins are concerned with information processing or storage has subsequently become an accepted idea in neurobiology (Agranoff, 1970; Ungar, 1970).

One of the best procedures for analysing brain proteins (and nucleic acids) is that of polyacrylamide gel electrophoresis. This is partly due to the fact that gels can be prepared easily from analytically pure starting materials, and by varying the proportions of the starting materials gels of different density and pore diameter can be prepared. Moreover, the gel is transparent, mechanically very stable, can be used over a wide range of pH, is insoluble in most of the solvents used for electrophoresis and other substances can also be copolymerised into the gels. The principle of polyacrylamide electrophoresis is simple. The gel is formed by polymerisation of acrylamide in a tube or slab mould and the gel tube or slab mould is positioned between two buffer chambers. The sample is then placed on the gel and an electric field applied across them. Most of the components then migrate as individual bands each with a characteristic electrophoretic migration rate depending upon its size and net charge (see Maurer, 1971; Rodbard and Chrambach, 1971). Fixation and staining or slicing and assay of the gel subsequent to electrophoresis reveals characteristic band positions. The possibility of applying polyacrylamide gel electrophoresis to separate very small amounts of materials was soon recognised and in 1964 Pun and Lombrozo developed the first microprocedure. This consisted of gel moulds of vertical glass channels (microcells of 1.8 × 1.0 × 70 mm) cut into a glass plate. Thus the cross-section of the gel strips is smaller than 1/10th of the standard gel cylinder (5–7 mm diameter). This made it possible to analyse from 0.8 to 2 mg of nervous tissue and the concentrating effect of the system allowed relatively large volumes of extracts to be used. The actual idea of using Drummond capillaries for polyacrylamide electrophoresis was conceived by Grossbach (1965), who separated nanogram quantities of albumen. In Grossbach's original publication he used 5-μl Drummond capillaries, in which he prepared separation and spacer gels according to the original method of Davis and Ornstein (1959), though later he used 1-μl Drummond capillaries to fractionate picogram quantities of proteins (Grossbach, 1971).

Similarly Hydén *et al.* (1966), used Drummond capillaries to separate picogram quantities of proteins and then analysed the Amido black-stained protein gels with a microdensitometer which, due to the high amplification capacity, permitted nanogram quantities of material to be detected (see also Hydén and Lange, 1968). In 1968 Neuhoff modified the latter method specifically to fractionate water-soluble brain proteins in the microscale. The modification consisted essentially of adding 0.5% hydantoin for the preparation of 20% polyacrylamide gels. The effect of hydantoin is to increase the total ion concentration of the gel mixture which results in a great improvement in the quality of protein separation. One of the aims of this chapter is to describe fully the microdisc electrophoretic procedure as modified by Neuhoff (1968) for the separation of minute quantities of water-soluble brain proteins.

A number of other microelectrophoretic procedures also exists for the separation of brain proteins. Maurer and Dati (1972) have described a polyacrylamide gel electrophoresis system on microslabs. This has the advantage over cylindrical gels in that the gels may be conveniently evaluated by densitometry and photography without risk of optical artefacts. Moreover, they can easily be dried and subjected to autoradiography and other contact print methods, e.g. enzyme detection. Similarly, Been and Rasch (1972) have described a microprocedure for analysing protein using thin sheets of polyacrylamide gels, thus allowing nanogram quantities of water-soluble proteins to be detected. Hazama and Uchimura (1972) also developed a microprocedure based on a continuous buffer system of mixed agarose acrylamide to permit fast and slow components of S-100 brain protein to be separated.

It has, however, become clear in recent years that less than 40% of brain proteins can be solubilised with classical aqueous solvents. The other proteins are membrane-bound and can only be extracted by other systems, e.g. acidic phenol system and sodium dodecyl sulphate (SDS) system. It is therefore necessary to have microprocedures capable of fractionating proteins extracted by these systems, since smaller amounts of starting material are required. Moreover, neuronal membranes have a number of very important functions, one of which is that of a permeability barrier, which would mean that the membranes must transfer matter and information across the boundary between the inner and outer phases, and another, excitability, which enables the propagation of the nerve impulse and support of catalytic functions. Clearly membrane proteins of individual neurons have to be studied in order to understand these phenomena better. Wilson (1971) was the first to describe a micromethod for separating protein species on SDS polyacrylamide, which allowed him to study protein synthesis for the first time in single neurons of *Aplysia*. Later Gainer (1971) also developed a microprocedure for fractionating nervous tissue proteins extracted by use of SDS. It has an advantage over Wilson's procedure in that a steady-state stacking step in the disc electrophoresis is used which preconcentrates the sample into very thin starting zones and hence increases the resolution of the system. This microdisc polyacrylamide gel system in SDS detects as little as 5×10^{-10} g protein and has been successful in the examination of differences in the patterns of protein synthesis in individual pulmonate neurons (Gainer, 1972a), and in the relation of electrophysiological activity of characterised neurons to their proteins (Gainer, 1972b).

The microdisc electrophoresis developed by Neuhoff (1968) and described in this chapter cannot be easily adapted to separate proteins extracted in an SDS system. This is because the Triton X-100 contained in the gel mixtures is not compatible with SDS, and cannot merely be omitted, since the Triton X-100 content allows the gel to be easily pushed out of the Drummond capillary by water pressure. Water pressure alone is not sufficient to extrude gels containing no Triton X-100, and the use of steel wires for the extruding process often destroys the gel. This is partly why microelectrophoresis in gradient gels has been developed in this laboratory (Rüchel *et al.*, 1973a; Dames and Maurer, 1974). Microelectrophoresis in gradient gels have a number of advantages in that, (a) it is possible to fractionate SDS extracts, (b) it is possible to measure the molecular weights of proteins, and (c) proteins of molecular weight greater than about 1.5 million and smaller than 10,000 (e.g. peptides) can be fractionated which cannot be done in Neuhoff's procedure because of the limits imposed by the 5% stacking gel and 20% separating gel respectively. The polyacrylamide gradients do not contain Triton X-100 and the high acrylamide concentration at the lower end of the gel means that a steel wire can be used to extrude the gel without causing any damage. Attention will also be paid to the different ways in which these procedures have been adapted to study enzymes and RNA species as well as proteins in microquantities of tissue.

2. General Principles of Polyacrylamide Electrophoresis

Fractionation of proteins can be based primarily on molecular size (e.g. gel filtration) or on molecular net charge (e.g. free electrophoresis, ion exchange chromatography). Polyacrylamide gel electrophoresis, in contrast, exploits differences both in molecular size and charge for purposes of fractionation, which in itself demonstrates its usefulness. Raymond and Weintraub (1959) were the first to recognise the applicability of polyacrylamide gels as a supporting medium for electrophoretic separations. The synthetic polymer, polyacrylamide, is simply produced by polymerising acrylamide with N,N′-methylenebisacrylamide or ethylene diacrylate as the cross-linking component. Usually certain catalyst redox systems which yield free radicals are used to initiate polymerisation, e.g. ammonium peroxydisulphate and N,N,N′,N′-tetramethylethylenediamine (TEMED). By varying the concentrations of reagents in the polymerisation reaction, various pore sizes (0.5–3 μ) in the polyacrylamide can be achieved.

Electrophoresis is defined as the movement of a particle of effective electrical charge through a medium caused by the action of an electrical field, though the actual electrophoretic migration of the particle in reality depends on a number of parameters, viz. size, shape, concentration, electrical charge, viscosity, pH value, temperature and ionic strength of the medium, electric field strength, migration time and degree of hydration and dissociation of the particle. In support electrophoresis, e.g. electrophoresis on polyacrylamide gels, as opposed to free-flow electrophoresis, additional factors become important and these may influence the mobility and sharpness of the separation. These are adsorption, inhomogeneity, electro-osmosis, suction effects, heating and evaporating influences and exchange capacity of the support material. Polyacrylamide gels have the great advantage of being almost completely homogeneous, and adsorption and electro-osmosis do not play a major role in the electrophoresis (see e.g. Maurer, 1971).

Electrophoresis on polyacrylamide gels, as in all other methods of electrophoresis, can be either 'continuous' (Raymond and Weintraub, 1959) or 'discontinuous' (Ornstein, 1964; Davis, 1964; Poulik, 1957). The idea of continuous electrophoresis is based on homogeneous buffer systems and a fixed pH, while a discontinuous system is based on different buffers and different pHs. However, some authors also include variation in pore size as a discontinuous system, so that electrophoresis in gradient gels will be placed under the heading of a discontinuous system. The definitions of this system need not be discussed here, though it must be pointed out that the term 'disc-electrophoresis' comes from the phrase '*dis*continuities in the electrophoretic matrix' and only coincidentally from the high resolution power which results in the separating zones appearing 'discoid' in shape (Ornstein, 1964). Disc-electrophoresis or discontinuous separating systems with varying buffer and pH composition as well as gel layers of different pore size create a discontinuous voltage and pH gradient, and these discontinuities produce very thin but highly concentrated starting zones immediately prior to separation, thus determining the sharpness of the fractionation. The theory of discontinuous electrophoresis with numerous examples of its use has been fully described by Maurer (1971), Jovin (1973a, b, c) and Ornstein (1964).

3. Microelectrophoresis (partly homogeneous polyacrylamide separating gel)

(a) GENERAL PRINCIPLE

The general principle of the microdisc-electrophoresis method as described by Neuhoff (1968) is shown in Fig. 79. Though the separating 20% gel appears to be homogeneous,

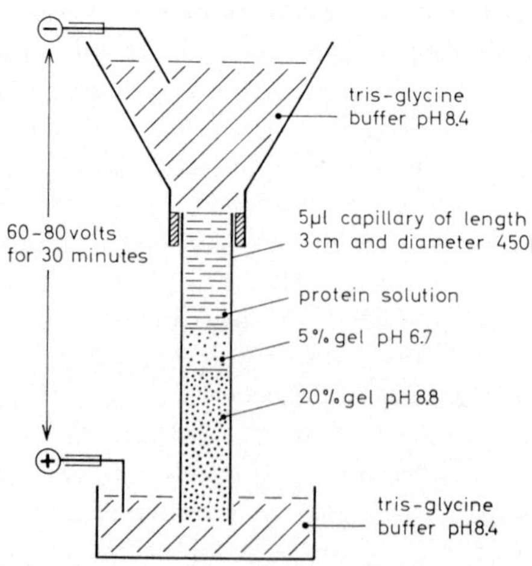

FIG. 79. Schematic diagram of Neuhoff's procedure for microdisc-electrophoresis in 5-μl capillaries. During the electrophoresis the current falls from an initial 60–80 μA to 15–20 μA.

in reality it is only partly so because the first few millimetres of the gel top are in fact a gel gradient caused simply by the method of casting the gel. This is demonstrated in Fig. 80, where it can be seen that the gel top is in fact a gradient and it is this property that explains why high molecular weight proteins still migrate in an apparent 20% homogeneous separating gel.

Polyacrylamide gels are usually made in 5-μl Drummond capillaries, though 1-, 2- and 10-μl capillaries can also be used. The protein solution (optimal concentration 1–3 mg/ml) to be fractionated cannot be applied directly to the separating gel, because it would tend to block the pores at the top of the gel, so permitting only the smaller sized proteins to enter the separating gel. Moreover, small proteins entering the separating gel would be

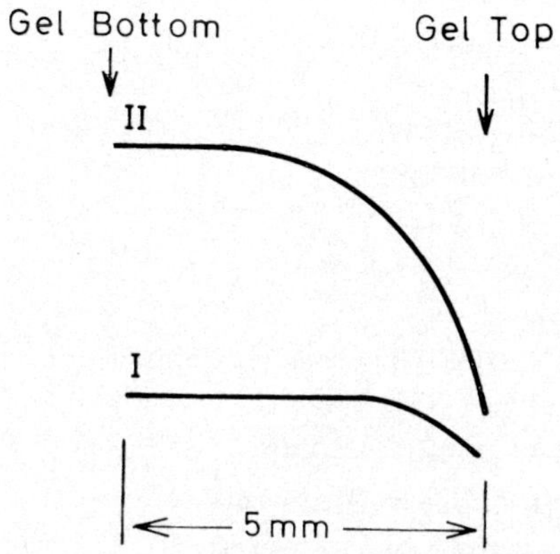

FIG. 80. Densitometric tracings of polymerised 20% gels prepared by the procedure described by Neuhoff (1968). The gels contained globulin and after polymerisation were stained with Amido black. It can be seen that the gels are not homogeneous, but have a slight gradient at the gel top. In (I) the 20% gel mixture in the micro cap was correctly and carefully overlaid with water whilst in (II) the pipette capillary was allowed to penetrate slightly the upper 2–3 mm of gel solution, so permitting the gel mixture and water to mix. (Rüchel, unpublished results.)

absorbed by the capillary space between the gel top and the glass wall and would migrate without being fractionated. These two phenomena are overcome by placing a small amount of 5% stacking gel above the separating gel. The stacking gel effectively prefractionates the protein mixture and thus solves the problem of larger proteins blocking the separating gel top, and also eliminates the interfering side effects caused by absorption of proteins by the capillary space between the gel top and glass wall. In order to have a discontinuous system the pH of the stacking gel is different from that of the separating gels, as is the electrophoretic buffer. With this system water-soluble proteins are separated by applying a potential difference of 60 V for about 30 min. The current in this time drops from about 60 μA to 17 μA. In practice, a little bromophenol blue and fluorescein are placed in the electrophoretic buffer and used as tracker dyes. Gels are extruded with water pressure into 1% Amido black in 7.5% acetic acid and stained and fixed for 5 min. Excess dye is then removed by placing the gel in 7.5% acetic acid for about 30 min.

Should very dilute protein solutions have to be fractionated, the capillary volume can be increased simply by connecting a Drummond capillary to the gel capillary by means of polythene tubing.

(b) REQUIREMENTS

(i) *Gel preparation*

All stock solutions are stored at 4°C in brown bottles. All, with the exception of solution D, keep for at least 2 months under these conditions. Fresh solution D should be prepared fairly often.

Stock A: 860 mg Tris
+ 8 ml H_2O
+ 0.063 ml N,N,N',N'-tetramethylenediamine (TEMED)
+ 3.6 N H_2SO_4 to pH 8.8 (about 0.45 ml)
+ H_2O to make total of 10 ml

Stock B: 2.85 g Tris
+ 25 ml H_2O
+ 8.7 M H_3PO_4 to pH 7
+ 1 M H_3PO_4 to pH 6.7
+ H_2O to make total of 50 ml

Stock C: for up to 20% gel
20 g acrylamide
+ 200 mg N,N'-methylene-bis-acrylamide
+ 3.75 mg $K_3Fe(CN)_6$
+ H_2O to make total of 37.5 ml

Stock D: 70 mg ammonium peroxydisulphate
+ 25 ml 4% Triton X-100 in water
+ 25 ml H_2O

Stock E: 5.9 g Tris
+ 50 ml H_2O
+ 0.46 ml TEMED
+ 8.7 M H_3PO_4 to pH 7
+ 1 M H_3PO_4 to pH 6.7
+ H_2O to make total of 100 ml

Stock F: 200 mg ammonium peroxydisulphate
+ 5 ml 4% Triton X-100 in water
+ 5 ml H_2O

When preparing the above solutions it is important to adjust the pH after the addition of TEMED, since this reagent is strongly alkaline. Potassium ferricyanide is used in solution C both to slow down the polymerisation and to give a better quality of gel. Triton-X-100 is added to facilitate the removal of the gels from the capillaries by water pressure and has no effect upon the separating qualities of the gel. The acrylamide or bis-acrylamide does not have to be recrystallised before use.

Gels which are required to be dissolved (in ammonium hydroxide or hydrogen peroxide) when using radioactive material can be made by simply substituting ethylene diacrylate in place of bis-acrylamide (Choules and Zimm, 1965). The addition of ethylene diacrylate has no obvious effects upon the separating qualities of the gel, but has the disadvantage of being highly toxic.

(ii) *Other solutions*

(a) Electrophoresis buffer: Dissolve 3.0 g of Tris and 14.4 g of glycine in 500 ml of water. Check the pH and adjust to 8.4 if necessary.
(b) Bromophenol blue solution: Dissolve 100 mg of bromophenol blue in 5 ml water.
(c) Acetic acid solution: Dilute 75 ml of glacial acetic acid to 1 litre with water.
(d) Amido black solution: Dissolve 1 g Amido black 10 B in 100 ml 7.5% acetic acid.
(e) Fluorescein solution: Saturate about 20 ml water with fluorescein.

(iii) *Voltage generator*

An electronically controlled voltage generator (Neuhoff *et al.*, 1967) which supplies a constant potential or a steady current is required. The constant voltage is adjustable between 0 and 450 V and the variation is about $\pm 1\%$. The apparatus (see Fig. 81) obtainable from Ernst Schütt jun. (Göttingen, Germany) is so constructed that a total current

FIG. 81. Voltage generator and stands required for microelectrophoresis. (Photograph by courtesy of V. Neuhoff.)

of 5 mA is supplied, making it possible to carry out ten electrophoretic separations simultaneously. A noise diode switch has been built into the apparatus to solve the problem of the regulation and stability of the very small currents in the microamp range. The current can be regulated continuously and held over a range of 0–2000 μA via a Helipot precision potentiometer. The voltage is measured with the aid of a built-in very high resistance tube-voltmeter, so that conditions are attained under which the current flow through the capillary varies by a maximum of $\pm 2\%$. This, in reality, means that the current is not appreciably influenced by the measurement of the effective voltage and is almost independent

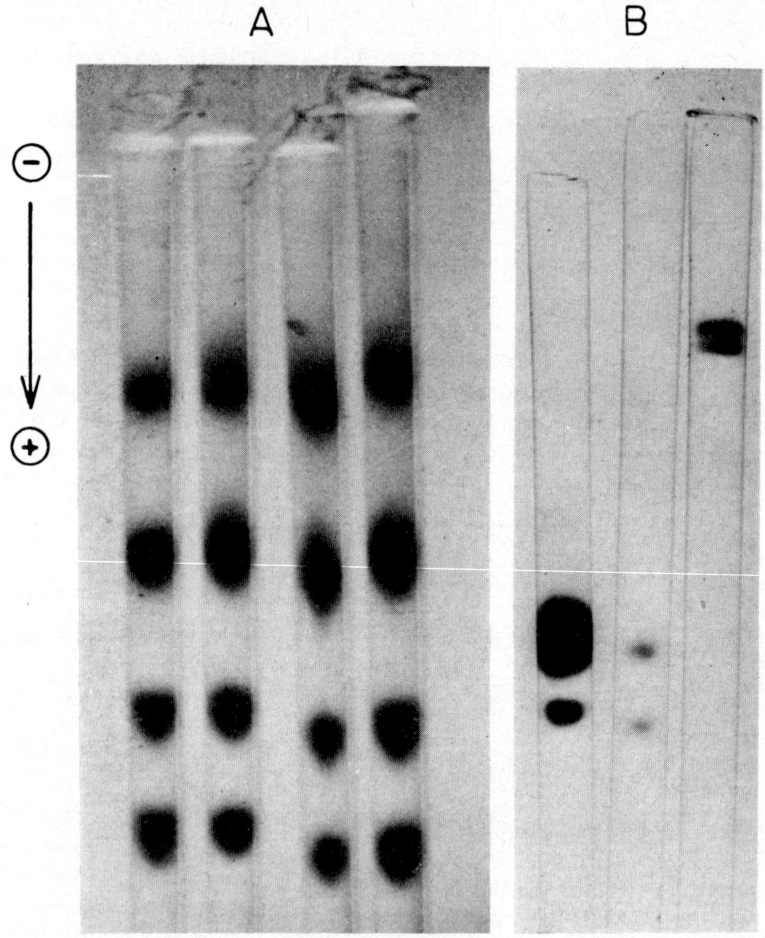

FIG. 82. Fractionation of dehydrogenases in 5-μl capillaries using microdisc electrophoresis. A shows the clear separation of four LDHs in hippocampus (the two gels on the left) and frontal lobe (the two on the right) from rat. B in contrast shows the separation of variants of Glc-6-P dehydrogenase from extracts of tissue culture cells from myeloma of man and mouse fibroblasts. In each instance the variant from man is the upper dehydrogenase. The gel on the right in B shows that an electrophoresis time of 40 min is sufficient for a good separation of the two variants, while an electrophoresis time of two hours (middle and left gel) produces a clear separation of the two species. In the gel on the left of B, the extract of human myeloma was ten times more concentrated than the extract of mouse fibroblasts, whilst the concentrations of both extracts were about the same in the other two gels. (Photographs by courtesy of W. Dames.)

of the resistance in the gel, which varies during electrophoresis. There are dials on the instrument to monitor not only the total current for one to ten microelectrophoreses connected in parallel, but also the current flow through the individual electrophoresis. It is thus possible to check the current at the beginning of the electrophoresis and detect any sources of error, e.g. the presence of air bubbles, or bad contacts. All values can be continuously measured and registered on dials throughout the electrophoresis. The polarity of all the electrophoresis columns can be simultaneously reversed by a switch on the voltage apparatus.

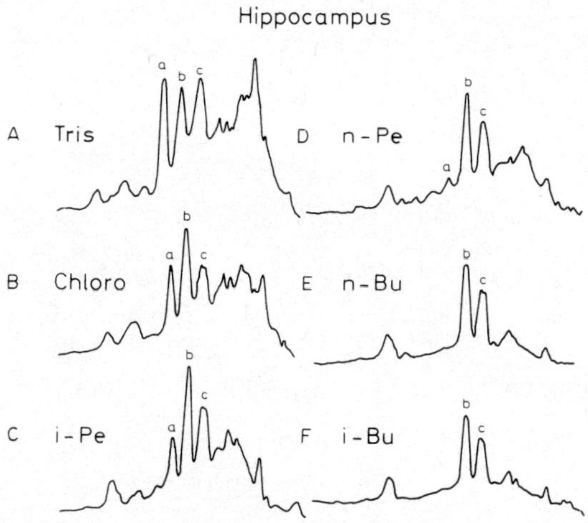

FIG. 83. Typical microdensitometric traces of proteins selectively extracted from the hippocampus of the rat using (A) 0.01 M Tris/HCl pH 7.4, (B) chloroform, (C) iso-pentanol, (D) n-pentanol, (E) n-butanol and (F) iso-butanol. (From Ansorg *et al.*, 1971.)

(iv) *Electrophoresis stands*

Also obtainable from Ernst Schütt jun. (Göttingen, Germany) is a convenient, adjustable electrophoresis stand (shown in Fig. 81) with the funnel containing the upper electrophoresis buffer attached to the capillary by a rubber cap. The other end of the capillary is located in a small jar containing the same buffer. The position of the funnel can easily be adjusted in the vertical direction, which is important when the length of the capillary is increased by the addition of another capillary. This is often necessary when dilute protein solutions are being fractionated and a greater volume is required.

(v) *Microdensitometer*

In order to analyse the protein bands in a protein gel, a suitable densitometer is needed. In this laboratory a Joyce-Loeble double-beam microdensitometer Model Mk III has proved to be a good choice.

(vi) *Some other requirements*

Drummond microcapillaries 1, 2, 5 and 10 μl
Special 'soft' forceps for handling gels (see Fig. 84)
Dissecting microscope
Pasteur pipettes
5-ml beakers (many) for staining and destaining
Measuring cylinders
Pyronine (for measuring R_f values)
Micropipette capillaries (see page 18)
Rubber tubing with mouthpiece for micropipettes (see Fig. 17)
Plasticine to hold capillaries

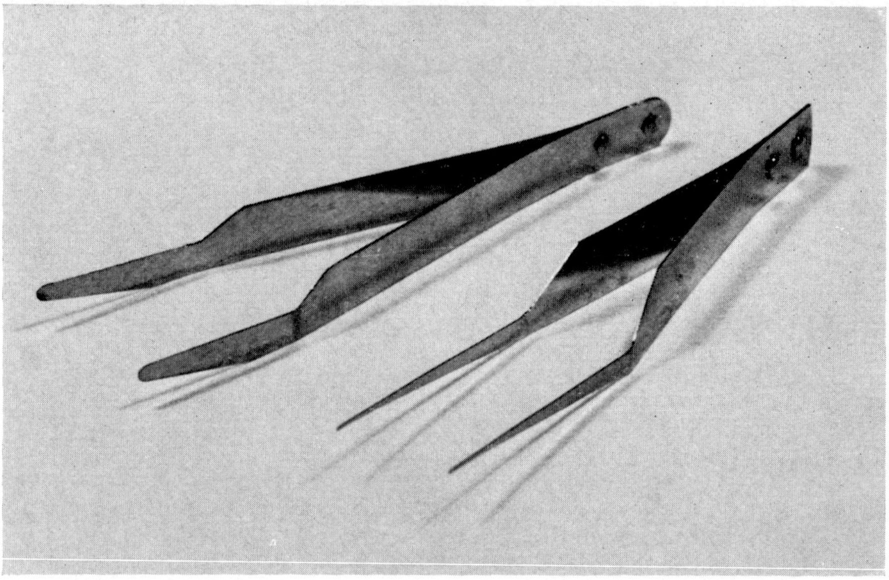

FIG. 84. Two different types of soft forceps needed for handling microgels.

Diamond knife
Stop clock
Parafilm
Chamber (see Fig. 85) for recording gel pattern under microdensitometer
Electrodes (0.5 mm diameter platinum wires)
Pipettes—various sizes
Syringe (2 ml) used for extruding gels (see Fig. 86)

FIG. 85. Chamber (6 × 6 cm) used for scanning stained microgels. The gels are placed within the grooves which contain 7.5% acetic acid and a cover-glass is laid over them. It is important to make sure that no air bubbles or dirt are trapped between the gels and the cover-glass.

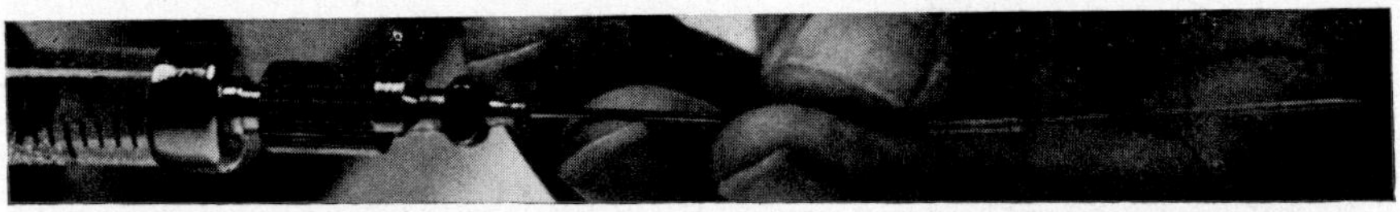

FIG. 86. This figure shows a 2-ml syringe filled with water and connected via a piece of polythene tubing to a capillary containing a polyacrylamide gel. By applying a little pressure, the gel, which contains Triton X-100, is easily extruded from the capillary.

FIG. 87. A simple saturation chamber set-up required to store capillaries containing microgels.

Polythene tubing of suitable size for attaching Drummond capillaries to each other, for example
Felt-tipped pen
Saturation chamber for gels, e.g. petri dish bottom with 50-ml beaker (see Fig. 87)
Hydantoin
Light box for viewing gels (see Fig. 88)
Chamber for photographing gels (see Fig. 88)
Ruler.

(c) PROCEDURE

The initial step is that of preparing the separating gel, most often 20%, which is done in the following way: mix carefully together 0.5 ml solution A and 1.5 ml solution C, then take 1 ml of this mixture and add 1 ml solution D. Should a higher percentage gel be required, e.g. 25%, an appropriate amount of acrylamide (i.e. 100 mg for a 25% gel) is simply added to the 2 ml 20% gel mixture. Gel concentrations lower than 20% are obtained by replacing a corresponding aliquot of solution C by water (see Table 36). It is worthwhile

FIG. 88. Chamber used for examining the quality of the microgels. This chamber has a transparent bottom, and by placing it on a light-box, the stained gels can be viewed easily.

noting that the smaller the capillary diameter, the higher the concentration of polyacrylamide required for similar separations of proteins. To the 2 ml 20% gel mixture is then added 10 mg of hydantoin. The added hydantoin increases the total ion concentration of the gel mixture, so improving the separating properties.

The next stage is to fill the capillaries partially (Drummond microcaps) with separating gels. This is simply done by dipping a capillary into the freshly made separating gel mixture and allowing about half to two-thirds of it to fill by capillary attraction. A number of capillaries are thus filled and then pressed into a plasticine cushion. This effectively seals the bottom of the capillaries and also keeps them upright. It is important to place the capillaries in a vertical position (see Fig. 87). The upper parts of the capillaries are then filled with water using a capillary pipette of suitable diameter, making absolutely certain that the pipette capillary tip only just touches the meniscus of the gel solution. It must be stressed that the formation of a sharp interphase between gel and water is of utmost importance, since the actual quality of the gel top predetermines the sharpness of the fractionated protein bands (see Fig. 80).

The gel mixture polymerises in the capillaries within 1 hr, depending on the concentrations of ammonium peroxidisulphate and TEMED. However, the polymerised gels have to be stored for at least 10 hr at room temperature in a simple saturation chamber (see Fig. 87), to ensure that the water layer does not evaporate, before use, otherwise the separated protein bands are poorly defined. Microgels prepared in this way can be used for up to 3 days after polymerisation, when stored at 4°C.

The third step in the process is to place the stacking gel over the separating gel. As pointed out earlier, the stacking gel is required to seal the capillary gap between the top of the gel and glass wall, thus overcoming the interfering effects which could occur in this region and also giving some prefractionation to the protein mixture. The stacking gel

TABLE 36. AMOUNT OF SOLUTIONS A, D AND C REQUIRED TO PREPARE GELS OF CONCENTRATIONS LESS THAN 20%

20% gel	0.75 ml solution C 0.25 ml solution A 1 ml solution D
16% gel	0.6 ml solution C 0.25 ml solution A 1 ml solution D 0.15 ml water
13% gel	0.5 ml solution C 0.25 ml solution A 1 ml solution D 0.25 ml water
8% gel	0.3 ml solution C 0.25 ml solution A 1 ml solution D 0.45 ml water
6.6% gel	0.25 ml solution C 0.25 ml solution A 1 ml solution D 0.5 ml water

also holds back some of the impurities in the protein solutions and retards very large protein molecules which could easily block the gels top. In practice the 5% stacking gel is placed above the separating gel 1–2 hr before electrophoresis. The 5% gel mixture is prepared in the following way: 0.5 ml solution E and 1.5 ml solution C are mixed together, and 1 ml of this mixture is then diluted with 1 ml water. 1 ml of this mixture is added to 1 ml solution E and 1 ml of this combination mixed with 0.8 ml water and 0.2 ml solution F. Polymerisation occurs within 5–10 min and can be prolonged slightly by adding 0.9 ml water instead of 0.8 ml, and 0.1 ml solution F instead of 0.2 ml.

The 5% stacking gel solution is applied immediately after removing the water above the polymerised separating gel. This water is carefully removed with a suitable pipette capillary lest the separating gel top be damaged by the pipette tip. (At this stage it is often advisable to mark the position of the gel top on the outside of the capillary by using a felt-tipped pen. This makes it easier to follow the migration distance of the protein into the separating gel (see page 166).) The 5% stacking gel mixture is now prepared and 3–5 mm of this is applied to the top of the capillary gel using a capillary pipette. Once again care must be taken not to damage the separating gel top, nor to allow any air bubbles to form between the separating and the stacking gel. Finally, the 5% stacking gel is overlayed with water. Again a capillary pipette is used, though in this instance it is not of such critical importance that the top of the 5% stacking gel is not touched.

After at least a 15-min wait for the stacking gels to polymerise, they can now be loaded with protein solution. Brain proteins are normally extracted with 0.01 M Tris-chloride buffer pH 7.4, though other buffer systems can be used. The optimum concentration of solution for fractionation should be 1–3 mg protein/ml. More concentrated solutions have to be diluted since it is very easy to overload the gels and produce bad separations. If the solution is more dilute than 1–3 mg protein/ml so that the volume of solution which has to be applied is greater than that above the stacking gel, a second capillary is simply attached by means of a piece of polythene tubing.

Normally a protein solution volume corresponding to 2–5 mm of capillary is placed above the stacking gel (in a 5-μl capillary 1 mm = 0.154 μl), depending upon the concentration of protein. In order to load the gel with protein solution, the water or solution B (dilute 1:3) above the 5% stacking gel is first removed with a pipette capillary. The protein solution is then applied, free of air bubbles, above the 5% gel. Any free space in the capillary not filled by the protein solution is filled with buffer solution B, diluted 1:3.

The next and fifth step in the process is the electrophoresis. Once filled, the capillary is removed from the plasticine cushion and a diamond knife is used to scratch the bottom of the capillary at the level where the plasticine has penetrated it. The small piece of capillary is then carefully broken off in such a way that the separated gel is also cleanly severed. It is important not to pull the separating gel in the process. The upper end of the capillary is then inserted through a 2-mm thick perforated rubber cap which fits over the lower end of the funnel, and the funnel is immediately filled with electrode buffer. The capillary with the funnel is then hung in the electrophoresis stand (Fig. 81) and the stand adjusted so that the lower end of the capillary is located in a small petri dish containing the same electrophoresis buffer. The electrodes made of platinum wire and attached to the voltage generator are placed in the electrode buffers. The anode is connected to the lower, and the cathode to the upper electrophoresis buffer.

The voltage generator (Neuhoff *et al.*, 1967) is now switched on. The actual voltage used depends upon the concentration of the separating gel and the type of protein solution to be fractionated. For water-soluble brain proteins and a 20% separating gel, a constant voltage of between 60 and 80 V is most suitable. At the start of the electrophoresis for this system, between 60 and 80 μA flows through the gel system and falls off slowly during the electrophoresis. The current depends on the ionic strength of the buffers used and a current which appears too low or unstable generally indicates that air bubbles are trapped in the capillary. If the voltage applied is too high, the proteins migrate as convex bands (owing to the formation of a temperature gradient across the gel) which are unsuitable for microdensitometer analysis.

The progress of the electrophoresis can now be followed by observing the current dial on the voltage generator. Normally when the current drops from 60–80 μA to 15–20 μA over a period of 30–40 min, the electrophoresis is completed. However, it is best to add a little bromophenol blue and fluorescein to the upper electrophoresis buffer, so as to produce a light coloration, and then to follow their electrophoretic migration. The bromophenol blue binds tightly to the protein mixture during electrophoresis and on entering the stacking gel is seen as a sharp blue band, but gradually becomes diffuse after having entered the separating gel. When the lower end of this diffuse band has migrated a distance of 4–6 mm into the separating gel (the marked position of the separating gel top on the outside of the capillary now becomes very useful), the electrophoresis is complete. Alternatively the fluorescein, which can be viewed as a yellow band and is clearly visible under ultraviolet light, migrates with the buffer front and can be detected entering the separating gel with the bromophenol blue. When the fluorescein band has migrated a distance of 5–7 mm into the separating gel, the electrophoresis is accomplished.

The next stage is to stain the gel. Immediately following electrophoresis the gel is expelled from the capillary directly into 1% Amido black solution (in 7.5% acetic acid) by water pressure. This is carried out by using a 2-ml syringe, the end of which is connected to the bottom part of the capillary by means of a tightly fitting polythene tube (see Fig. 86). It is advisable to expel the gel slowly so that it is not distorted. Five minutes is generally sufficient to stain the protein in Amido black, a shorter time only stains the periphery of the gel which makes it difficult to analyse it densitometrically. The stained gels are then transferred to 7.5% acetic acid which washes out all stain unbound to the proteins within 30 min. The proteins are thus simultaneously fixed and coloured by this procedure.

Coomassie blue (0.1% in 7.5% acetic acid) instead of Amido black can be used as a stain, though in this laboratory Amido black is preferred.

Finally, the stained gels can either be photographed, simply by placing them in a glass chamber containing 7.5% acetic acid (see Fig. 88), or analysed microdensitometrically. To obtain pherograms of the gels, they are transferred to a special glass trough (see Fig. 85) containing 7.5% acetic acid. The floor of this trough is made of optical glass. A single gel is positioned to lie in the middle of a groove and a cover glass placed over it. After focusing the gel in the light path of the instrument and making sure that the optical system of the microdensitometer is clean, pherograms of the gel can be recorded (see Fig. 83). Recording with the Joyce-Loeble double-beam microdensitometer is generally carried out with a gearing ratio 1:20 or 1:50.

(d) DETERMINATION OF R_f VALUES OF PROTEINS

To compare the migration rates of different protein bands from different fractionations, it is necessary to have a reference point that is dependent on the conditions of the electrophoresis. The best way of doing this in the microscale is to treat the gels with 1% pyronine solution in 7.5% acetic acid before Amido black staining (Neuhoff *et al.*, 1970). Pyronin forms a complex with the fluorescein band which is registered as a peak in microdensitometer records (see Fig. 89), and this can be used as a reference value for the determination

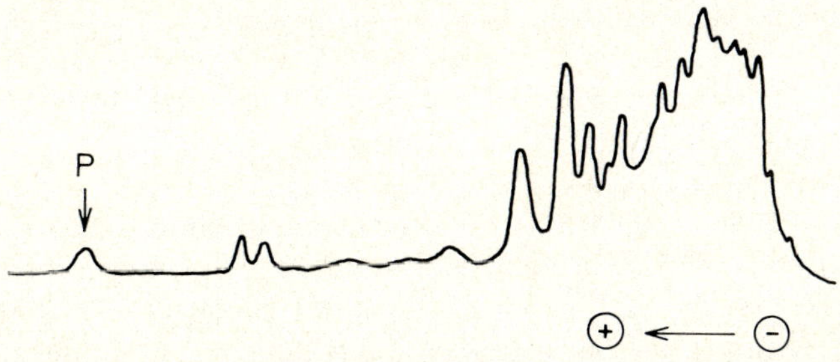

FIG. 89. Pherogram of microgel after fractionation and staining first in 1% pyronine in 7.5% acetic acid (for 10–30 sec) followed by Amido black staining of the proteins. The pyronine produces an insoluble complex with fluorescein (P) which migrates with the buffer front. The pyronine complex can then be used as a reference point to calculate the R_f values of the individual protein-stained bands.

of R_f values of individual bands. In practice the gels are stained in 1% pyronin for a very short period (10–30 sec), since some proteins are denatured by this process. Alternatively only the bottom part of the gel containing the fluorescein band is pushed out of the capillary and the part stained quickly in 1% pyronine in 7.5% acetic acid.

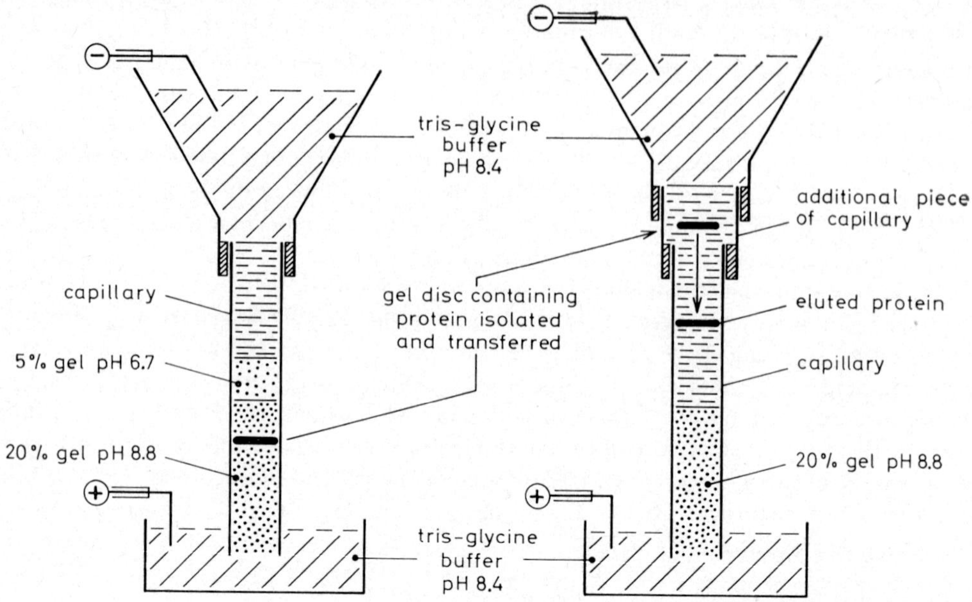

FIG. 90. Procedure used for the electrophoretic elution of proteins from isolated gel discs.

(e) ISOLATION OF INDIVIDUAL PROTEINS

Individual proteins separated by microdisc-electrophoresis can be re-isolated in the way shown in Fig. 90. The second capillary (receiver capillary) required for the electrophoretic elution of the protein band contains only a small amount of a 20% gel and is overlayed by solution B diluted 1:3. Moreover, a short piece of capillary tubing with an inner diameter only slightly larger than the outer diameter of the gel capillary is attached to the top of the gel capillary with Harvard glass wax (Harvard Dental Ges., Berlin, Germany).

To ensure that the proteins to be isolated are not denatured, the fractionated proteins on the gel should only be stained for a short time (about 1 min) with buffered 0.1% bromophenol blue solution. The staining produced is weak and only the protein bands in high concentrations are easily located. The gel is then transferred to a glass slide covered with a layer of paraffin wax and the protein disc cut out of the gel under a stereomicroscope. During the process the gel is held by soft forceps and a stainless steel razor blade is used for the cutting. One or more gel slices from several gels can then be transferred to the receiver capillary (see Fig. 90). After making sure that the receiver has no air bubbles and is filled to the top with buffer B (diluted 1:3), it is attached to the perforated rubber cap of the funnel. The capillary is then subjected to microelectrophoresis as described previously. It is best not to apply too high a voltage for elution. After electrophoresis has proceeded for a while, the eluted proteins which appear as a blue band can be seen migrating in the buffer above the 20% gel. At this stage the electrophoresis is stopped and the larger diameter piece of capillary removed. The solution containing the proteins above the 20% gel can now be removed with a capillary pipette. To check for complete elution, the eluted gel discs can be stained in Amido black.

The above method of electrophoretic elution of proteins is only effective for small proteins (molecular weight up to 100,000). Larger proteins are best eluted from gel slices by first placing the gel slice in a small amount of pure glycerine and then freezing in liquid air. The glycerine infiltrates the gel fairly quickly and protects the protein molecules, while the freezing procedure effectively destroys the internal gel structure. After thawing, the gel slice is transferred to the receiver capillary and the proteins eluted as described.

A similar system of elution can be used to check the electrophoresis of an isolated protein band so as to test the homogeneity of the band. In this instance the receiver capillary contains a separating and stacking gel (see Fig. 91). The isolated gel slice is placed in the receiver capillary and the second electrophoresis can be carried out in the gel immediately following elution. In the analysis of the homogeneity of a protein band, the separating gel of the receiver capillary can have a different concentration from that of the separating gel from the first electrophoresis.

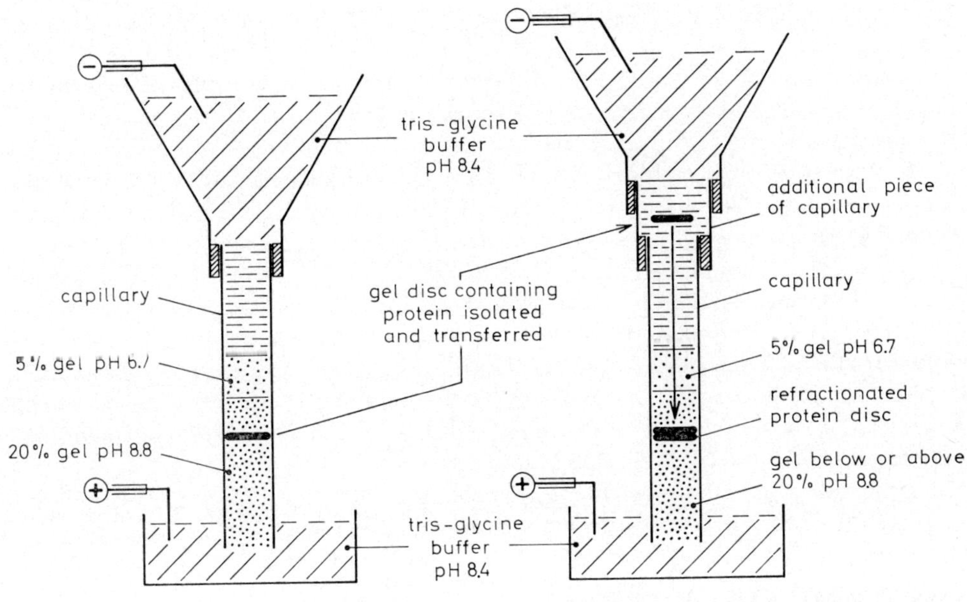

FIG. 91. Procedure used for checking the homogenity of protein(s) from an isolated gel disc.

(f) AUTORADIOGRAPHY OF MICROGELS

An analysis of gels containing separated proteins into which radioactivity has been incorporated can easily be carried out by autoradiography (see Neuhoff, 1973). The gel is initially stained with Amido black and after the unbound dye has been washed out, it is transferred to a small petri dish containing a little water. The dish and its contents are then frozen in liquid nitrogen and placed in a freeze-dryer. The embedding of the gel in ice ensures that the gel retains its external form on freeze-drying and so is virtually unable to shrink. After freeze-drying the gel is placed on a 3 × 3 cm glass plate and fixed to it at each end with a small drop of celloidin (1 % in 50 % ether/ethanol). The autoradiogram film is placed on top of the gel and another 3 × 3 cm glass plate placed on top. (X-ray films of the type employed for making autoradiograms of microchromatograms containing ^{14}C–dansyl compounds are used (see page 75).) This last procedure is of course carried out in the dark. After a suitable length of time the autoradiogram is developed, and the gel, when placed in 7.5% acetic acid, swells up and assumes its original shape. Densitometric traces of the stained gel and the autoradiogram can then be compared to determine which protein bands contain the radioactivity. The protein discs from the gel can be isolated and evaluated by liquid scintillation counting to give some quantitative data.

4. Microelectrophoresis on Gradient Gels

In gradient gels, the pore sizes of the separating gels are not homogeneous but range from small (at the high concentration gel part) to large (at the low concentration gel part). Such a gel allows for a great separation of different sizes of proteins since molecules are basically separated by molecular sieving, with the result that more information can be obtained on the protein composition of a sample than when it is fractionated on a system where a homogeneous pore size separating gel is used. It is partly for this reason that microelectrophoresis on gradient gels has been developed in this laboratory. Microcapillaries (Drummond microcaps) with linear gel gradients are produced either in large quantities by using a special apparatus (Dames and Maurer, 1974) or in single capillaries by using capillary attraction (Rüchel *et al.*, 1973a).

It has subsequently been found that microelectrophoresis on gradient gels often gives a sharper separation of proteins compared with the microdisc electrophoresis method of Neuhoff (1968) where homogeneous pore sizes separating gels are used. Moreover, the actual procedure is simpler and does not demand the skill necessary for microdisc electrophoresis. For example, the 5% stacking gel required for the one procedure is not required in microgradient electrophoresis. Other advantages of microgradient electrophoresis include (a) the buffer-systems employed, which can be either discontinuous or continuous, (b) the slope of the gradient can be varied from 2 to 40% gel concentration, so allowing separation of various sizes of proteins and which also (c) can easily be adapted to fractionate proteins extracted with SDS.

The procedure to be described thus appears more suitable for routine analysis of proteins than the microdisc-electrophoresis method of Neuhoff (1968).

(a) PREPARATION OF GRADIENT GELS

(i) *Washing the capillaries* (*Rüchel, 1973*)

The new Drummond capillaries used to make gradient gels must first be cleaned, otherwise the protein fractionation is affected. This can be done effectively by washing the capillaries in chromic acid and then rinsing in distilled water, absolute ethanol, and finally acetone. To ensure that they are completely dry, the capillaries are left overnight in an oven at 37°C. A batch of capillaries can easily be filled with chromic acid if the pressure above them is reduced by creating a vacuum with a water-pump. The capillaries are then rinsed with water and transferred to a suitable sintered-glass filter-funnel over which a separating funnel is placed. The separating funnel is fitted with a stopper so that it fits tightly onto the top of the sintered funnel, which is connected to a water-pump (see Fig. 92). By placing water, ethanol or acetone in the separating funnel and controlling the release of the solution, the capillaries are thoroughly cleaned.

Capillaries which have already been used for electrophoresis should first be treated

FIG. 92. This photograph shows the equipment required for cleaning capillaries. The solution in the separating funnel is allowed to flow into the lower chamber containing the capillaries which rest on a sintered glass filter. The lower chamber is connected to a water-pump system, thus enabling all solution to be removed from the capillaries. This simple equipment allows a number of capillaries to be rinsed with different solutions.

with Javelle's solution (undiluted aqueous potassium hypochlorite) to dissolve any polyacrylamide, before washing by the above procedure. It is best to transfer the dirty capillaries to a suction flask half-filled with Javelle's solution and connected to a vacuum pump. The suction flask should have a perforated stopper with a valve which is operated by a motor so that the vacuum is interrupted at intervals. This allows the gas, formed by the action of potassium hypochlorite, to be removed immediately from the capillaries. If this is not done, gas bubbles remain in the capillaries and prevent the potassium hypochlorite from acting on the polyacrylamide.

(ii) *General requirements*

These are the same as for microelectrophoresis on homogeneous gels.

(iii) *Gel solutions*

The following solutions are used for the production of gradients:

Solution 1: 35 mg ammonium peroxydisulphate is dissolved in 50 ml glass distilled water.

Solution 2: 3.39 g Tris are mixed with 6 ml 1 N H_2SO_4 and 0.1 ml TEMED. This solution is then adjusted to pH 8.8 by the addition of approximately 0.6 ml 6 N H_2SO_4, and then glass distilled water added to make a total of 10 ml.

Solution 3: 20 g Acrylamide, 400 mg bisacrylamide and 3.75 mg $K_3Fe(CN)_6$ are dissolved in 37.5 ml glass distilled water.

All these solutions can be stored for prolonged periods at 4°C, except for solution 1, which should not be kept for longer than a month. Before use the solutions should be at room temperature.

(iv) *Preparation of gel gradients*

(a) Method of Rüchel *et al.* (1973a). In this procedure, individual capillaries are filled by capillarity, first with some solution 1, and then with a mixture of solutions 2 and 3. By placing a crystal of bromophenol blue into the mixture of solutions 2 and 3, the formation of the gradient can be observed. The lower mixture initially penetrates the upper solution in a cone-shaped manner (see Fig. 93) which depends on the diameter of the capillary,

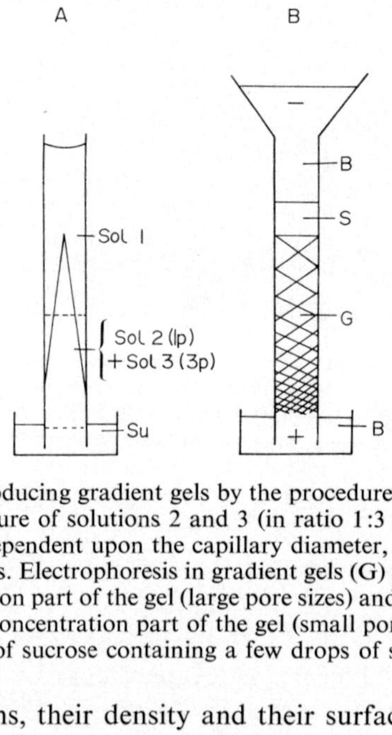

FIG. 93. A schematic view of producing gradient gels by the procedure of Rüchel *et al.* (1973a) is shown in A. It can be seen that a mixture of solutions 2 and 3 (in ratio 1:3 by vol) penetrates solution 1 in a cone-shaped manner which is dependent upon the capillary diameter, the viscosity of the solutions and their density and surface tensions. Electrophoresis in gradient gels (G) is shown in B, with the sample (S) placed above the low concentration part of the gel (large pore sizes) and electrode buffer (B) placed above the sample and below the high concentration part of the gel (small pore sizes). (Photograph by courtesy of R. Rüchel.) Su is a solution of sucrose containing a few drops of solution 1.

the viscosity of the solutions, their density and their surface tension. The shape of the gradient in the vertical direction is thus achieved when the capillary is filled with the gel solutions. However, the distribution of the solutions in the horizontal direction takes a while and depends upon diffusion. This diffusion process in a 5-μl capillary requires at least 10 min and it is important that the polymerisation of the gel mixture does not begin within this time. For the gel mixtures used, polymerisation only begins after a period of 10 min and proceeds from the top (low concentration) to the bottom of the gradient. It is essential in the formation of the gradient that turbulence does not upset the polyacrylamide before polymerisation. Turbulence is often the result of heat generated in the polymerisation process, so it is necessary that the gel polymerises from the top and proceeds to the bottom. This is achieved in this method because the peroxydisulphate also forms a gradient from the top of the gel to the bottom in the same manner as the polymerisation process.

In practice it is best to hold a capillary with a pair of forceps and to dip it into solution 1 so that half of the capillary is filled by capillarity. It is advisable to hold a finger over the top of the capillary to control the movement of the capillary contents and prevent air bubbles from entering (see Fig. 94). The capillary is then immediately dipped into a mixture

FIG. 94. This photograph demonstrates the procedure for filling capillaries with the appropriate amounts of solution when preparing gradient gels by the method of Rüchel *et al.* (1973a).

of solutions 2 and 3 (one part solution 2 and three parts solution 3) and filled to the upper edge of the capillary by capillarity. The filled capillary is then placed vertically in a small beaker containing a 2–4 mm layer of 50% sucrose solution (see Fig. 95). The adhesive force produced by the sucrose solution and the glass walls of the capillary and beaker holds the capillary firmly in a vertical position. If capillaries are only partly filled with the gel solutions, they should be placed vertically in a cushion of plasticine and not in a sucrose solution, otherwise some of this solution will enter the capillary by capillarity and interfere with the formation of the gradient.

For routine purposes it is advisable to place a crystal of bromophenol blue into the freshly prepared mixture of solutions 2 and 3 in order to visualise the formation of the gradient. This also has the advantage of revealing any failures, besides making it easy to see the high and low concentration parts of the gradient gel. The dye has no effect upon the fractionation of the proteins. After the gels have completely polymerised, a small quantity of excess solution 1 and a mixture of solutions 2 and 3 remain respectively above and below the gel. The polymerised gels can be stored in a glass container of solution 2 (diluted 1:8 with water) for at least a week at 4°C.

The following relationship exists for the formed gradient gels using the above stock

FIG. 95. The procedure used for temporarily storing capillaries filled with polyacrylamide gradients. The filled capillaries are placed around the periphery of a small beaker containing about 5 mm of 50% sucrose containing a little solution 1. This picture shows polymerised gradient gels where the upper layer of excess solution has been removed.

solutions. In a 5-μl capillary (0.45 mm diameter) the gradient has a length of about 15 mm. As the capillary diameter increases, the gel gradient becomes shorter and steeper, while a decrease in diameter results in a longer and less steep gradient (see Fig. 96A).

The length of steepness of gel gradients in a particular size of capillary can also be varied by either adding sucrose to stock solution 1 (see Fig. 96B), or by varying the concentration of acrylamide in solution 2/3.

(b) Method of Dames and Maurer (1974). Another way of preparing gradient microgels is that described by Dames and Maurer (1974). This procedure allows more than 100 5-μl Drummond capillaries to be filled simultaneously with gel solutions, so that a number of gels with identical gradients are prepared. This technique, however, necessitates the use of mixing chambers suited to the volumes required for filling the capillaries. Figure 98 shows the apparatus for preparing gradient gels. It consists of three chambers (diameter 12 mm, height 45 mm) drilled into a perspex block. The chamber containing the high concentration gel (A) is connected to the chamber containing the low concentration gel (B) via a small groove (G) which can be closed by the valve (E). The third chamber (C) contains sucrose solution. The outflow of solutions from chambers B and C is controlled by the valves H and I respectively. The other major part of the apparatus is the receiving chamber (R). For this part a cellulose nitrate tube, measuring 14 mm in diameter and 9.5 cm long, has been found most convenient because a hole can easily be drilled through the bottom. It is through this hole that the gel solution enters. The flow of the gel solution is controlled not only by valve H but also by valve F. The capillaries are stacked within the cellulose nitrate tube, and are first stood on a nylon mesh (N) which is held upright by a stainless steel holder (H). To keep the capillaries in position, a small glass cylinder (R) is also placed

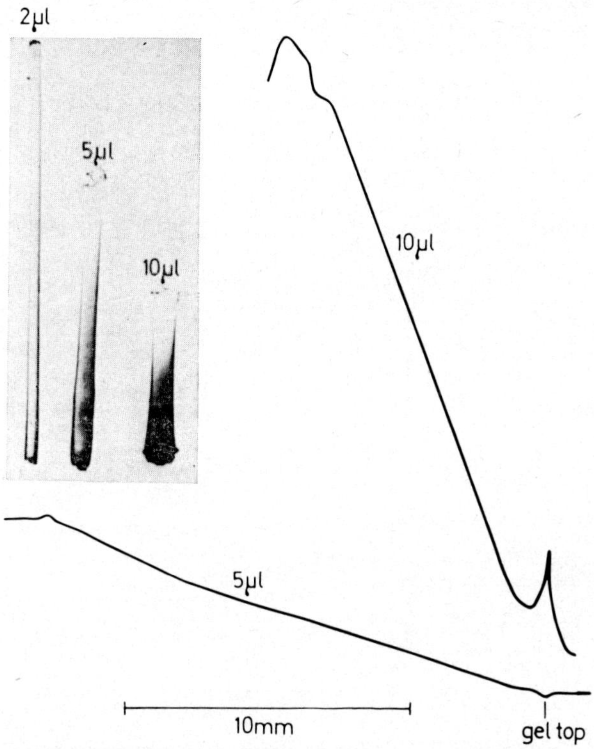

FIG. 96A. The relationship between the slope and length of a polyacrylamide gradient and the inner diameter of capillaries. This was obtained by mixing globulin with the polyacrylamide solutions and then staining the gels after polymerisation with Amido black followed by microdensitometry. (Photograph by courtesy of R. Rüchel.)

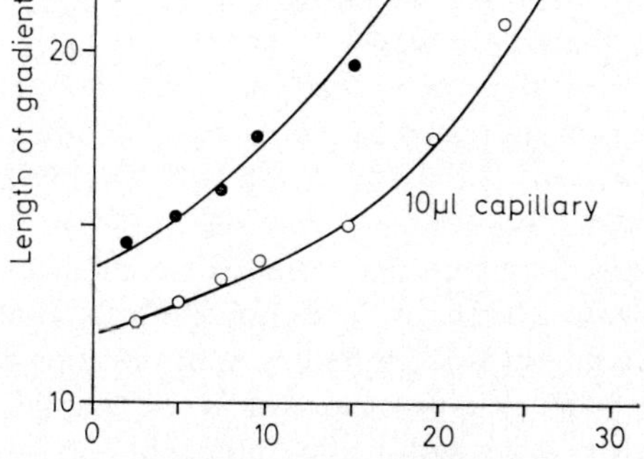

FIG. 96B. The effect of adding sucrose to the gel mixture on the length of polyamide gradients in 5-μl and 10-μl capillaries. (After Rüchel *et al.*, 1973a.)

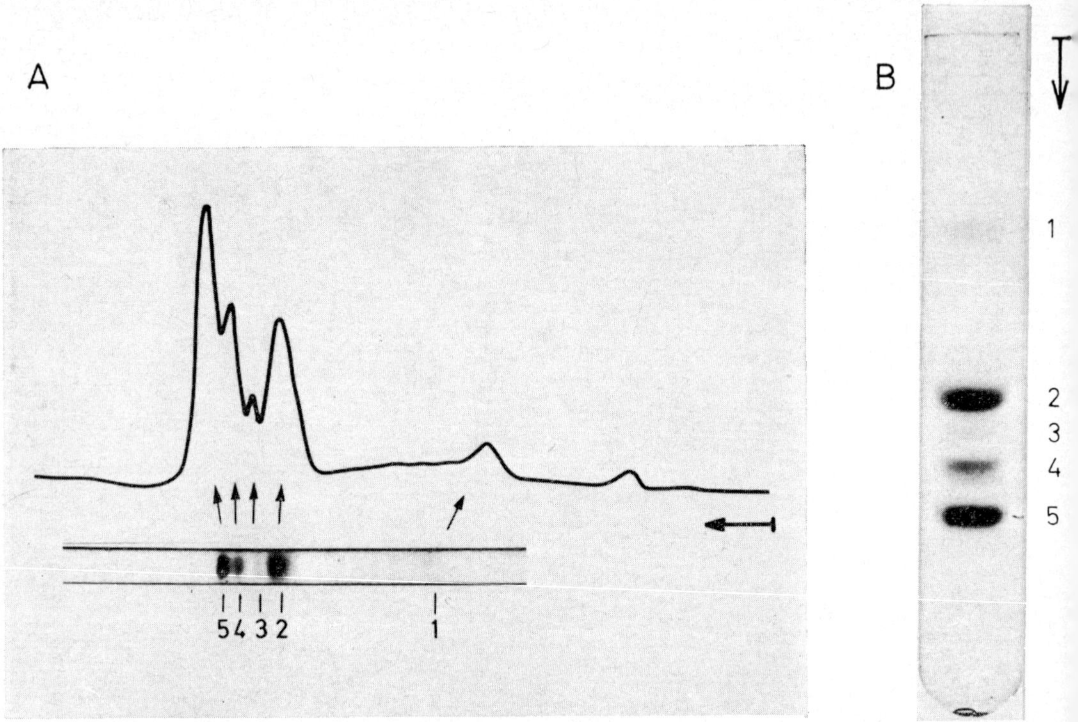

FIG. 97. A comparison of the fractionation of purified myelin protein of mouse brain in SDS-systems using a micropolyamide gradient gel in a 5-μl capillary (A) and a normal 5 mm diameter gel (B). A also shows a microdensitometer scan of the gel to demonstrate all the protein bands. It can be seen that the same protein pattern is obtained when employing either dimension of gel. The proteins occurring in myelin can be identified as follows: 1, Wolfgram's protein; 2, proteolipid protein; 3, protein band DM-20, defined by Agrawal *et al.* (1972); 4 and 5, basic proteins. (A by courtesy of R. Rüchel and B by courtesy of T. V. Waehneldt.)

on the nylon mesh, so that the capillaries stand in an upright position between the glass cylinder and the stainless steel holder. To ensure that the capillaries are upright in the cellulose nitrate tube, a little T-piece (T) is placed on the top of the stainless steel holder, which guarantees that the holder is in a vertical position within the tube.

Gradients are prepared in the following way. The apparatus is assembled as shown in Fig. 98, without any solutions present and with all the valves closed. Note that the receiving chamber is situated lower than the other chambers. Some solution 2, diluted 1:8 by volume with water (i.e. buffer solution), is then placed in chamber B and sucrose (50%) solution in chamber C. Valve I is slowly opened so that sucrose solution fills the polythene tubing space (free of bubbles) between the chamber and the valve, and the valve is then closed. Next, valves H and F are slowly opened to allow the buffer to run evenly without any air bubbles into the receiving chamber (R). When the capillaries are totally immersed in buffer solution, the two valves (H and F) are closed, after making sure that all the tubing between the receiving chamber (R) and the low concentration chamber (B) is filled with buffer and free from air bubbles. Any excess buffer in chamber B is now removed. This chamber (B) is then partly filled with freshly prepared low concentration (see Table 37) gel mixture (about 1 ml) and chamber A with the same amount of freshly prepared high concentration gel mixture. A little magnetic stirrer (M) is also placed in chamber B. Valves

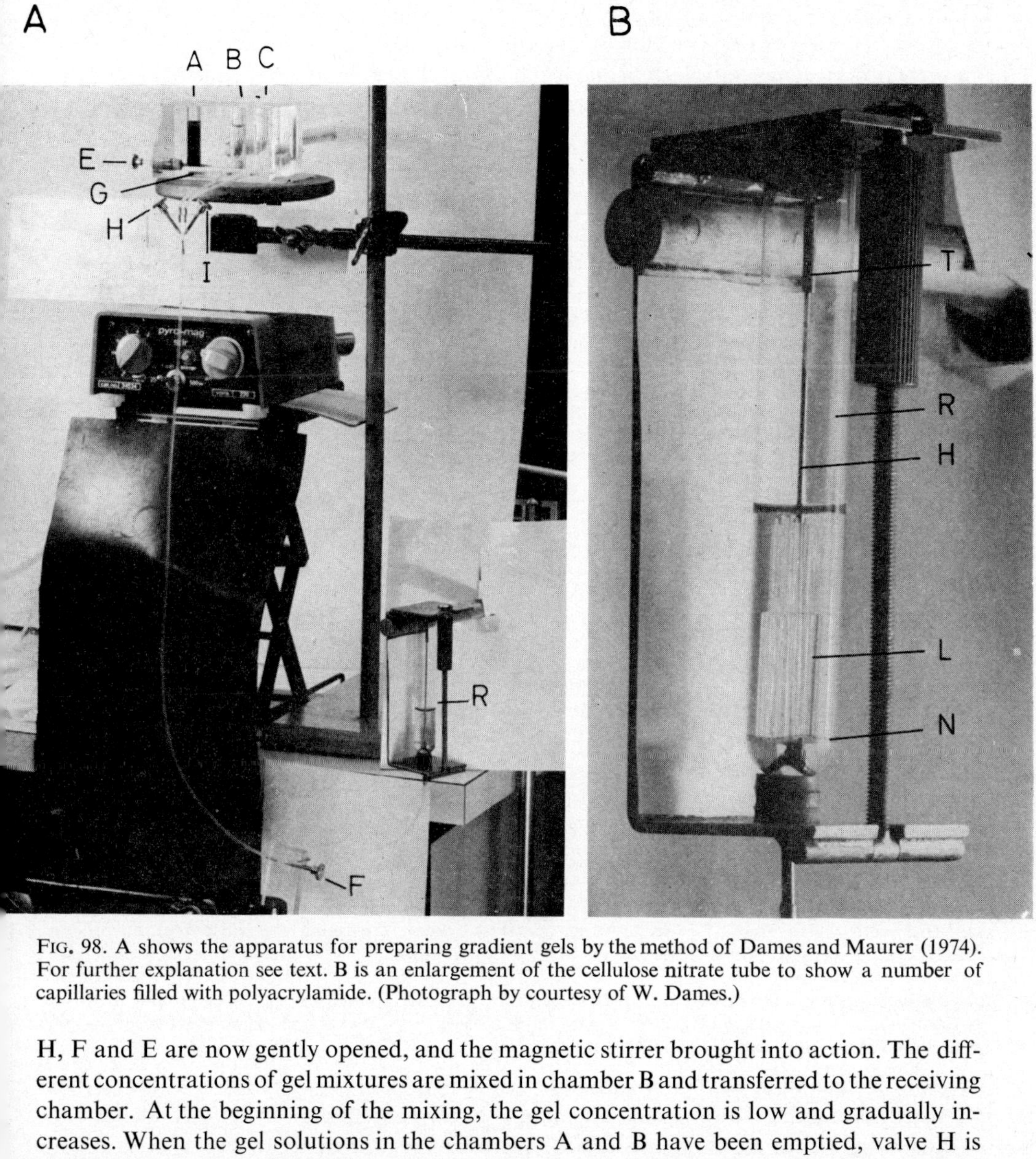

FIG. 98. A shows the apparatus for preparing gradient gels by the method of Dames and Maurer (1974). For further explanation see text. B is an enlargement of the cellulose nitrate tube to show a number of capillaries filled with polyacrylamide. (Photograph by courtesy of W. Dames.)

H, F and E are now gently opened, and the magnetic stirrer brought into action. The different concentrations of gel mixtures are mixed in chamber B and transferred to the receiving chamber. At the beginning of the mixing, the gel concentration is low and gradually increases. When the gel solutions in the chambers A and B have been emptied, valve H is closed, and again no air bubbles are allowed to enter the lower part of the tubing. Valve I is now opened and the sucrose solution is allowed to push the remaining gel solution from the tubing into chamber R. The flow of sucrose is continued for a while, until the amount of sucrose in the receiving chamber is sufficient to allow thc capillaries to balance on the sucrose meniscus. All the valves are now closed. After the gels have polymerised, the stainless steel holder and the capillaries are removed from chamber R. The outer ring of polymerised polyacrylamide is peeled off and the capillaries are simply freed.

The whole procedure of filling the capillaries to form a gradient should not exceed 10–15 min, since polymerisation begins after about 25 min. Because of the higher potassium ferricyanide content in the acrylamide-bis-acrylamide stock solution (see Table 37) the polymerisation process in this procedure is slightly and essentially prolonged. It is important for polymerisation to begin at the top of the gel to avoid turbulence (see page 172). The formation of the gradient can be visualised by adding a crystal of the bromophenol blue to the concentrated polyacrylamide gel solution.

TABLE 37. SUMMARY OF THE AMOUNTS OF STOCK SOLUTIONS REQUIRED TO PREPARE DEFINED POLYACRYLAMIDE GRADIENT GELS BY THE PROCEDURE OF DAMES AND MAURER (1974) In each instance 0.25 ml of stock solution N and 0.25 ml of stock solution O have to be added to the shown amounts of stock solution P and water to give 2 ml gel solution

% Acrylamide conc.	Amount of stock solution P in ml	Amount of water in ml	% Acrylamide conc.	Amount of stock solution P in ml	Amount of water in ml
1	0.0375	1.4625	21	0.7875	0.7125
2	0.075	1.425	22	0.825	0.675
3	0.1125	1.3875	23	0.8625	0.6375
4	0.15	1.35	24	0.900	0.5925
5	0.1875	1.3125	25	0.9375	0.5625
6	0.225	1.275	26	0.975	0.525
7	0.2625	1.2375	27	1.0125	0.4875
8	0.30	1.20	28	1.05	0.45
9	0.3375	1.1625	29	1.0875	0.4125
10	0.375	1.125	30	1.125	0.375
11	0.4125	1.0875	31	1.1625	0.3375
12	0.45	1.05	32	1.20	0.30
13	0.4875	1.0125	33	1.2375	0.2625
14	0.525	0.975	34	1.275	0.225
15	0.5625	0.9375	35	1.3125	0.1875
16	0.60	0.90	36	1.35	0.15
17	0.637	0.863	37	1.3875	0.1125
18	0.675	0.825	38	1.425	0.075
19	0.7125	0.7875	39	1.4625	0.0375
20	0.75	0.75	40	1.5	0.000

The compositions of stock solutions are slightly different from those described in the text as it is impor tant to have a higher amount of ferricyanide present in order to prolong the polymerisation process. The following stock solutions as according to Dames and Maurer (1974) are therefore recommended Stock solution N is 48.5 g Tris-sulphate, 3 ml 98% H_2SO_4, and 0.65 ml TEMED dissolved in 90 ml water and adjusted to pH 8.9 with H_2SO_4 Stock solution O is 580 mg ammonium peroxydisulphate dissolved in 100 ml water. Stock solution P is 26.14 g acrylamide, 0.54 g N,N′-methylene-bis-acrylamide and 6 mg potassium ferricyanide dissolved in 50 ml water.

With this technique of making gradient gels it is possible to alter the steepness of the gradient, the length of the gradient, etc., simply by altering the concentration and amounts of the two gel mixtures. Gradients of different steepness, for example, can be achieved by altering the concentration of the freshly prepared high concentration gel mixture (Fig. 99 and Table 37). This figure also demonstrates that gradients produced by this method are almost perfectly linear. Since it is essential that turbulence does not upset the polyacrylamide before polymerisation (see page 172) it is absolutely necessary to add potassium ferricyanide to the acrylamide-bisacrylamide stock solution, i.e. solution P. This slows down the polymerisation of polyacrylamide in low concentrations to a lesser extent than in high concentrations, even though its relation to acrylamide in high or low concentrations is constant. The method of producing gel mixtures of various concentrations is shown in Table 37.

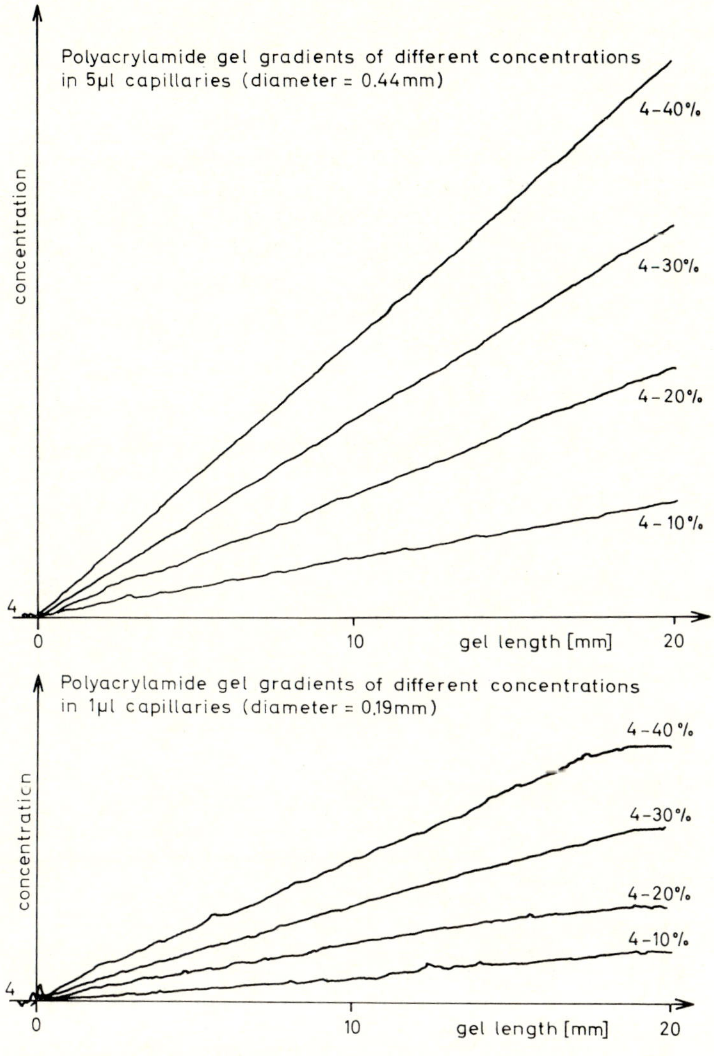

FIG. 99. Densitometric traces of different gradient gels containing Blue Dextran 2000 prepared by the procedure of Dames and Maurer (1974). The upper trace shows gradients of different concentrations produced in 5-μl capillaries and the lower trace those produced in 1-μl capillaries. Only the first 20 mm of each gel was scanned because of the technical limits of the densitometer. (From Dames and Maurer, 1974.)

It is of interest to note that Smeds and Björkman (1972), writing independently, have also described a procedure for preparing micropolyacrylamide gels which is based on the same idea as the method of Dames and Maurer (1974). In their method the gradient mixing device consists of two cylindrical chambers, the 'dense chamber' and the 'mixing chamber', connected to one another via a channel opening controlled by a valve. Once the gradient solution is prepared, the capillaries are filled by dipping them into the solution by means of an excentric plate connected to a motor with stepwise controls. This procedure appears to be unsuitable for the production of large quantities of gradient gels.

(b) FRACTIONATION OF PROTEINS

Fractionation of proteins on gradient gels in the microscale can either be carried out with the discontinuous buffer system of Ornstein (1964) and Davis (1964), or the discontinuous system of Allen *et al.* (1969) as shown in Figs. 100 and 101 respectively. In the first procedure about 4 mm Tris/phosphate buffer pH 6.7 (stock solution B of disc-electrophoresis on homogeneous gels) containing 20% sucrose is placed above the gel immediately after removing the overlaying liquid. This layer of solution is used instead of a 5% stacking gel. The sample (in Tris/phosphate buffer, i.e. solution B diluted 1:3) containing a little sucrose (10%) is then applied to the top of the sucrose-buffer solution and the remaining capillary space filled with Tris/phosphate buffer (i.e. solution B diluted 1:3). The same electrode buffers as for microdisc electrophoresis in homogeneous gels are used, viz. Tris/glycine pH 8.4–8.8.

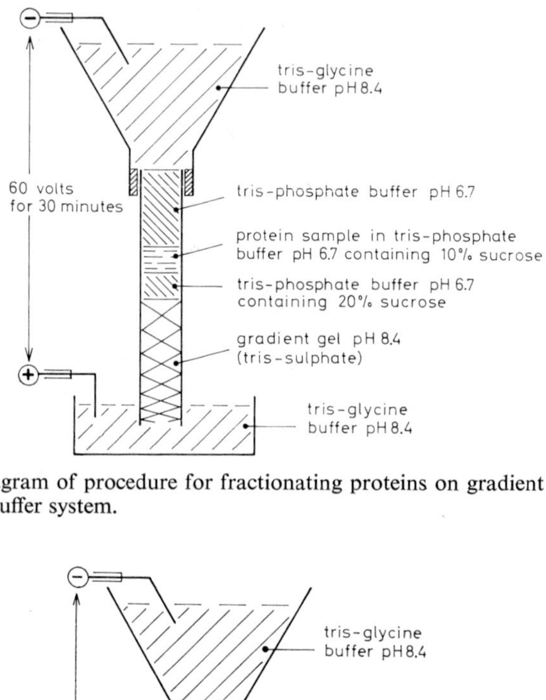

FIG. 100. Schematic diagram of procedure for fractionating proteins on gradient gels in the microscale using a discontinuous buffer system.

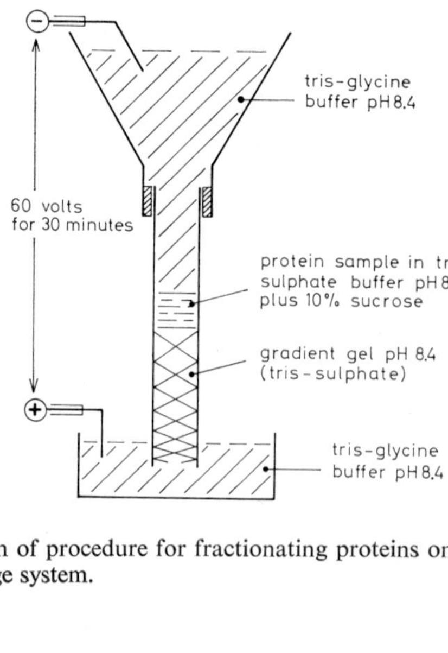

FIG. 101. Schematic diagram of procedure for fractionating proteins on gradient gels in the microscale using a discontinuous voltage system.

When using the discontinuous voltage system as described by Allen *et al.* (1969), the protein sample is prepared in diluted (1:20) Tris/sulphate buffer (i.e. stock solution 2). A little sucrose (10%) is added to the sample and some of this solution is placed directly above the gel after removing the overlaying liquid. The remaining capillary space can either be filled directly with electrode buffer (Tris/glycine pH 8.4–8.8) or with diluted buffer of the same composition and pH as the gel buffer but without TEMED. The latter functions as the 'cap-gel' in the system of Allen *et al.* (1969).

Since the protein molecules are separated by the molecular sieving effect of the gradient gels, the duration of electrophoresis has little effect upon the sharpness of the bands. In practice it is found that the best fractionations are obtained after 20–30 min at a constant potential of 60 V. The electrophoresis can be completed in a couple of minutes by using voltages approaching 300 V, though this is not advocated because of possible protein degradation. If the electrophoresis is allowed to continue for more than 30 min at 60 V and up to 3 hr, the relative positions of the bands are unaltered, though additional bands due to breakdown of some proteins may appear. The migration of the proteins can be observed by adding a little bromophenol blue or bromophenol red to the upper electrode buffer. After electrophoresis the gel is extruded from the capillary by placing a steel wire into the bottom of the capillary and pushing the highly concentrated part of the gel so that the low concentration part is the first to emerge (see Fig. 102). The gels are then stained and analysed in the manner described for electrophoresis of homogeneous gels.

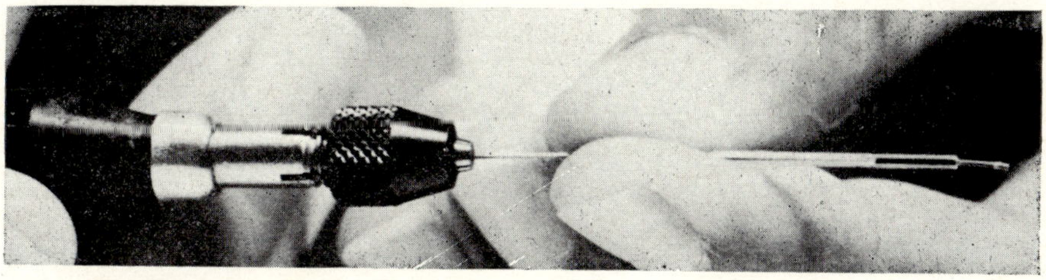

FIG. 102. A photograph to show the procedure for removing gradient gels (containing no Triton X 100) in capillaries. A steel wire is used to push the gel (at the highly concentrated tip of the gel) gradually out of the capillary.

One advantage of using gradient gels to separate proteins in shown in Fig. 103. In this experiment brain proteins from different areas were extracted with Tris/buffer and then subjected to microelectrophoresis, employing both homogeneous and gradient gels. It can clearly be seen that a better fractionation of proteins occurs when gradient gels are used.

(c) FRACTIONATION OF PROTEINS CONTAINING SDS

When SDS extracts are being fractionated on gradient gels, the presence of SDS in the gel is not required (see Fig. 104). Moreover, Triton X-100 must not be included in the gel mixture. The SDS concentration in the sample to be fractionated should be more than twice the protein concentration. The electrode buffer remains unchanged apart from the addition of 0.1% SDS to the upper electrode buffer. The fractionation of a purified sample of myelin by electrophoresis in the micro- and normal scales is shown in Fig. 97. It can be seen that in each of the gels the pattern of myelin proteins is the same.

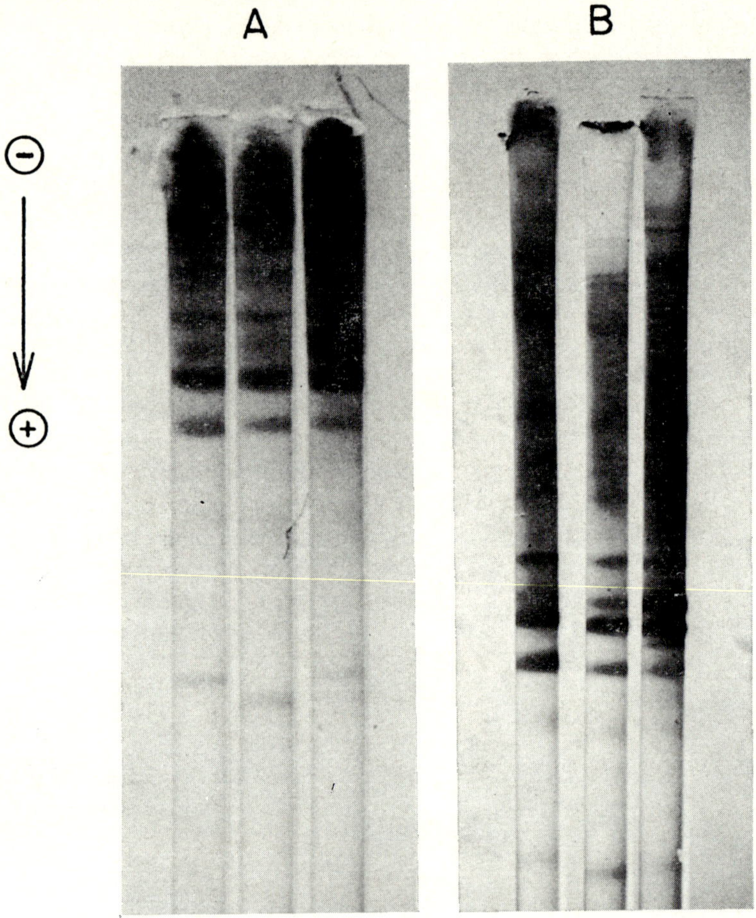

FIG. 103. Fractionation of water-soluble proteins by microelectrophoresis in 5-μl capillaries using (A) homogeneous and (B) gradient polyacrylamide gels. In each instance the gels on the left, right and in the middle show the separation of proteins extracted from rat grey matter, rat pineal gland and rat white matter respectively. The magnification of B is slightly lower than of A. (Photograph by courtesy of W. Dames.)

Rüchel *et al.* (1973b) have studied critically the effect of SDS on the electrophoretic patterns of proteins, using the microgradient polyacrylamide gels. The general conclusions formed were as follows. Despite the greater sharpness of the bands and the shorter separation times which can be achieved with discontinuous electrophoresis buffer systems, preference should be given to the continuous buffer systems for the SDS electrophoresis. In continuous buffer systems the labile SDS protein combinations (Maizel *et al.*, 1970) are less 'stressed' in constant decrease in the electrical field (electrical field strength = V/cm) than in changing field strength, which is produced by a discontinuous system and can easily cause unloading artefacts in the proteins. In a continuous system, the amount of SDS present does not therefore play such an important role and can be increased so as to stabilise the labile complexes. In contrast, the SDS content in a discontinuous system has strict limits, and even then interferes with the system. These points are exemplified by studies on human albumen (see Fig. 105), which contains four polymeric forms as well as a monomer. The albumen solution was incubated with 10 mg SDS per 1 mg protein for

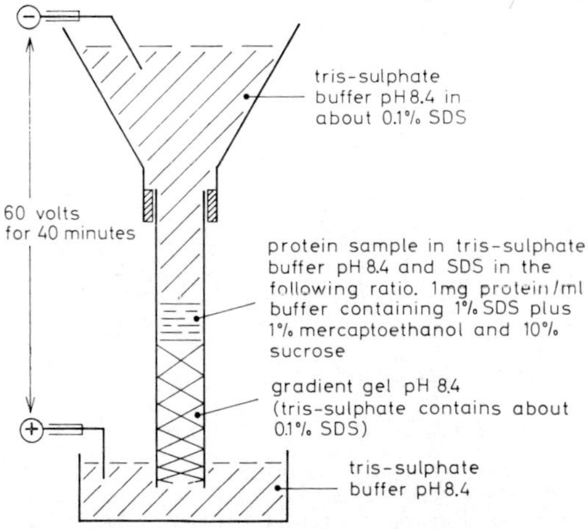

FIG. 104A. Schematic diagram of procedure for fractionating SDS-treated proteins on gradient gels in the microscale using a continuous buffer system. The amount of SDS used is critical and should be adjusted in the upper electrode buffer and the gel to approach the amount in the sample solution.

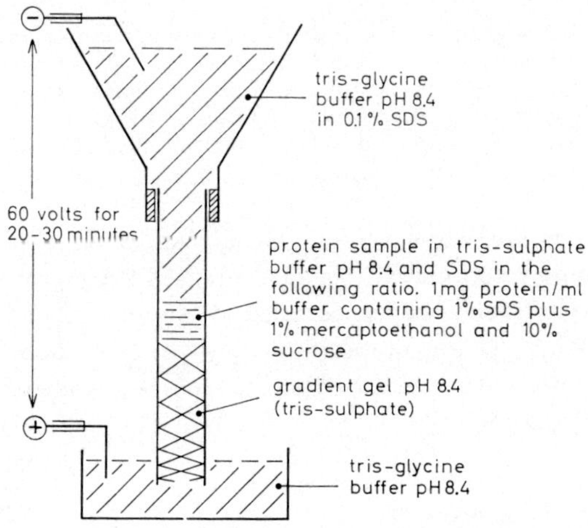

FIG. 104B. Schematic diagram of procedure for fractionating SDS-treated proteins on gradient gels in the microscale using a discontinuous buffer system. The amount of SDS used is critical in order to obtain consistent data.

5 min at 100°C and 0.3 μl samples were electrophoresed on gradient gels in 5 μl capillaries. Fractionation of the SDS-treated albumen was then carried out with a continuous buffer system (Fig. 105A) and a discontinuous buffer system (Fig. 105B). It can clearly be seen that use of the continuous buffer system produced the anticipated pattern of albumen and its polymers (total of five bands), while the discontinuous buffer system produced at least twelve different bands.

From the above observations it is obvious that great care has to be taken when separating proteins containing SDS, and that the type of electrophoresis system used must be critically assessed. Generally, the continuous SDS gel electrophoresis system (Fig. 104A) is preferred to the discontinuous system (Fig. 104B) for 'reliable' protein separations, as shown

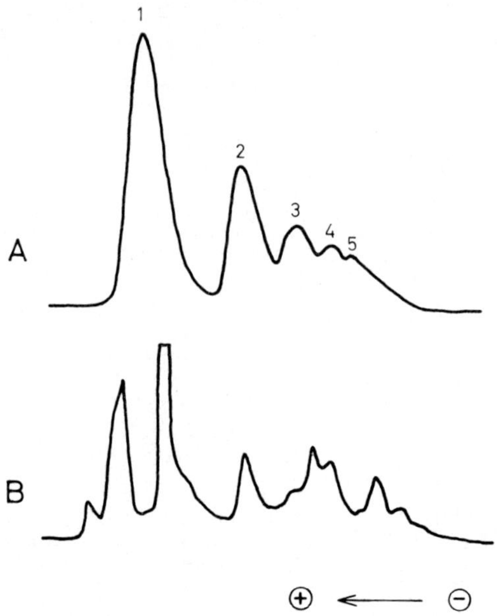

FIG. 105. Use of continuous (A) and discontinuous (B) systems for the fractionation of SDS-treated albumen in gradient gels. In the continuous electrophoresis system (A), 0.1 % SDS in sodium phosphate buffer was used (for upper buffer and to suspend sample), while in the discontinuous system (B), 0.1 % SDS in Tris/glycine buffer was used. It can be seen that with the continuous system the predicted five albumen polymers are formed, whilst when applying the discontinuous system a great number of bands are formed. (Photograph by courtesy of R. Rüchel.)

in the case of human albumen. However, even when using a continuous buffer system, the number of SDS proteins fractionated can be increased simply by allowing the electrophoresis to proceed for a much longer duration. It should perhaps also be pointed out that electrophoresis in a continuous buffer system (with or without SDS) requires appreciably longer electrophoresis time and the bands are not as sharp as those obtained with discontinuous buffer system.

5. Adaptation of Microelectrophoresis

(a) RIBONUCLEIC ACID (RNA) SPECIES

Various authors have adapted micro- or semimicroelectrophoresis procedures to study the distribution of RNA species in nanogram quantities of starting materials (Ringborg *et al.*, 1968; Naumova, 1971). Using the same method of making gradient gels as Rüchel *et al.* (1973a), Wolfrum *et al.* (1974) developed a micromethod to fractionate RNA species. In this method the same procedures as described on page 170 are used for preparing the gradient gels, except that in solution 1 the amount of ammonium peroxydisulphate is increased from 35 to 50 mg, and it is essential that it is fresh. The actual preparation of the gradient gels is otherwise as described on page 172. By diluting the stock solutions, various concentrations of gradient gels can be obtained. For example, to produce a gradient gel from about 1 % to 10 %, which is often necessary for the fractionation of certain RNAs, e.g. ribosomal, gel solutions 2 and 3 are made up by mixing together five parts solution 2, three parts solution 3 and eight parts water, while solution 1 is used as normal.

Since RNA is rapidly degradated by ribonucleases which have a ubiquitous distribution, fairly sterile conditions are required for the fractionation of an extract of RNA. Drummond (or Brand) capillaries must be cleaned in Javelle's solution, treated with alcohol and acetone (see page 170), and then dried at 80°C. It is also very important when preparing extracts of RNA to use microtubes which have been siliconised by dipping them in 2% dimethyldichlorosilane in benzene and then drying them for 2 hr at 80°C.

Extracts of RNA from nervous tissue can be obtained in the following way. Tissues (0.1–5 mg), which may be fresh or fixed (often fixed tissue in for example, 70% ethanol is preferred because the fixation process prevents the breakdown of the RNAs), are homogenised in siliconised microtubes at 4°C in 50 mM Tris/sulphate pH 8 plus 2 mM EDTA and 1% SDS as described on page 53. Since it is important not to damage the RNAs during the homogenisation, the latter must be gentle and no faster than 1000 rpm. Some 0.5% proteinase K can also be added to the solution to help in the digestion of proteins, even in the presence of SDS and EDTA (Wiegers and Hilz, 1971, 1972; Gross-Bellard *et al.*, 1973). After homogenisation, the solution is allowed to stand at room temperature for 1 hr to enable the proteinase K to digest the proteins and then centrifuge for 10 min at 10,000 *g*. Should the extract contain a high concentration of RNA, as for example, in a ribosomal preparation, the supernatant can be used immediately for electrophoresis. When proteinase K is omitted, a modification of the extraction method of Kirby (1965) can be applied to the supernatant in order to remove the remainder of the proteins (and proteinase K, if used). This involves making a phenol–cresol mixture saturated with 100 mM Tris/sulphate pH 8.4. The supernatant containing the RNA is mixed thoroughly with the phenol–cresol mixture and the sample is then centrifuged for 10 min at 20,000 *g*. The upper layer containing the RNA is subsequently removed. The recovery of RNA through this step is about 80%, and this procedure can be repeated if required. The solution containing the RNA is now thoroughly mixed with chloroform/isoamylalcohol (24:1 v/v) and then centrifuged for 10 min at 20,000 *g*. The supernatant is removed and the RNA precipitated out by adding 1/10 vol of 20% sodium chloride and 2.5 vol ethanol, and is left overnight at −20°C. After centrifugation at 20,000 *g* for 10 min at 0°C, the upper alcohol phase is removed with a pipette and the remainder evaporated under vacuum. If necessary, at this stage any DNA is removed from the RNA extract by treatment with DNAse. The RNA sample is now dissolved in either 200 mM Tris/sulphate buffer pH 7.5 or 200 mM phosphate buffer pH 6.5 containing 0.1% SDS, 1 mM EDTA and 10% sucrose. The EDTA and SDS partly inhibit the nucleases. Using the above extraction procedure, RNA from brain tissue can be extracted in a relatively pure state for microelectrophoresis (see Fig. 106).

Fractionation of the different RNA species is carried out by a system similar to that described by Allen *et al.* (1969), as shown in Fig. 107. In order to obtain well-fractionated RNAs, in particular of high molecular weight species, and to prevent RNA aggregation, the concentrating effect of the discontinuous system must be avoided, and the sieving properties of the electrophoresis system relied upon. This can be achieved partly by (a) having a low electric field strength, (b) having a high ionic strength in the sample buffer, and (c) by the addition of fast-moving ions to the system, viz. EDTA and SDS (Wolfrum *et al.*, 1974). Just before applying the sample (dissolved in 200 mM Tris/sulphate buffer pH 7.5 or 200 mM phosphate buffer pH 6.5 plus 0.1% SDS, 1 mM EDTA and 10% sucrose), the upper aqueous phase is carefully removed from the polymerised gel with a pipette capillary. The free space is then rinsed with distilled water and 2 mm 50% sucrose in Tris/sulphate buffer (solution 2 without TEMED, diluted 1 to 8 with water (see page 172)) pH 8.4 containing 1 mM EDTA and a little bromophenol blue is applied. Special care must be taken not to damage the gel top. About 3 mm (or more) of solution containing the RNA is then added (see Fig. 107). When the gel is overloaded with RNA solution, only the lower molecular weight RNAs are adequately fractionated. The remainder of the free space in the capillary is now filled with electrode buffer as shown in Fig. 107, and the capillary attached to the electrophoresis apparatus as described on page 158. When very

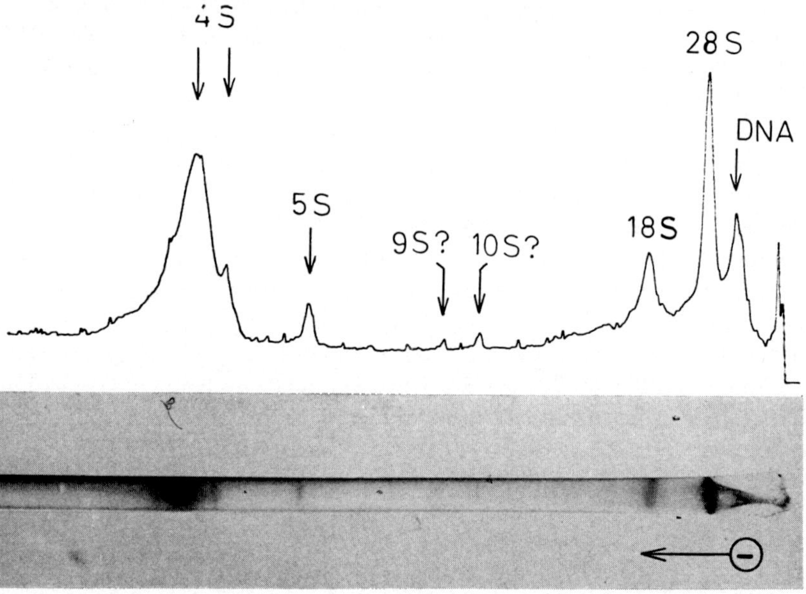

FIG. 106. This photograph shows the fractionation of 0.5 μg RNA in a gradient gel cast in a 5 μl Drummond capillary, using a discontinuous buffer system (see Fig. 107). The electrophoresis was carried out at 35 V and terminated after 120 min. The RNA was extracted from the cortex of a cat's brain as described in the text. A microdensitometric trace of the gel is also shown, so that the various RNAs (28S, 18S, 5S and 4S) can be clearly identified. A typical accumulation of DNA on the gel head is also visible, and probably some messenger RNA (9S and 10S) too. (Photograph by courtesy of D. Wolfrum.)

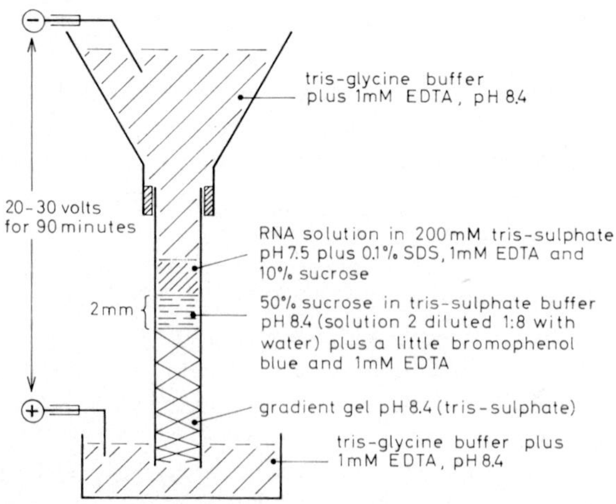

FIG. 107. Schematic diagram of procedure used to fractionate RNAs on gradient gels in the microscale using a discontinuous buffer system.

dilute solutions of RNA are to be fractionated, another capillary can be attached to the gel capillary (see page 168). Electrophoresis is carried out at 20–40 V and terminated after 90–120 min (see Fig. 108). The anode is below the gel and the cathode above. The actual voltage used is of critical importance for sharp and well-defined bands. Should higher voltages be used, i.e. between 50 and 60 V, a number of RNAs are precipitated (see Fig. 108), probably because the concentrating effect is too rapid.

As soon as the bromophenol blue and SDS in the sample has migrated through the gel, the electrophoresis is terminated, and the gel is extruded from the capillary with a steel wire. The RNAs are then localised by staining the gels with either 0.2% methylene blue in 10% acetic acid (Peacock and Dingman, 1967) or 0.5% Toluidine blue in 10% acetic acid (Konings and Bloemendal, 1969) and destained in 7.5% acetic acid. The methylene blue solution is particularly good for revealing 4S and 5S RNA, though the latter is better as a general stain. A problem with all staining methods is that, owing to the small diameter of the gels, stain associated with low molecular weight RNAs diffuses out of the gel very rapidly (about 1–2 hr). This means that all densitometer traces must be carried out immediately after staining and the gel must always be bathed and stored in 7.5% acetic acid, weakly coloured with dye. Staining with 0.1% Toluidine blue in water followed by destaining in water is also recommended since in this system the dye binds ionically with the different RNAS.

Though the above described electrophoresis is suitable for fractionating most RNAs, and in particular, low molecular weight species, it can be adapted with a continuous buffer system (see e.g. Loening, 1969) to separate ribosomal RNA (see Figs. 109 and 110). In this procedure, however, the fractionation of low molecular weight RNAs is not as sharp as with the discontinuous system (see Figs. 108 and 110). It is important to terminate the electrophoresis when the 4S and 5S RNAs migrate into the lower end of the gradient (40%) gel. This is achieved by ending the electrophoresis 30–60 min after the colour dye has migrated through the gel.

In these methods it is possible to introduce SDS into the gradient gels. In this case, 0.1–0.2% SDS is added to the gel mixture when the gradient gels are prepared, and the electrophoresis buffers also contain the same amount of SDS. After staining SDS-containing gels with dye solution (methylene blue or Toluidine blue in 10% acetic acid) for 5 min, a destaining time of at least 5–10 hr in 7.5% acetic acid is required because of the formation of a SDS complex with the stain.

(b) ANALYSIS OF DIFFERENT DEHYDROGENASES

The described microprocedure for disc-electrophoresis in homogeneous separating gels can be easily adapted to assay different dehydrogenases. After electrophoresis the gel is simply incubated in the appropriate staining mixture so that dehydrogenases like glucose-6-phosphate, 6-phosphogluconate, alcohol dehydrogenase, melate dehydrogenase, lactate dehydrogenase, etc., can be studied. The actual staining of the dehydrogenase makes use of a colourless tetrazolium salt which is reduced to a coloured compound (see Fig. 111). The chemistry of the staining method for lactate dehydrogenase (LDH), for example, is as follows (see Fig. 111). With lactate as the substrate, LDH forms pyruvate and NADH in the presence of the coenzyme NAD. Phenazine methosulphate (PMS) acts as an intermediate electron carrier, converting the colourless tetrazolium salt to a purple precipitating formazan and oxidising NADH to NAD. This 'cycling' reaction (see page 33) allows small amounts of LDH to be easily observed.

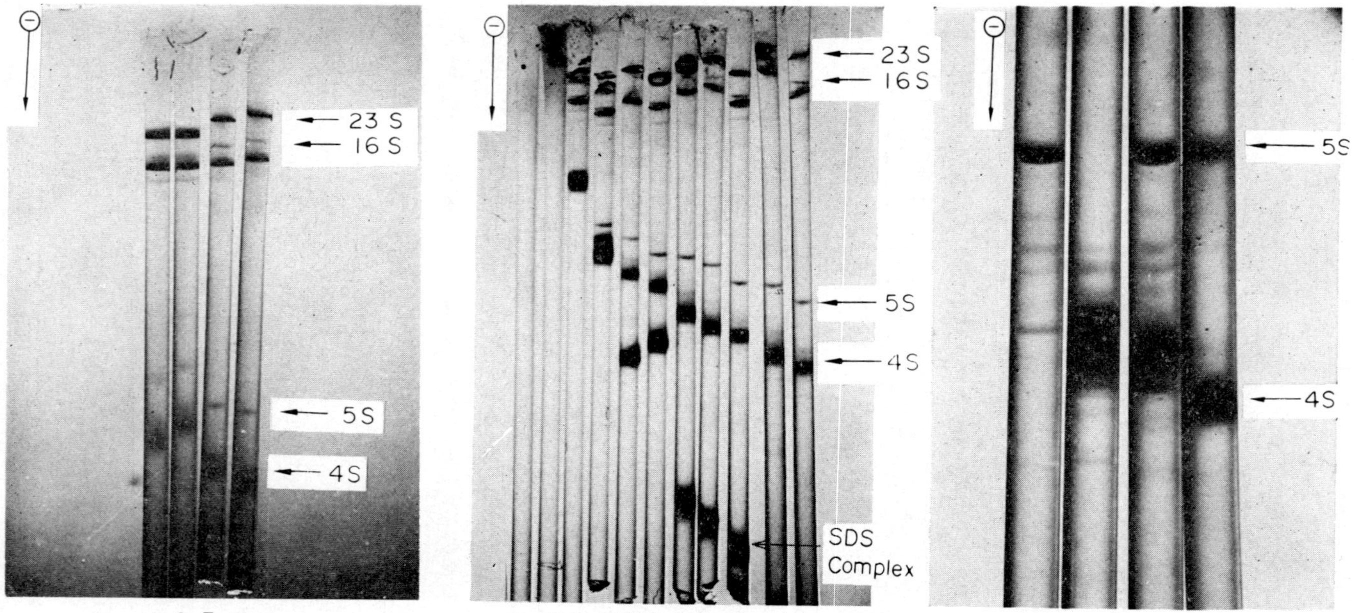

FIG. 108. The gradient gels (1–40%) in (a) show the fractionation of 0.5 μg RNA from *E. coli* in 5-μl Drummond capillaries, using continuous (gels 1 and 2) and discontinuous (gels 3 and 4) buffer systems (for further details see text and Figs. 107 and 109). In each case the voltage was 30 V and the electrophoresis was terminated after 30 min. It can be seen that at 30 V the separation of RNAs in both systems is approximately the same. (b) shows a number of gradient gels (1–40% polyacrylamide gradients in 5-μl Drummond capillaries) where 0.5 μg RNA from *E. coli* was fractionated in a discontinuous system at 60 V and the electrophoresis terminated at various intervals of time (1, 2.5 min; 2, 5 min; 3, 10 min; 4, 15 min; 5, 20 min; 6, 25 min; 7, 30 min; 8, 35 min; 9, 40 min; 10, 45 min; 11, 50 min). In this series of gels 0.1% SDS was added to the sample buffer. It can be seen that at 60 V the ribosomal RNAs (23S and 16S) form aggregations, while the low molecular RNAs (4S and 5S) are clearly fractionated. It can also be seen that an electrophoresis time of more than 20 min is required for the clear separation of 4S and 5S RNAs, and that as the duration of electrophoresis is increased, so the SDS monomer (visible as a complex formed between SDS and the dye toluidine blue) is gradually 'electrophoresed out' of the gel. (c) is a photograph of parts of a number of gradient gels (1–53%) in 5-μl Drummond capillaries (the cross-linkage of the gel system was 4% and not 2% as in all other cases quoted) showing the fractionation of 5S (gel 1), 4S (gel 2) and a mixture of 4S and 5S (gels 3 and 4) RNAs from *E. coli* (kindly provided by Dr. Sprintzel). Both discontinuous (gels 1–3) and continuous (gel 4) buffer systems were used. In these experiments the ionic strength in the sample buffers was lower than normal (20 mM Tris-sulphate buffer). The electrophoresis was carried out at 80 V and terminated after 30 min for the discontinuous system and 60 min for the continuous system. It can be seen that by employing the discontinuous system a number of minor bands occur, due to the concentrating and sieving effects. In the continuous system, where the concentrating effects are negligible, these minor bands are not visible. It would seem that the minor bands are not artefacts, though their actual significance is not clear. (Photographs by courtesy of

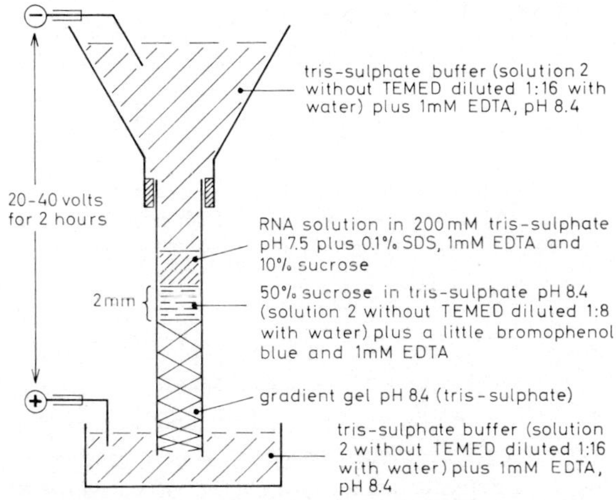

FIG. 109. Schematic diagram of the procedure for fractionating RNAs on gradient gels in the microscale using a continuous buffer system.

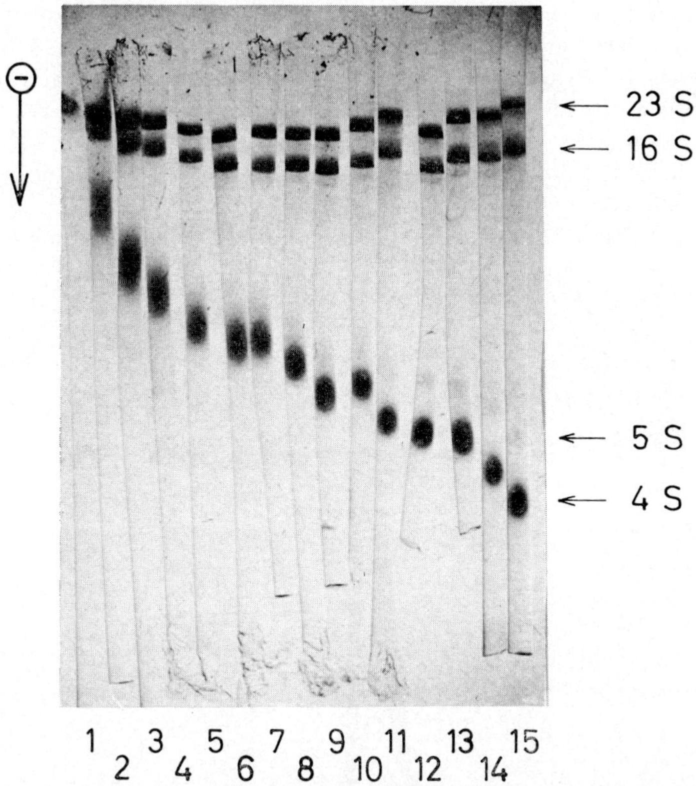

FIG. 110. Fractionation of RNA from *E. coli* in polyacrylamide gradient gels (1–40%) cast in 5-μl Drummond capillaries, using a continuous buffer system (see Fig. 109 and text). The photograph shows a number of gels (0.5 μg was fractionated for each), and in each case the electrophoresis was carried out at 30 V, although it was terminated at various intervals, ranging from 0 to 120 min (1, 5 min; 2, 10 min; 3, 15 min; 4, 20 min; 5, 25 min; 6, 30 min; 7, 40 min; 8, 50 min; 9, 60 min; 10, 70 min; 11, 80 min; 12, 90 min; 13, 100 min; 14, 110 min; 15, 120 min). It can be seen that a good separation of ribosomal RNAs (23S and 16S) occurs within 20 min of electrophoresis, whilst a longer period (between 70 and 120 min) is required for a good separation of 5S and 4S RNA. (Photograph by courtesy of D. Wolfrum.)

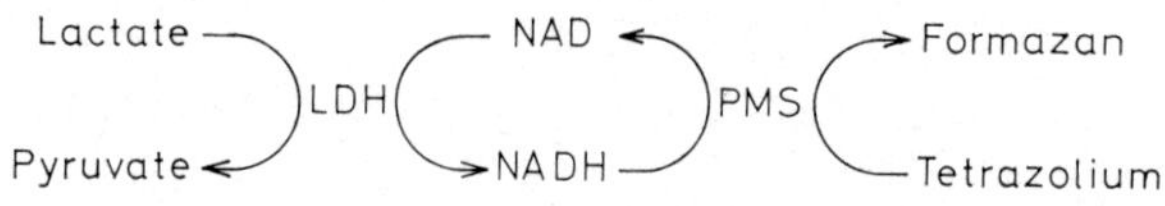

FIG. 111. The reaction involved in the staining for LDH isoenzymes. With lactate as the substrate, LDH in the presence of the coenzyme NAD, forms pyruvate and NADH, while PMS (phenazine methosulphate) acts as an intermediate electron carrier, converting the colourless tetrazolium salt to a purple-blue formazan and oxidising NADH to NAD.

Using the above procedure for analysing glucose-6-phosphate dehydrogenase (Cremer *et al.*, 1972), it was found that the system is at least 1000 times more sensitive than photometric methods. Moreover, the tetrazolium procedure has a higher sensitivity (than Amido black) for staining bands of dehydrogenase enzymes. For example, the minimum amount of pure glucose-6-phosphate dehydrogenase from yeast, detected by Amido black staining in gels formed in 5-μl capillaries, is approximately 1 ng, while the tetrazolium procedure detects about 1 pg.

Since the diameter of the microgels is so small, the substrate diffuses into the gels almost instantaneously, when the latter are placed in the incubation mixture. This means that all the enzyme in a gel is exposed to the substrate. It is for this reason that a linear increase in optical density at the site of enzymatic activity in the gel can be observed (see Fig. 112) and followed for periods of 15–80 min, depending on the enzyme concentration. It is therefore possible to consider the microgel as being a 'microcuvette' and containing the enzyme in a volume of 0.003–0.015 μl since a single protein band in a 5-μl gel measures from 20 to 100 μm. Because of this 'microcuvette' property, the formation of the blue dye

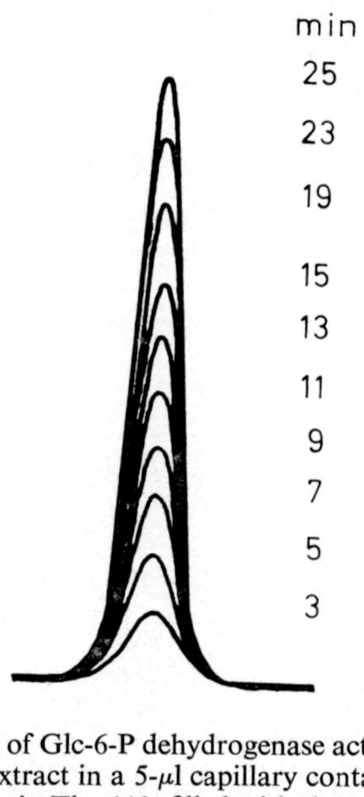

FIG. 112. Microdensitometric tracings of Glc-6-P dehydrogenase activity at different intervals in the same microgel. After fractionating a liver extract in a 5-μl capillary containing a 20% polyacrylamide gel, the gel was placed in a chamber as shown in Fig. 113, filled with the tetrazolium incubation mixture and at intervals of two minutes was scanned with a microdensitometer in the dark room. The figure shows that the formazan peak increases linearly with the time of incubation. (After Cremer *et al.*, 1972.)

at the enzymatically active site of this gel can be recorded at different time intervals, so allowing the kinetics of the enzyme to be studied.

Another advantage of the gel electrophoresis procedure is the further purification of the enzyme. Thus the occurrence of other NAD-reducing enzymes in an extract under investigation does not interfere with the assay of Glc-6-P DH, for example, 6-phosphogluconate dehydrogenase migrates much more slowly than Glc-6-P DH in the separating gel. Furthermore, substrates and inhibitors are removed by electrophoresis, unless they are tightly bound to an enzyme. And although there is a reversible partial inactivation of Glc-6-P DH by dilution which is NADP independent (Yoshida, 1966), this phenomenon is avoided in this procedure because the enzyme, even when in very dilute solutions, is concentrated in the stacking gel.

(i) *Assay of glucose-6-phosphate and lactate dehydrogenase*

Tissue is homogenised in 0.01 M phosphate buffer pH 7.0 (1 mg/10 μl) at 0°C and then centrifuged. The clear supernatant may be stored at −20°C until required for microelectrophoresis. Repeated thawing and freezing of the supernatant destroys the dehydrogenase activities.

Just before subjecting part of the solution to microelectrophoresis, it is diluted 60–100-fold with phosphate buffer (0.01 M pH 7). When analysing glucose-6-phosphate dehydrogenase (Glc-6-P DH), 0.1 ng/ml NADP (but not NAD) must be added to this buffer to stabilise the enzyme. NADP is not required for the analysis of lactate dehydrogenase (LDH), though NAD is.

Microdisc electrophoresis of the diluted sample is then carried out in 5-μl capillaries on 20% polyacrylamide homogeneous gels as described on page 156. Other buffer systems, e.g. Tris/HCl, Tris/phosphate, triethanolamine/phosphate or triethanolamine/sulphate, can be used to prepare the fractionating gel, provided a pH of 8.8 is maintained. Moreover, 1-μl or 2-μl capillaries can be used and the concentration of acrylamide can be increased to 25%. Normally, upper and lower electrode buffers are identical (Tris-glycine pH 8.4), though in the analysis of Glc-6-P DH the upper electrode buffer contains 0.1 mg NADP/ml. A little fluorescein is also added to the upper electrode buffer in order to follow the buffer front during electrophoresis. Microelectrophoresis is carried out at room temperature at a constant voltage (50 V for Glc-6-P DH and 100 V for LDH), and is stopped when the fluorescein marker has migrated a certain distance, depending on the type of separation (e.g. Glc-6-P DH variants), into the separating gel. Electrophoresis normally takes from 35 to 50 min, after which the gels are extruded onto a watch glass by pressure from a syringe filled with triethanolamine buffer and then transferred to a tetrazolium incubation mixture.

The incubation mixture for detecting Glc-6-P DH is as follows:

4 mg NADP
5 mg glucose-6-phosphate
2.5 mg *p*-nitro-blue tetrazolium chloride
1 mg phenazine methosulphate
20 ml 0.05 M Tris/HCl buffer pH 8

For detecting LDH, a similar staining medium is used. However, sodium lactate and NAD instead of glucose-6-phosphate and NADP are present in the medium. The contents for the staining mixture of LDH are given on page 208. The staining mixtures must be kept in the dark because the tetrazolium system is light-sensitive. Thus it is advisable to use freshly prepared mixtures. The microelectrophoretic separation of glucose-6-phosphate varients (Fig. 82B) and LDH (Fig. 82A) are demonstrated in the figures.

(ii) *Analysis of enzyme kinetics*

Cremer *et al.* (1972) analysed the enzyme kinetics of Glc-6-P DH by following closely the formation of formazan at the actual site in microgels. The chamber shown in Fig. 113

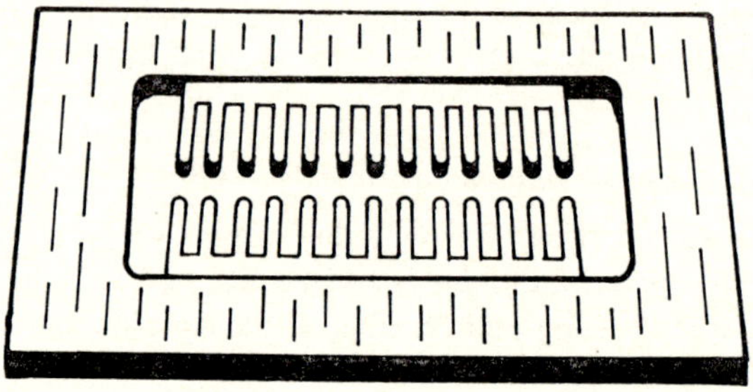

FIG. 113. Chamber used by Cremer *et al.* (1972) to record continuously the enzyme activity in a microgel. The floor of the chamber is made from optical glass and the two combs from Perspex. After filling the chamber with assay mixture the gel is placed along the centre and then covered with a cover-glass.

was used to record the enzymatic activity in an individual gel. This allowed any excess of incubation mixture to bathe the gel during the recording. The formation of the blue dye at the site of the enzyme activity was recorded in the dark at various time intervals with a densitometer. As shown in Figs. 112 and 114 there is a linear correlation between the area of the peaks and the time of incubation. Slope 'a' of the curve in Fig. 114 is a determination for the measure of enzyme activity. From Fig. 115 it can be seen that the linear increase in the area under each peak with time depends upon the concentration of enzyme present at the active site of the gel. When this linear increase in area under each peak is plotted against the glucose-6-phosphate concentration in the assay mixture, it can be used to determine the K_m value of the enzyme (see Fig. 116). The binding constant of Glc-6-P DH for glucose-6-phosphate can then be calculated by plotting the corresponding Lineweaver-Burk diagram (see Cremer *et al.*, 1972). The K_m value of 2.3×10^{-4} M calculated for yeast Glc-6-P DH with glucose-6-phosphate as substrate is higher than that calculated by Glaser and Brown (1955), Lowry *et al.* (1961) and Rose (1961). This may be due to the embedding of the enzyme in the gel matrix, as Neuhoff *et al.* (1969) have shown that the activity of DNA-dependent RNA-polymerase from *E. coli* is lower in the gel than after elution from the gel.

6. Microanalysis of Brain Proteins

Microelectrophoresis for studying brain proteins has been used to study two major problems in particular: firstly, to detect any neuron-specific proteins and secondly, to see whether any variation in characterised proteins can be correlated with behaviour. Apart from differences in the levels of specific enzymes, there are very few facts about the individuality of proteins in neurons or glia. The two most specific brain proteins studied

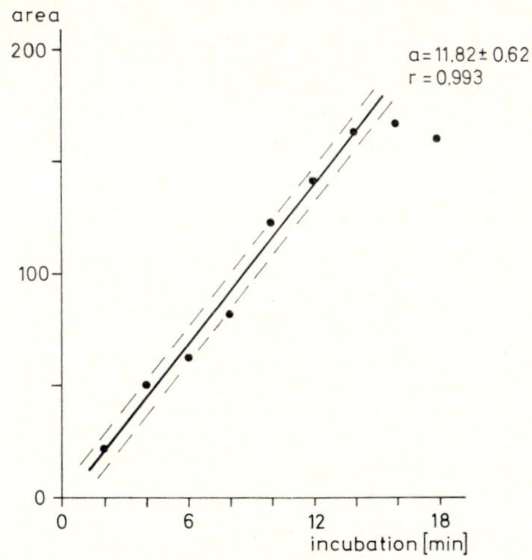

FIG. 114. Glc-6-P dehydrogenase of rat kidney extract fractionated by microelectrophoresis followed by incubation for different periods in assay mixtures containing no $MgCl_2$. A linear correlation occurs (r area time = 0.993) between the area of the peaks and the time of incubation from 2 to 14 min. The slope a of the curve is a measure of the increase in optical density per unit time and can therefore be taken as a measure of enzyme activity. The dashed line indicates the standard deviation. The equations of the curve were calculated according to the method of linear regression, i.e.

$y = b + ax$. Therefore $y = [(-2.57 \pm 5.56) + (11.82 \pm 0.62)]\, x$.
$b = \pm 6.58$, $r = 0.993$.

(From Cremer *et al.*, 1972.)

are the so-called 14-3-2 protein, found predominantly in neurons, and the S-100 protein, which is localised in glial cells (Moore and Perez, 1968). The S-100 protein, which is enriched in glutamic acid, also exists in small amounts in neuronal nuclei (Hydén and McEwen, 1966), and its role, though not yet definitely resolved, is thought to be associated with learning.

(a) PROTEINS RELATED TO LEARNING IN NERVE CELLS

Learning can be loosely defined as the capacity to store information which can readily be retrieved. In such a definition no distinction is made between 'instinct learning' and 'experienced learning'. The question that arises is: How do animals learn and what is the connection between learning and memory? For comprehensive reviews on these subjects see Glassman (1969), Rose (1971) and Ungar (1970). Among the macromolecules in nervous tissue, proteins seem to be involved in learning and memory mechanisms. This is because protein molecules undergo conformational changes, they have sites for absolute recognition, and they can be incorporated into the membranes of nerve cell bodies, synapses and glial, thus enabling simultaneous recognition in a vast number of neurons. Moreover, because proteins react to changes in the environment, e.g. the electrical field, they may constitute the mechanism whereby electrical changes in the neuron are transformed into molecular and cellular effects. It is partly for these reasons that Hydén and collaborators in Göteborg analysed protein changes in specific nerve cells (and defined brain areas) following conditioning and learning. The procedure used by Hydén and collaborators to fractionate the small amounts of proteins was similar to that described on page 156.

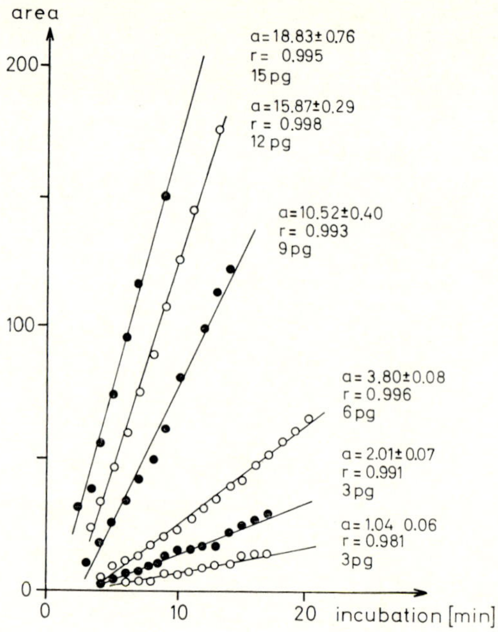

FIG. 115. Enzyme kinetics of Glc-6-P dehydrogenase from kidney cortex of rat. Extracts of kidney cortex of rat corresponding to 6.7, 13.4 and 20 μg wet weight of tissue were fractionated on 20% polyacrylamide microgels until the fluorescein marker had migrated a distance of 7 mm into the separation gel. The peak areas are plotted against incubation time at 1-min intervals. The calculations of the corresponding regression lines are:

$$6.7\ \mu g:\ y = [(-3.975 \pm 1.213) + (3.485 \pm 0.073)]\,x$$
$$= \pm 2.014;\ r = 0.996$$
$$y = [(-1.830 \pm 0.529) + (3.793 \pm 0.053)]\,x$$
$$= \pm 0.695;\ r = 0.999$$
$$13.4\ \mu g:\ y = [(-6.476 \pm 1.666) + (8.756 \pm 0.242)]\,x$$
$$= \pm 1.556;\ r = 0.997$$
$$y = [(-5.604 \pm 1.338) + (9.787 \pm 0.146)]\,x$$
$$= \pm 1.744;\ r = 0.998$$
$$20.0\ \mu g:\ y = [(-23.164 \pm 2.000) + (14.191 \pm 0.233)]\,x$$
$$= \pm 2.434;\ r = 0.998$$
$$y = [(-16.464 \pm 2.461) + (14.030 \pm 0.252)]\,x$$
$$= \pm 3.406;\ r = 0.998.$$

(After Cremer *et al.*, 1972.)

Right-handed rats can be trained to be left-handed (Hydén and Egyhazi, 1964). Once learned, this behaviour remains stable for more than 300 days. To trace protein synthesis during this learning process, rats were injected intraventricularly in each hemisphere with H^3-leucine half an hour before the final training period. Fifteen minutes after the last training period the animals were killed and nerve cells dissected free-hand from the hippocampal region. The hippocampus, which is subdivided into different parts called CA 1 to CA 4, constantly influences the hypothalamus and it is known that the hippocampus must be intact for formation of long-term memory (Lashley, 1964), though memories do not appear to be stored within the hippocampus itself. About 0.5 μg of proteins extracted from dissected nerve cells in the different hippocampal regions were subsequently subjected to microdisc electrophoresis. There was a consistent stimulation of leucine incorporation into total proteins in the CA 3 region throughout the course of learning, while

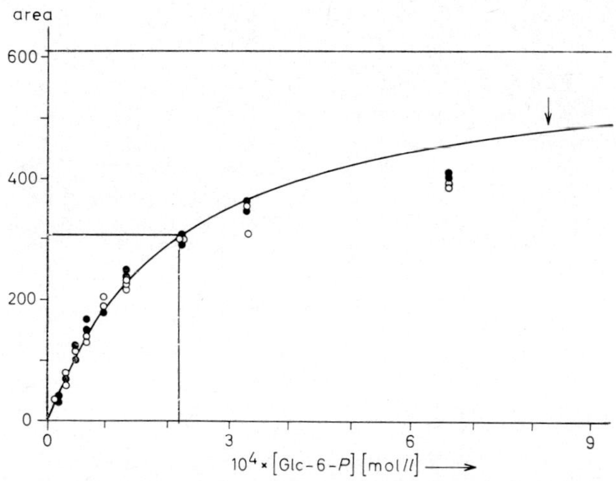

FIG. 116. Determination of the K_m value for Glc-6-P dehydrogenase from yeast in a microgel (after Cremer *et al.*, 1972). The open circles represent results of experiments with 18 different gels and the closed circles results of experiments with 19 different gels. After fractionation each gel was incubated for 20 min at 25°C in assay mixtures with concentrations of Glc-6-P as indicated on the abscissa, and then scanned with a microdensitometer. The peak areas of the individual curves were then plotted against the concentration of Glc-6-P and the data required to produce the above curve were calculated from the values of the regression line. The lines parallel to the abscissa represent the peak areas corresponding to V and $V/2$ respectively. The concentration of substrate corresponding to $V/2$ represents the K_m value for Glc-6-P under the conditions of the experiments and the arrow shows the concentration of Glc-6-P used in subsequent experiments.

incorporation in the CA 4 region was increased only during the early period of training. No clear stimulation could be detected in the CA 1 region. Since the S-100 and 14-3-2 proteins are specifically brain proteins (Moore and Perez, 1968; Hydén and McEwen, 1966), Hydén and his collaborators paid particular attention to the incorporation of labelled leucine into these two proteins in nerve cells of the hippocampal region. Close analyses of densitometric traces from gels of extracts of cells from the CA 3 region showed two peaks for S-100 protein in trained animals and a single peak in controls (Fig. 117). That the two peaks are components of the S-100 protein was revealed by the fact that both peaks gave a positive immunological reaction when treated with S-100 antiserum (Hydén and Lange, 1970a, b). Moreover, it is known that the S-100 protein can be electrophoretically fractionated into a number of components (as many as seven), some of which can be separated by microdisc electrophoresis (see Ansorg and Neuhoff, 1971). It was therefore concluded from the training experiments that an increase in the amount of S-100 occurred in the trained animals because the sum of the integrated curves of the trained material of the two S-100 fractions was always greater than the same value for the control material for the S-100 fraction. Since it is known that calcium causes conformational changes in the S-100, the calcium content of hippocampal nerve cells from the CA 3 region from trained and control rats was also studied (Haljamäe and Lange, 1972). It was found that more calcium was present in the CA 3 region from trained animals than in material from control animals (see Table 38), while the sodium and potassium contents remained constant. Moreover, the calcium content was greater in the faster migrating band than in the slower migrating band of the S-100 complex in trained animals. It may thus be concluded that

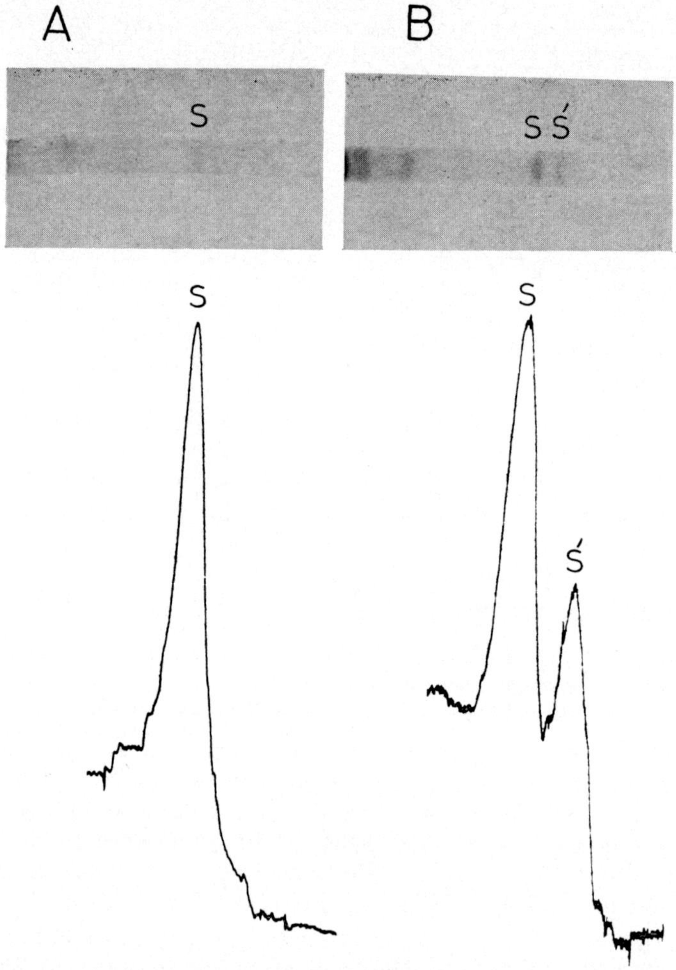

FIG. 117. This photograph shows the separation of water soluble nerve cell proteins (from hippocampus, CA 3 region) from control rats using one paw (A) and from rats trained to use a non-preferred paw (B). In each case part of a micropolyacrylamide gel and a microdensitometric trace of the gel are demonstrated to show precisely the electrophoretic position of the acidic S-100 brain specific protein (designated S and S′). It can be seen that the effect of training is not only to increase the overall content of S-100, but also to produce another conformation of the protein (S′). For further explanation see text. (From Hydén and Lange, 1970; photographs by courtesy of H. Hydén.)

training produced an increase of S-100 in the hippocampal nerve cells and that part of the S-100 produced during training has a different conformation from that of the S-100 from controls. This difference in conformation of the S-100 produced in trained animals is related to the calcium content. It may well be that during the training there is an increase in intracellular calcium in the hippocampus due to the increased activity, and this causes a conformational change in the S-100. Alternatively, an S-100 of different conformation may be formed during the training.

The effect of learning on the other brain specific protein 14-3-2 has also been analysed by microelectrophoresis (Hydén and Lange, 1970b). The same type of reversal of handedness experiment was carried out to show that there was an increase in incorporation of

TABLE 38. POTASSIUM, SODIUM, CALCIUM AND WATER CONTENT OF BRAIN SAMPLES FROM THE CA3 REGION IN THE HIPPOCAMPUS OF TRAINED AND CONTROL RATS.
Electrolyte values obtained both on wet weight and dry weight basis
(from Haljamäe and Lange, 1972)

	Calcium mg/100 g		Sodium mEq/100 g		Potassium mEq/100 g		
	dry weight	wet weight	dry weight	wet weight	dry weight	wet weight	% water
Trained (n = 20) S.E.M.±	44.0* 1.8	8.04† 0.19	20.04 0.43	4.04 0.10	50.42 1.50	9.99 0.34	81.0 0.5
Control (*n* = 18) S.E.M.±	37.0 1.9	7.36 0.27	20.43 0.35	3.87 0.07	51.20 1.60	10.17 0.38	81.4 0.4

* $p < 0.001$. † $p < 0.05$.

labelled leucine into two protein fractions (which contain the 14-3-2 protein) from nerve cells located in the CA 3 hippocampal region in trained rats. When the data were examined after repeated training or after training which had been resumed 30 days later, the response was found in no way to vary from that in the control material.

(b) ANALYSIS OF PROTEINS IN INVERTEBRATE NEURONS

In an effort to discover proteins specific to neurons, extracts from GSCs from *Helix pomatia* were subjected to microdisc electrophoresis where the separating gel was homogeneous (Osborne *et al.*, 1971). Only the water-soluble proteins (extracted in 0.01 M Tris/HCl buffer) of the cells were fractionated and compared with the same proteins present in other tissues of the snail. It can be seen from Fig. 118 that the electrophoretic patterns of all tissues examined showed a number of differences regarding both slow- and fast-running proteins. Of special interest is the protein band which moves faster than any other proteins and which is concentrated particularly in the GSCs (Fig. 118A), brain (Fig. 118B) and optic tentacle (Fig. 118C). This protein band, which appears to consist of three components, is present to a lesser extent in the penis retractor muscle (Fig. 118D), oesophagus (Fig. 118E), heart (Fig. 118F), liver (Fig. 118G) and kidney (Fig. 118H). Results from experiments like this indicate that the fast-moving band is specifically associated with nervous tissue, since it is most highly concentrated in isolated nerve cells and decreases in tissues containing lesser amounts of nervous tissue. The liver and kidney, which are very poorly innervated (see Bullock and Horridge, 1965), are almost completely free of this fast-running protein band. This (these) nervous specific protein(s) of the snail may be identical or similar to the S-100 and/or 14-3-2 brain-specific proteins of vertebrates (see Shooter and Roboz-Einstein, 1971). Initial studies in our laboratory indicated that they are unlikely to be the S-100 protein, since the fast-moving band cannot be dissolved in saturated ammonium sulphate as can S-100 (Hydén and Lange, 1970b).

In an attempt to study the molecular weight distribution of proteins synthesised in identified neurons, Wilson (1971) used polyacrylamide gel electrophoresis in the presence of sodium dodecyl sulphate (SDS). The parietovisceral ganglia of *Aplysia californica* were incubated in a medium containing labelled leucine, single identified neurons were dissected

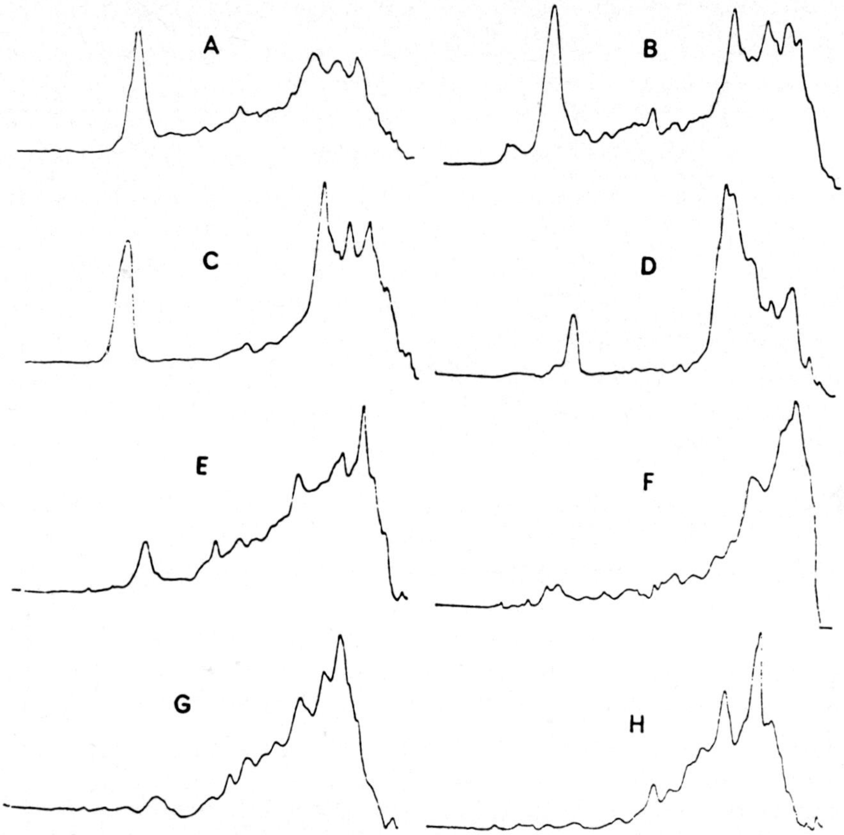

FIG. 118. Pherograms of microgels after fractionation and Amido black staining of soluble proteins from identified neurons (GSCs) of the snail *Helix pomatia*. A, brain; B, optic tentacle; C, penis retractor muscle; D, oesophagus; E, heart; F, liver; H, kidney. Gels were scanned with a Joyce-Loeble double beam microdensitometer. The migration of proteins is from right (cathode) to left (anode). (After Osborne *et al.*, 1971.)

and their proteins separated by electrophoresis on 5% SDS-polyacrylamide gels. The distribution of total or newly synthesised proteins from single neurons was determined by staining or slicing and liquid scintillation counting of the gels. The results showed that a number of proteins were continually synthesised in nerve cells, the most significant incorporation occurring into material of high (275,000) molecular weight, and that different identified neurons revealed different labelling patterns.

An improved SDS-gel micromethod described by Gainer (1971) has enabled him to study protein synthesis in identifiable neurons of *Aplysia*. This method makes use of a steady-state stacking step of disc electrophoresis which preconcentrates the sample into very thin starting zones. This, together with the higher concentration of SDS-polyacrylamide gel (7%) used, enabled him to increase the resolution of proteins, allowing polypeptides with molecular weight less than 12,000 to be observed. Gainer was thus able to note a difference between two identifiable cells of *Aplysia* (R_2 and R_{15}) regarding the occurrence of a low molecular weight polypeptide (5000 daltons) in cell R_{15}, which was not detected by Wilson (1971).

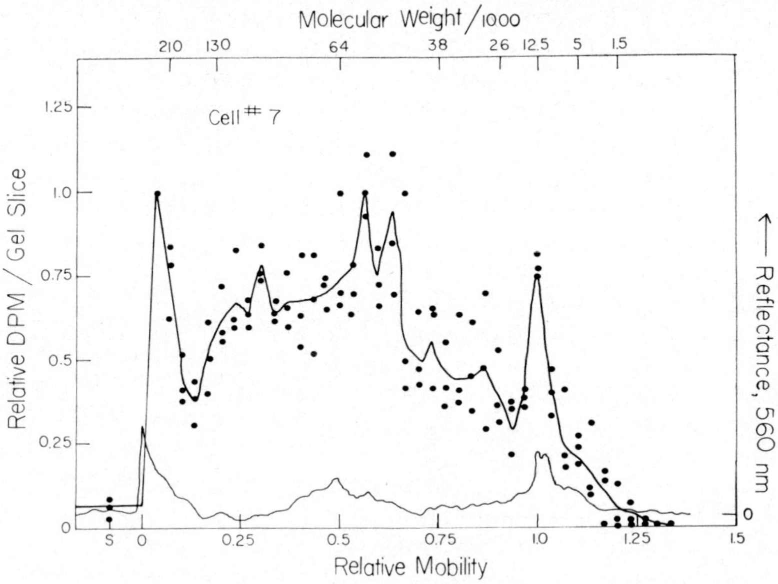

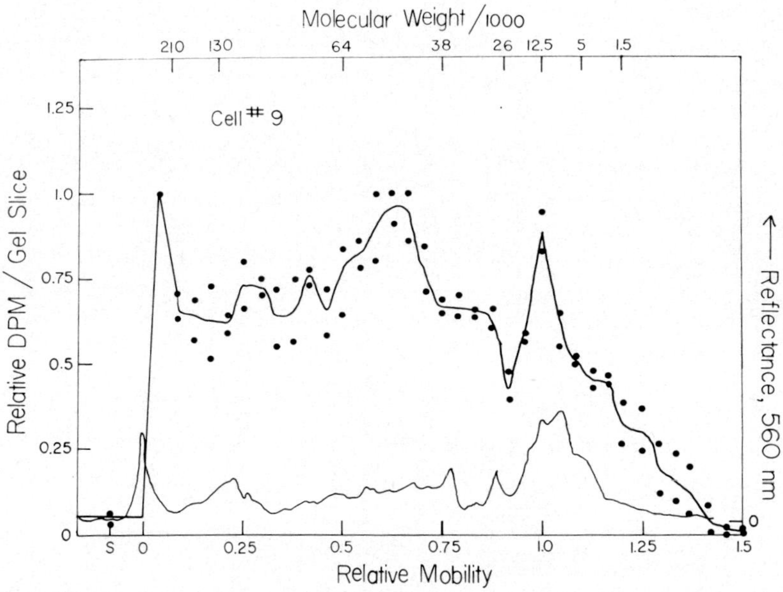

FIG. 119. Electrophoretic patterns on 7.5% acrylamide microgels from extracts of the cell bodies of neurons 7 and 9 (above), 10 and 12 (see p. 200) and 11 (see page 201) isolated from snails after 20 hr incubation in media containing L-(^{3}H) leucine. The heavy solid line represents the spectrum of labelled proteins and is drawn through the average value of the points at each relative mobility unit. The radioactivity (in dist./min minus background) in each gel is plotted on the ordinate, relative to radioactivity in the first gel slice after the origin, O, from the same microgel. The radioactivity of the total upper, or stacking, gel is plotted over S on the abscissa. The lower abscissa indicates the mobility relative to cytochrome C and the upper abscissa shows the molecular weight corresponding to the relative mobilities in the lower abscissa. The thin solid lines in each instance represent the reflectance scan of the microgels at 560 nm. It can be seen that only cell 11 contains a unique protein (4000–5000 daltons). (From Gainer, 1972a.)

Gainer (1972a) also attempted to characterise the rates of incorporation of isotopic leucine into individual identified neurons of the snail *Otala lactea* to see whether there were different patterns of protein synthesis in the various neuron types. Incorporation rates appeared to bear an inverse relationship to the spontaneous electrical activity of the cells. Moreover, individual neurons showed distinct molecular weight distribution of synthesised proteins, and one neuron (cell 11) synthesised low molecular weight proteins (5000 daltons) which were not synthesised by the other four neurons (see Fig. 119). The effects of experi-

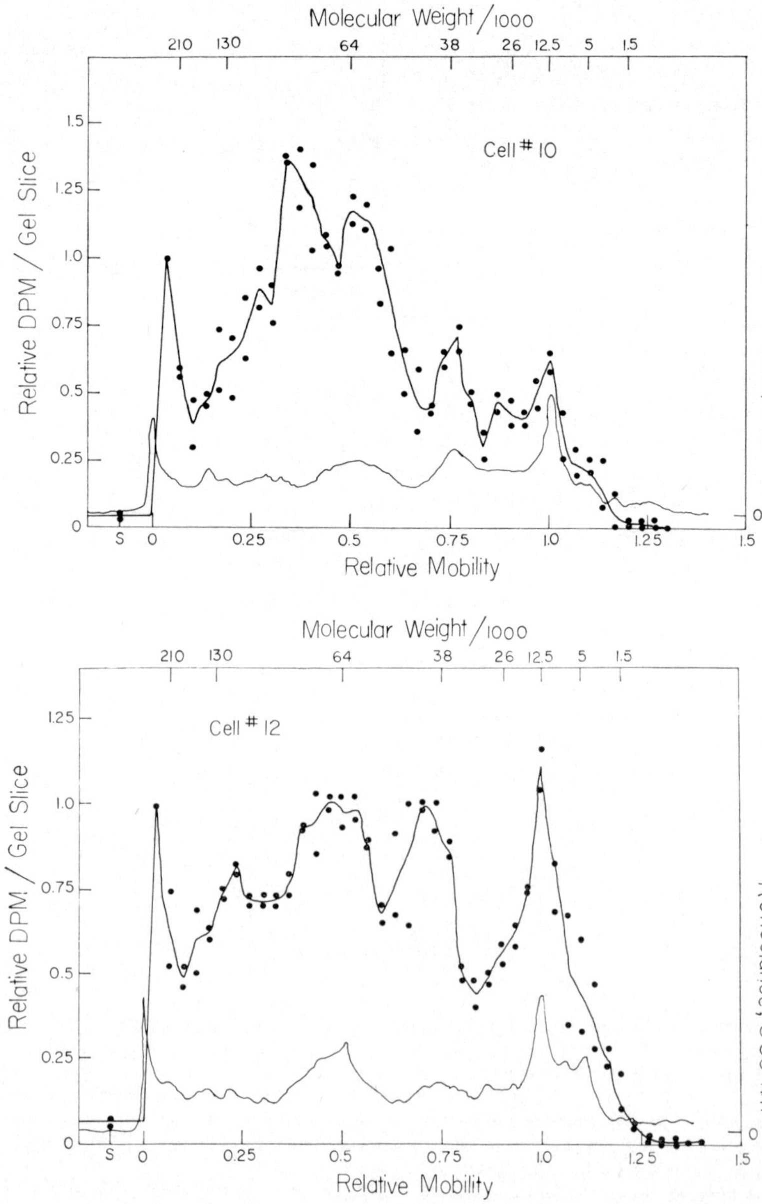

FIG. 119 (*cont.*)

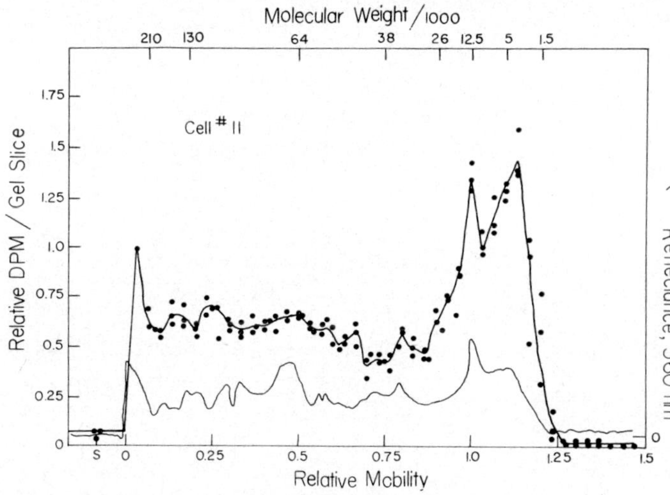

FIG. 119 (*cont.*)

mentally induced diapause on the protein synthesis patterns of the five identified neurons were also studied (Gainer, 1972b). The results showed that the unique low molecular weight protein of 5000 daltons occurring in cell 11 was profoundly affected by diapause (see Fig. 120), although as in all the cells studied, the incorporation of radioactivity into the other protein species was also slightly affected. From the specificity of the low molecular weight protein synthesised in cell 11 under different experimental conditions, and from the electrophysiology of this cell, it was concluded by Gainer (1972b) that cell 11 is a neurosecretory-type cell.

(c) DEVELOPMENTAL CHANGES

During development a great number of changes occur in the brain. For example, with the onset of myelination, both the lipid content and composition change. There are also alterations in the extracellular space, water and protein content and in the level of a number of other substances, many of which participate in protein metabolism, such as energy rich phosphates and amino acids (see e.g. Lajtha, 1964). In fact it appears as if the brain undergoes three direct changes during development: (1) the formation of the adult number of neurons; (2) a period of accelerated growth of the brain during which extensive morphological maturation takes place, (3) a final period of consolidation which is of varying duration and finally leads to the mature adult nervous system. With all these alterations, parallel changes in the protein metabolism during development can be expected and have indeed been found through studies correlating enzyme composition (Cohn and Richter, 1956) and protein distribution (Kiyota, 1959; Grossfeld and Shooter, 1971; Moore and Perez, 1968; Waehneldt and Neuhoff, 1974) in the nervous system with development.

Microelectrophoresis, employing gradient gels, has been used to see whether any obvious changes occurred in the pattern of water-soluble proteins in the brains of rats of

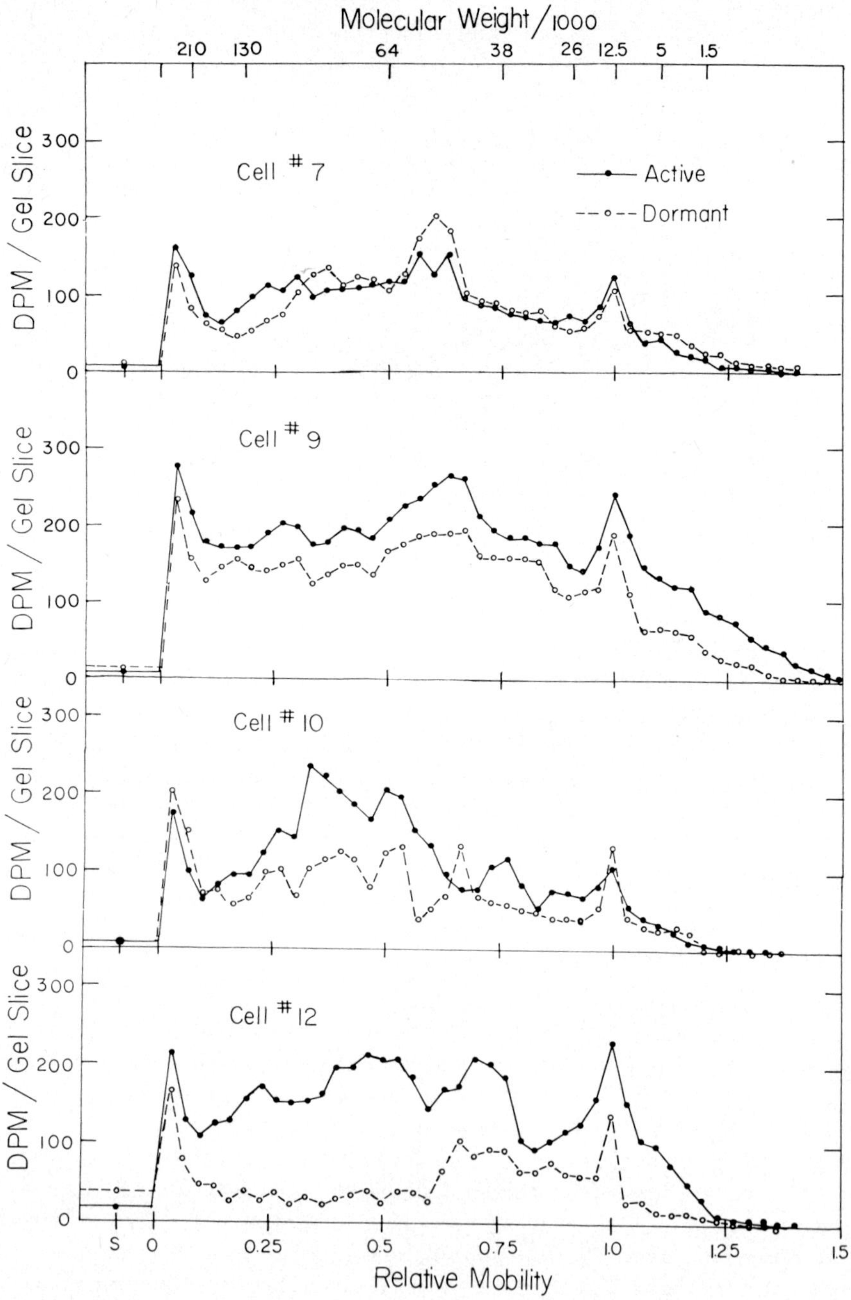

FIG. 120. Electrophoretic patterns of proteins on 7.5% acrylamide SDS microgels from extracts of cell bodies of neurons 7, 9, 10, 12 (above) and 11 (see p. 203) isolated from active (filled circles, continuous line) and dormant (open circles, broken line) snails after 20 hr incubation of the ganglion in media containing L-(^{3}H) leucine. The lines drawn through the points in the graph represent the spectrum of labelled proteins in a single identified cell. The points on the graph were plotted as the average dis./min (minus background) per slice against the relative mobility of the proteins in the gel slice (lower abscissa) and the molecular weight represented at the specific mobility unit (upper abscissa). The letter O denotes the origin of the gel and S the total radioactivity of the stacking gel. It can be seen that the specific protein in cell 11 (cf. Fig. 119) is inhibited by diapause. (From Gainer, 1972b.)

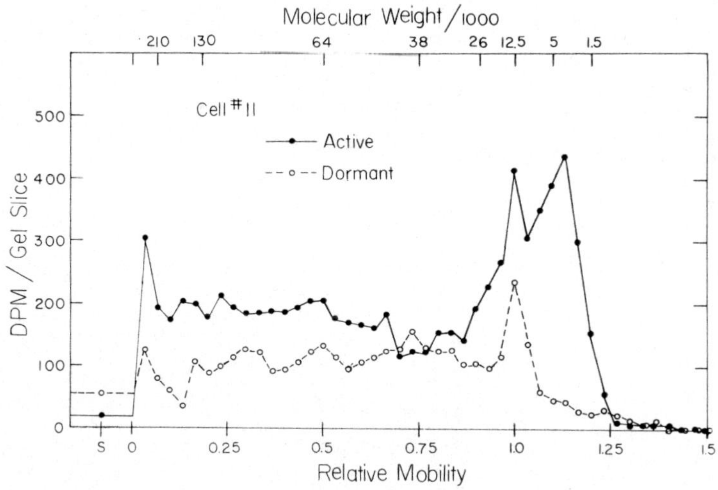

FIG. 120 (*cont.*)

various ages. Though a detailed study is still being carried out, it can be seen from the pherograms in Fig. 121 that generally the pattern of proteins is similar in brains from animals ranging from 2 to 93 days, with the exception of bands A and B. It is evident that at the 2-day stage the height of bands A and B is similar, and that as the animal develops the content of protein associated with band B increase in proportion to band A. Examination of fractionated water-soluble brain proteins when utilising normal size electrophoresis reveals similar results (see Fig. 122). It can be seen again that as the animal develops, band B increases in relation to band A.

Microelectrophoresis, after the procedure of Neuhoff (1968), has been used by Nir *et al.* (1973) to study the pattern of water-soluble proteins in the pineals of rats of different ages by night and day. From the pherograms obtained (see Fig. 123 and Table 39), a constant pattern of proteins was observed in all the ages studied. However differences existed in content, the very mobile proteins marked 14–22 (see Fig. 123) were present to a greater extent in the immature than the mature animals. In contrast, the slow-moving fractions 3–10, decreased relative to the other components in the immature rats. Table 39 shows that these differences in protein content were similar in pineals from all animals, whether dissected at night or by day, although variations among the contents of the same proteins from animals of the same age were caused by the time at which the pineals were isolated.

7. **Microisoelectric Focusing** (LDH isoenzymes)

The basic principle of isoelectric focusing is as follows. The net charge of a protein molecule in an acidic solution is positive because most amino groups carry a net positive charge and carboxy groups are protonated and electrically uncharged. Should the pH be increased slowly, the number of carboxylic groups which carry a negative charge will increase and the number of positively charged groups will decrease. At a certain pH value,

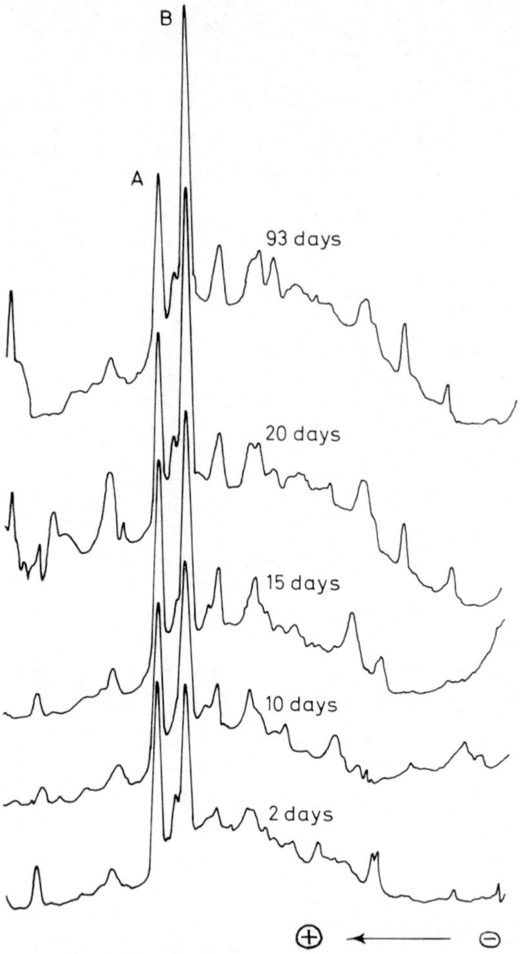

FIG. 121. Densitometric traces of micropolyacrylamide gradient gels showing the separation of water soluble brain proteins from rats of different ages, using a discontinuous buffer system and staining with Amido black. It can be clearly seen that the content of band B increases with age compared with band A. Similar results are also observed when fractionating water soluble brain proteins on normal sized polyacrylamide gels (see Fig. 122).

the isoionic point, the net charge of the protein molecule will be zero. The isoionic point of a molecule is thus determined by its proteolytic groups and their dissociation constants and can vary within the pH range of 3–11. In conventional electrophoresis there is a constant pH between the anode and the cathode so that any positively charged ions, e.g. proteins, migrate to the cathode, while negatively charged ions migrate to the anode. In electrofocusing, a stable pH gradient is arranged, i.e. pH increases successively from the anode to the cathode. If a protein is put into this system at a pH lower than the isoionic point, the net charge of the molecule will be positive and will migrate in the direction of the cathode. Due to the presence of the pH gradient, the protein will migrate to an environment of higher pH values which in turn will influence the ionisation and overall charge of the molecule. Eventually the protein will reach a pH where its net charge is zero, and so

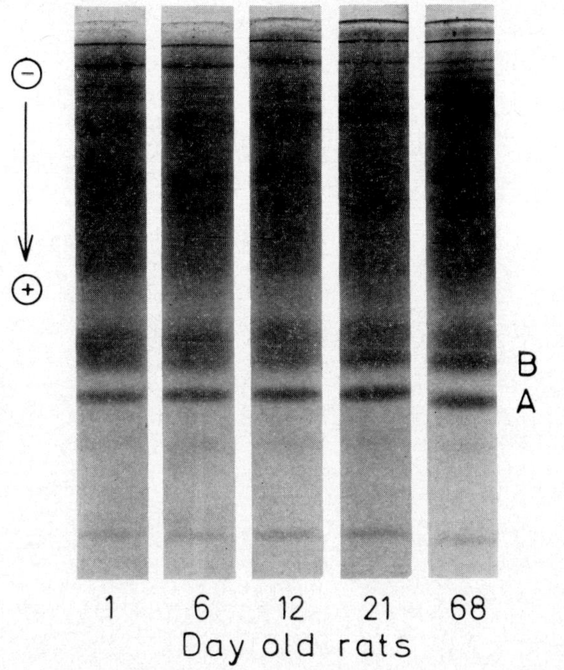

FIG. 122. Fractionation of water-soluble brain proteins from rats of different ages on normal sized polyacrylamide gels. It can be clearly seen that the content of protein associated with band B increases with age in comparison to band A (cf. Fig. 121) (Photograph by courtesy of T. V. Waehneldt.)

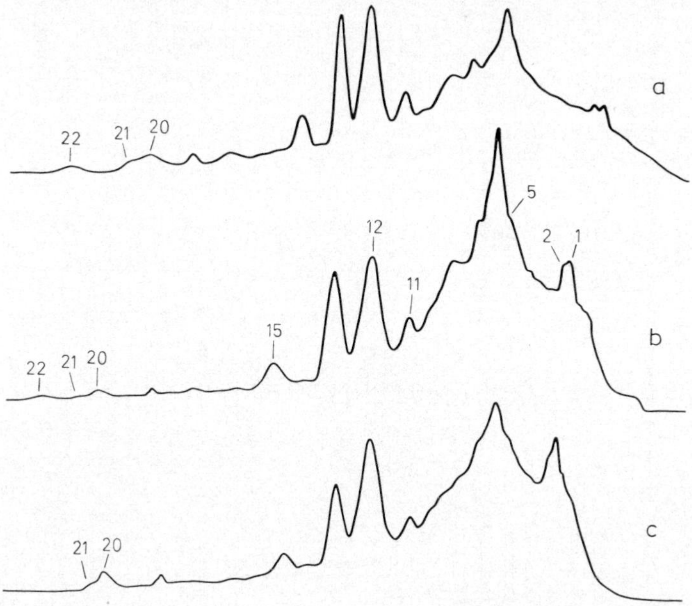

FIG. 123. Pherograms of water-soluble proteins of pineal glands from rats of different ages after fractionation by microdisc electrophoresis and staining with Amido black. (a) 31 days, (b) 10 weeks, (c) 1 year. (From Nir *et al.*, 1973.)

TABLE 39. DAY AND NIGHT COMPOSITION OF VARIOUS PROTEIN FRACTIONS (see Fig. 121) IN THE PINEAL GLAND OF RATS AT DIFFERENT AGES (from Nir *et al.*, 1973)

Protein fractions Band no.	Day			Night		
	Immature (31 days) A	Mature (10 weeks) B	Adult (1 year) C	Immature (31 days) A	Mature (10 weeks) B	Adult (1 year) C
1/2	4.36 ± 0.57	4.62 ± 0.62	5.84 ± 0.53*	3.83 ± 0.51	5.08 ± 0.78	5.63 ± 0.74*
3–10	45.80 ± 0.90*	53.64 ± 1.89*	50.07 ± 0.87*	52.30 ± 0.55	57.39 ± 2.80	53.25 ± 0.79
11	6.54 ± 0.47*	5.16 ± 0.46	5.78 ± 0.47	5.74 ± 0.39	5.16 ± 0.51	5.79 ± 0.34
12	14.64 ± 1.21	12.39 ± 1.00	12.88 ± 1.27	14.15 ± 1.52	13.14 ± 1.57	12.67 ± 1.71
13	8.95 ± 0.86	9.10 ± 0.84	8.52 ± 0.57	8.86 ± 0.36	8.12 ± 0.93	8.51 ± 0.57
14	2.35 ± 0.14	1.84 ± 0.25	1.67 ± 0.06*	1.26 ± 0.11	1.32 ± 0.22	1.20 ± 0.09
15	3.74 ± 0.29	3.77 ± 0.28	4.04 ± 0.35	3.49 ± 0.24*	2.85 ± 0.17*	3.71 ± 0.21
16	3.98 ± 0.28*	2.97 ± 0.43	3.03 ± 0.46	2.84 ± 0.25	2.12 ± 0.49	2.87 ± 0.39
17	3.54 ± 0.14*	2.53 ± 0.18	2.36 ± 0.22*	2.77 ± 0.29*	1.96 ± 0.09	1.99 ± 0.14*
18/19	1.73 ± 0.12*	1.18 ± 0.12*	1.69 ± 0.23	1.38 ± 0.11*	0.73 ± 0.13	1.47 ± 0.45
20/21	3.51 ± 0.56	2.54 ± 0.26	4.12 ± 0.13	2.71 ± 0.29	2.09 ± 0.83	2.98 ± 0.41
22	0.89 ± 0.17*	0.27 ± 0.13*	not detectable*	0.64 ± 0.12	0.35 ± 0.08*	not detectable
14–22	19.75 ± 1.69*	15.11 ± 1.67	16.92 ± 1.45	15.10 ± 1.41	11.42 ± 1.91	14.22 ± 1.95

* $0.001 \leqslant p < 0.0027$.   † $0.0027 \leqslant p < 0.05$.

it will stop migrating. This is the isoelectric point of the protein. The consequence of this is that every protein will migrate to its respective isoelectric point irrespective of the apparatus, the current applied, or the point of application. This is contrary to conventional electrophoresis.

The pH gradients necessary for isoelectric focusing are obtained by special buffering substances called ampholytes (Kolin, 1954; Svensson, 1961) and since their discovery the isoelectric focusing principle has been used as a very powerful separating procedure.

The technique of isoelectric focusing was initially carried out in saccharose gradients (Vesterberg and Svensson, 1966) and later in polyacrylamide gels (Dale and Latner, 1968; Wrigley, 1968). The first authors to describe a microprocedure for isoelectric focusing were Riley and Coleman (1968) and Catsimpoolas (1968) who required 0.2–0.4 mg of a starting protein mixture. Later Grossbach (1971), Gainer (1973) and Quentin and Neuhoff (1972), working independently, all refined the microisoelectric focusing procedure so that smaller amounts of starting material were required. Grossbach (1971) carried out isoelectric focusing in 100-μm diameter tubes and was able to analyse bovine serum albumen and β-lactoglobin in the picogram range. Similarly Gainer (1973), using glass tubes with an inner diameter of 0.58 mm, found it possible to detect picogram quantities of α-casein, ovalbumen, bovine serum albumen, haemoglobin, chymotrypsin, α-chymotrypsinogen-A and bovine pancreatic ribonuclease by isoelectric focusing. Finally, a microprocedure for detecting LDH (lactate dehydrogenase) isoenzymes by isoelectric focusing in 5-μl Drummond capillaries was described by Quentin and Neuhoff (1972).

(a) PRINCIPLE OF THE MICROISOELECTRIC FOCUSING METHOD FOR DETECTING LDH ISOENZYMES

Since Nielands (1952) discovered that tissues LDH has more than one component (termed 'isoenzyme' by Markert and Möller, 1959), many methods have been developed for characterising the individual isoenzymes. It is now established that serum contains five LDH isoenzymes (Wieland and Pfleiderer, 1957), each of which is a tetramer composed of one or two different monomers.

In the microprocedure to be described, the LDH isoenzymes are separated by isoelectric focusing in micropolyacrylamide gels and the isoenzymes are located by staining with a modified van der Helm reagent (van der Helm, 1961) which contains a tetrazolium salt. The chemistry of the reaction is outlined in Fig. 111 and discussed on page 186.

Comparison of the fractionation of the isoenzymes of LDH by isoelectric focusing and by microelectrophoresis (see Fig. 82) has shown that the individual isoenzymes are better separated by isoelectric focusing. However, the limit of detection of LDH using the same tetrazolium incubation mixture is about 3 to 4 times greater in the microelectrophoresis procedure than in the microisoelectric focusing method. This is due to the presence of the ampholine mixture in gels used for isoelectric focusing.

(b) REAGENTS

(i) *Gel mixture*

4 g Acrylamide + 100 mg Bisacrylamide in 7.5 ml water	0.25 ml
0.063 ml TEMED in 10 ml water	0.25 ml
60% Saccharose	0.40 ml

40% Ampholine pH 3–10 (batch 34, LKB)	0.20 ml
10 mg Ammonium peroxydisulphate in 10 ml water	1.00 ml
	2.10 ml

(ii) *Staining solution*

1 mg phenazine methosulphate
2 mg NAD
3 mg *p*-nitro-blue-tetrazolium chloride
100 mg Sodium lactate
20 ml 0.05 M Tris/HCl buffer pH 8

The staining mixture should be mixed in a brown bottle shortly before use and kept in the dark because the tetrazolium system is sensitive to light. This solution is adequate for 20–40 microgels.

(iii) *Electrophoresis buffer*

Electrolytes are very dilute sulphuric acid of pH 2.95 containing sucrose (12%) at the anode side and dilute NaOH of pH 10.4 at the cathode side. The anode side is placed on top of the capillary and a little bromophenol blue can be put in the electrolyte to follow the migration.

(iv) *Equipment and other requirements*

These are the same as required for microelectrophoresis.

(c) EXTRACTS

Homogenise tissue samples in 0.01 M phosphate buffer pH 7.0 in a ratio of 1:10 (w/v) at 0°C and centrifuge for 60 min at about 24,000 *g*. These extracts can then be diluted prior to electrophoresis with a sucrose solution to give a final concentration of 12% sucrose in the sample.

(d) PROCEDURE

The method described below gives a reproducible migration of isoenzymes, whether samples are electrophoresed simultaneously or separately, a desirable feature for comparing samples and for detecting abnormal migratory rates.

(a) Fill 5-μl Drummond capillaries (scrupulously cleaned according to the procedure given on page 170) three-quarters full of freshly prepared acrylamide gel solution by immersing the lower end of the capillary into the gel mixture. The capillary is then placed into a cushion of plasticine 2 mm thick, so that it is in a vertical position. The gel mixture is overlayed by water with a suitable pipette capillary, while care is taken that air bubbles are not trapped between the two layers and that gross mixing between the ungelled acrylamide and the water does not occur. Polymerisation begins after about 15 min, though in practice it is advisable to store the gels for a little longer than 60 min before use. They can be used for a period of up to 48 hr after polymerisation if stored at room temperature in a moist chamber.

(b) After removing the water layer from the top of the gel (using a pipette capillary), 0.4–0.8 μl of the sucrose diluted protein sample is applied to the gel top and the remaining space between the sample and the top rim of the 5-μl capillary is filled with a solution containing 4% ampholine pH 3–10 and 12% sucrose.

(c) About 2 mm of the bottom of the 5-μl capillary is now scratched with a diamond knife and broken off so that all the plasticine is removed. The capillary is then attached to the rubber cap which fits over one end of a glass funnel, and is placed in the electrophoresis equipment as described on page 161.

(d) After putting the appropriate electrolyte in the anode (top of the capillary) and cathode compartments, isoelectric focusing is carried out at a constant voltage of 200 V at room temperature. The current at the beginning is about 80–100 μA per gel, and when it reaches 5 μA (about 50 min later) the electrophoresis is terminated.

(e) The gels are now extruded into a watch glass by means of water pressure. As the gel is pushed out of the capillary by pressure applied from the top end, the cathodic side of the gel can be marked by dipping the tip of the gel into 1% coomassie blue solution in water for a couple of seconds. The gels are transferred to the incubation mixture. Incubation takes place at 37°C in the dark.

(f) When the gels are adequately stained (*ca.* 90 min), the effect of the incubation mixture is terminated by transferring the gels to 7.5% acetic acid. The blue precipitate of formazan at the sites of the enzymic activity can now be quantified by microdensitometry of the gel followed by measuring the peak areas by planimetry.

Corrections for the total activity of LDH may be made by varying the sample size or the staining time. The bands should not develop for too long a time in the incubation mixture as there is a maximal intensity range which can be used for densitometry. With this in mind, the staining time can be varied from 30 to 160 min with good quantitative recovery, and thus staining time can be varied to compensate for misjudgement in sample size.

As for Glc-6-P DH (see page 191) there is a linear correlation between the enzyme activity and the time of incubation in microgels, so that it is also possible to analyse the enzyme kinetics of the different LDH isoenzymes.

(e) DEMONSTRATION OF LDH PATTERNS IN DIFFERENT DRAIN REGIONS

Quentin and Neuhoff (1972) used the above procedure to study the LDH distribution in different regions of the rat brain. Pherograms of microgels from different brain extracts after isoelectric focusing and incubation for 90 min in the substrate mixtures are shown in Fig. 125. It can be seen that there are striking variations in the patterns of the five isoenzymes in the various tissues. In order to obtain values for the content of the individual isoenzymes, densitometer tracings were evaluated by planimetry. Table 40 shows the percentage mean values of the LDH isoenzyme activities for the brain extracts of Fig. 125 and a number of other rabbit tissues too. It is evident that the variation in the contents of the individual isoenzymes in the different tissues is enormous. Most of the nervous tissues not directly involved in the optic function of the animal, with the exception of the hippocampus, have low amounts of LDH 5 and LDH 4 in comparison to the other three isoenzymes. Nervous

TABLE 40. PERCENTAGE COMPOSITION OF LDH ISOENZYMES IN DIFFERENT REGIONS OF THE RABBIT AS ANALYSED FROM PHEROGRAMS (from Quentin and Neuhoff, 1972)

	LDH^1	LDH^2	LDH^3	LDH^4	LDH^5
Lamina quadrigemina post	59.84	24.83	13.36	1.71	0.26
Hypothalamus	59.49	21.72	16.97	1.46	0.37
Cerebellum	51.17	21.91	24.47	1.92	0.53
Medulla	58.39	21.30	15.54	3.76	1.00
Hypophysis	49.32	35.40	12.38	1.74	1.16
Pons	48.73	24.47	19.47	6.96	0.63
Thalamus	40.00	20.36	26.55	9.82	3.27
Bulbus olfactorius	31.44	20.11	26.63	13.26	8.56
Hippocampus	30.25	16.33	20.75	19.14	13.52
Frontal lobe	34.93	20.60	26.00	11.05	7.41
Retina	22.09	14.34	27.23	17.31	19.04
Lamina quadrigemina ant.	38.17	15.94	16.19	8.48	21.21
N. opticus	18.77	15.15	22.07	17.40	26.61
Chiasma opt.	24.90	12.70	14.31	18.95	29.13
Tractus opt.	24.09	11.25	18.28	18.93	27.45
Heart	100.00	—	—	—	—
Kidney	55.20	24.26	14.11	6.44	—
Liver	0.94	0.94	11.92	46.08	40.13
Spleen	30.89	28.66	27.39	10.19	2.87

tissues involved in the optic function of the animal have, in contrast, similar amounts of LDH isoenzymes 1, 2, 3, 4 and 5. In heart tissue only LDH 1 is present, whereas LDH 4 is the predominant isoenzyme in liver and LDH 1 in the kidney and spleen.

(f) ALTERNATIVE BUFFER SYSTEM FOR MICROISOELECTRIC FOCUSING

Bispink *et al.* (1974) have found that by altering the electrophoresis buffer system, the above procedure can be simplified and made more sensitive. Gels are cast into 5-μl capillaries as described previously and overlayed after polymerisation with the extract to be analysed. Extracts are prepared in the same way, but diluted the second time with a solution

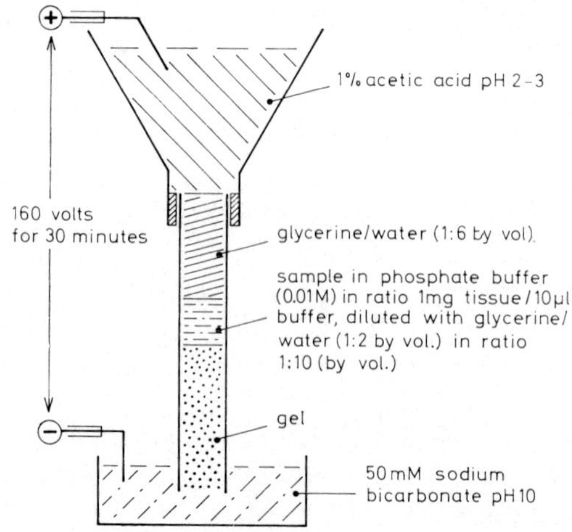

FIG. 124. Schematic diagram of isoelectric focusing procedure of Bispink *et al.* (1974).

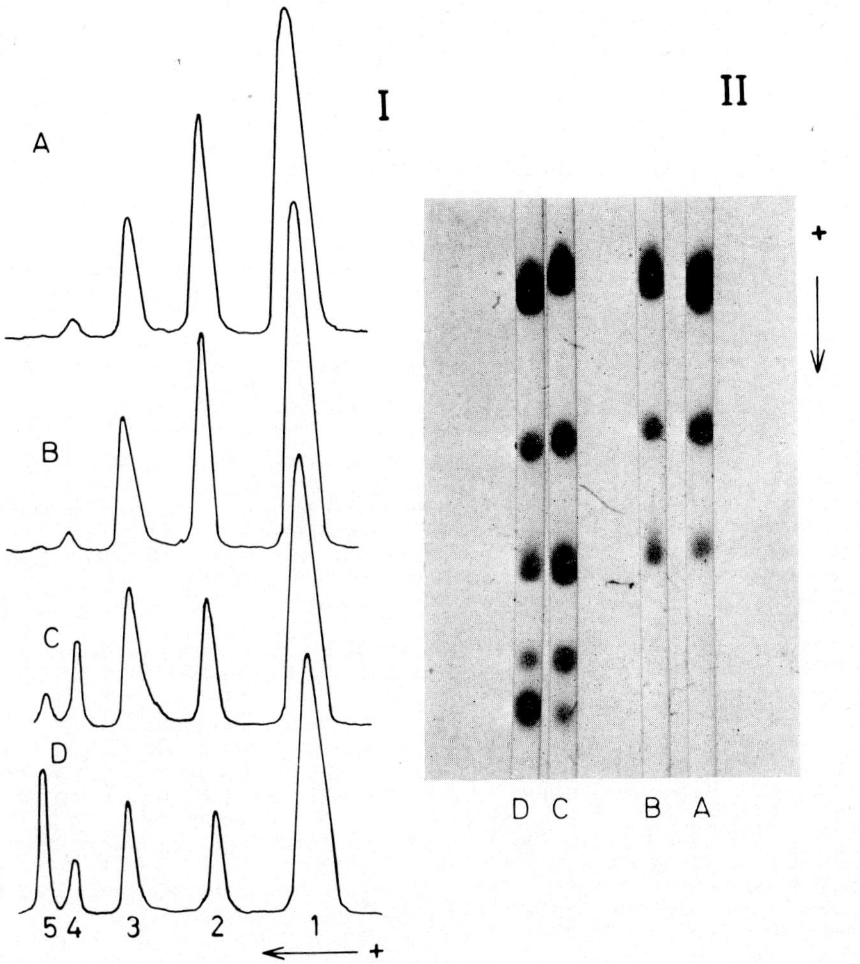

FIG. 125. Photograph (II) showing LDH isoenzymes in microgels cast in 5-μl capillaries after isoelectric focusing and incubation in a tetrazolium mixture. The following tissues from rabbit were fractionated: lamina quadrigemina posterior (A), hypothalamus (B), frontal lobe (C) and lamina quadrigemina anterior (D). Actual microdensitometric traces of the individual gels are shown for comparison (I), so that various LDHs (marked 1 to 5) can be easily identified. (From Quentin and Neuhoff, 1972.)

of glycerine and water (1:2 by vol) instead of with a sucrose solution. Free space above the extract is filled with a solution of glycerine and water (1:6 by vol). The upper electrode buffer is 1% acetic acid pH 2.8 and the lower 50 mM sodium bicarbonate pH 10. The anode is placed in the upper buffer, the cathode in the lower, and isoelectric focusing is allowed to occur at 160 V for 30 min. Figure 124 shows the system used by Bispink *et al.* (1974). It is simpler in that sucrose is not needed in the upper electrophoresis buffer, and ampholine is not required to be placed above the extract before isoelectric focusing. This procedure is slightly more sensitive than the one described previously.

References

AGRANOFF, B. W. (1970) Protein synthesis and memory formation. In: *Protein Metabolism of the Nervous System* (Ed. A. Lajtha), pp. 533–543. Plenum Press, New York.

ALLEN, R. C., MOORE, D. J. and DILWORTH, R. H. (1969) New rapid electrophoresis procedure employing pulsed power in gradient gels at a continuous pH: the effect of various discontinuous buffer systems on esterase zymograms. *J. Histochem. Cytochem.* **17**, 189–190.

ALVORD, E. C. Jr. (1968) *The Etiology and Pathogenesis of Experimental Allergic Encephalomyelitis in the Central Nervous System* (Eds. O. T. Bailey and D. E. Smith), pp. 52–70. Williams & Wilkins, Baltimore.

ANSORG, R. and NEUHOFF, V. (1971) Micro-disc electrophoresis of brain proteins. III. Heterogeneity of the nervous specific protein S-100. *Int. J. Neuroscience*, **2**, 151–160.

ANSORG, R., DAMES, W. and NEUHOFF, V. (1971) Micro-Disc-Elektrophorese von Hirnproteinen. II. Untersuchung verschiedener Extraktionsverfahren. *Arzheim.-Forsch.* (*Drug Research*), **21**, 699–710.

BEEN, A. C. and RASCH, E. M. (1972) A vertical microsystem for discontinuous electrophoresis of insect tissue proteins using thin sheets of polyacrylamide gel. *J. Histochem. Cytochem.* **20**, 368–384.

BISPINK, G., DAMES, W. and NEUHOFF, V. (1974) Unpublished data.

BULLOCK, T. H. and HORRIDGE, G. A. (1965) *Structure and Function of the Nervous System.* Freeman, San Francisco.

CATSIMPOOLAS, N. (1968) Micro-isoelectric focusing in polyacrylamide gel columns. *Analyt. Biochem.* **26**, 480–482.

CICERO, T. J., COWAN, W. M., MOORE, B. W. and SUNTZEFF, V. (1970) The cellular localisation of the two brain specific proteins, S-100 and 14-3-2. *Brain Research*, **18**, 25–34.

CHOULES, G. L. and ZIMM, B. H. (1965) An acrylamide gel soluble in scintillation fluid. Its application to electrophoresis at neutral and low pH. *Analyt. Biochem.* **13**, 336–344.

COHN, P. and RICHTER, D. (1956) Enzymatic development and maturation of the hypothalamus. *J. Neurochem.* **1**, 166–172.

CREMER, Th., DAMES, W. and NEUHOFF, V. (1972) Microdisc electrophoresis and quantitative assay of glucose-6-phosphate dehydrogenase at the cellular level. *Hoppe-Seyler's Z. Physiol. Chem.* **353**, 1317–1329.

DALE, G. and LATNER, A. (1968) Isoelectric focusing in polyacrylamide gels. *Lancet*, **I**, 847–848.

DAMES, W. and MAURER, H. R. (1974) Simultaneous preparation for electrophoresis of a large number of micropolyacrylamide gels with continuous concentration gradients In: *Electrophoresis and Isoelectric Focusing in Polyamide Gel* (Eds. R. C. Allen and H. R. Maurer), pp. 221–231. de Gryter, Berlin and New York.

DAVIS, B. J. (1964) Disc-electrophoresis. II. Method and application in human serum proteins. *Ann. N.Y. Acad. Sci.* **121**, 404–427.

DAVIS, B. J. and ORNSTEIN, L. (1959) A new high resolution electrophoresis method. Delivered at the Society for the Study of Blood at the New York Academy of Medicine.

D'MONTE, B., MARKS, N., DATTA, R. K. and LAJTHA, A. (1970) Protein turnover in membranous fractions. In: *Protein Metabolism of the Nervous System* (Ed. A. Lajtha), pp. 185–217. Plenum Press, New York.

GAINER, H. (1971) Microdisc electrophoresis in sodium dodecyl sulphate. An application to the study of protein synthesis in individual, identified neurons. *Analyt. Biochem.* **44**, 589–605.

GAINER, H. (1972a) Patterns of protein synthesis in individual, identified molluscan neurons. *Brain Research*, **39**, 369–385.

GAINER, H. (1972b) Effects of experimentally induced diapause on the electrophysiology and protein synthesis patterns of identified molluscan neurons. *Brain Research*, **39**, 387–402.

GAINER, H. (1973) Isoelectric focusing of proteins at the 10^{-10} and 10^{-9} g level. *Analyt. Biochem.* **51**, 646–650.

GLASER, L. and BROWN, D. H. (1955) Purification and properties of D-glucose-6-phosphate dehydrogenase. *J. Biol. Chem.* **216**, 67–69.

GLASSMAN, E. (1969) The biochemistry of learning: an evaluation of the role of RNA and protein. *Ann. Rev. Biochem.* **38**, 605–646.

GROSSBACH, U. (1965) Acrylamide gel electrophoresis in capillary columns. *Biochim. Biophys. Acta* (*Amst.*), **107**, 180–182.

GROSSBACH, U. (1971) *Cromosomen-Struktur und Zell-Funktion.* H2, pp. 93–108. Mitt. Max-Planck-Ges.

GROSS-BELLARD, M., ONDET, P. and CHAMBON, P. (1973) Isolation of high molecular weight DNA from mammalian cells. *Europ. J. Biochem.* **36**, 32–38.

GROSSFELD, R. M. and SHOOTER, E. M. (1971) A study of the changes in protein composition of mouse brain during ontogenetic development. *J. Neurochem.* **18**, 2265–2277.

HALJAMÄE, H. and LANGE, P. W. (1972) Calcium content and conformation changes of S-100 protein in the hippocampus during training. *Brain Research*, **38**, 131–142.

HAZAMA, H. and UCHIMURA, H. (1972) Separation of lactate dehydrogenase isoenzymes of nerve cells in the central nervous system by micro-disc electrophoresis on polyacrylamide gels. *Biochim. Biophys. Acta*, **200**, 414–417.

HYDÉN, H. and EGYHAZI, E. (1964) Changes in RNA content and base composition in cortical neurons of rats in a learning experiment involving transfer of handedness. *Proc. Nat. Acad. Sci.* **52**, 1030–1035.

HYDÉN, H. and LANGE, P. W. (1968) Micro-electrophoretic determination of protein and protein synthesis in the 10^{-9} to 10^{-7} gram range. *J. Chromat.* **35**, 336–351.

HYDÉN, H. and LANGE, P. W. (1970a) Protein changes in nerve cells related to learning and conditioning. In: *The Neurosciences, 2nd Study Program* (Ed. F. O. Schmitt), pp. 278–289. Rockefeller University Press, New York.

HYDÉN, H. and LANGE, P. W. (1970b) Brain-cell protein synthesis specifically related to learning. *Proc. Nat. Acad. Sci.* **65**, 898–904.

HYDÉN, H. and McEWEN, B. (1966) A glial protein specific for the nervous system. *Proc. Nat. Acad. Sci.* **55**, 354–358.

HYDÉN, H., BJURSTAM, K. and McEWEN, B. (1966) Protein separation at the cellular level by micro-disc electrophoresis. *Analyt. Biochem.* **17**, 1–15.

JOVIN, T. M. (1973a) Multiphase zone electrophoresis. I. Steady-state moving-boundary systems formed by different electrolyte combinations. *Biochemistry*, **12**, 871–879.

JOVIN, T. M. (1973b) Multiphase zone electrophoresis. II. Design of integrated discontinuous buffer systems for analytical preparative fractionation. *Biochemistry*, **12**, 879–890.

JOVIN, T. M. (1973c) Multiphase zone electrophoresis. III. Further analysis and new forms of discontinuous buffer systems. *Biochemistry*, **12**, 890–898.

KIRBY, K. S. (1965) Isolation and characterisation of ribonucleic acid. *Biochem. J.* **96**, 266–269.

KIYOTA, K. (1959) Electrophoretic protein fractions and the hydrophilic property of brain tissue. 1. The proteins of normal brain. *J. Neurochem.* **4**, 202–208.

KOLIN, A. (1954) Separation and concentration of proteins in a pH field combined with an electric field. *J. Chem. Phys.* **22**, 1628–1629.

KONINGS, R. N. H. and BLOEMENDAL, H. (1969) Synthesis of lens protein *in vitro*. 3. Ribonucleic acid with template activity isolated from calf lens tissue. *Europ. J. Biochem.* **7**, 165–173.

LAJTHA, A. (1964) Protein metabolism of the nervous system. In: *International Review of Neurobiology* (Eds. C. C. Pfeiffer and J. R. Smythies), Vol. 6, pp. 1–98. Academic Press, New York.

LASHLEY, K. S. (1964) *Brain Mechanism and Intelligence*. Hafner Publishing Company, New York.

LOENING, U. E. (1969) The determination of the molecular weight of ribonucleic acid by polyacrylamide gel electrophoresis. *Biochem. J.* **113**, 131–138.

LOWRY, O. H., PASSONNEAU, J. V., SCHULZ, D. W. and ROCK, M. K. (1961) The measurement of pyridine nucleotides by enzymatic cycling. *J. Biol. Chem.* **236**, 2746–2753.

MAIZEL, J. V., SUMMERS, D. F. and SCHARFF, M. D. (1970) SDS-acrylamide gel electrophoresis and its application to the proteins of polio-virus and adenonvirus-infected human cells. *J. Cell Physiol.* **76**, 273–288.

MARKERT, C. L. and MÖLLER, F. (1959) Multiple forms of enzymes: tissue, ontogenetic and species specific patterns. *Proc. Nat. Acad. Sci.* **45**, 753–763.

MAURER, H. R. (1971) *Disc Electrophoresis and Related Techniques of Polyacrylamide Gel Electrophoresis*. Walter de Gruyter, Berlin.

MAURER, H. R. and DATI, F. A. (1972) Polyacrylamide gel electrophoresis on microslabs. *Analyt. Biochem.* **46**, 19–32.

MOORE, B. W. and PEREZ, V. J. (1968) Specific acidic proteins of the nervous system. In: *Physiological and Biochemical Aspects of the Nervous System* (Ed. F. D. Carlson), pp. 343–360. Prentice-Hall, New York.

NAUMOVA, L. P. (1971) *Technika Mikroelektrophoresa na Smeshannych Blyakrylamid Agarosnych Geliach Shkola Seminar Mikrometody Analysa Nukleinowych Kislot*, pp. 17–23. Org. Chem. Novosibirsk/UdSSR.

NEUHOFF, V. (1968) Micro-Disc Electrophorese von Hirnproteinen. *Arneimittel-Forsch.* **18**, 35–39.

NEUHOFF, V. (1973) Microelectrophoresis on polyacrylamide gels. In: *Micromethods in Molecular Biology* (Ed. V. Neuhoff), Vol. 14, pp. 1–83. Springer, Berlin.

NEUHOFF, V., MÜHLBERG, B. and MEIER, J. (1967) Strom- und spannungs-konstantes Netzgerät für die Micro-Disc-Elektrophorese. *Arneimittel Forsch.* **17**, 649–651.

NEUHOFF, V., SCHILL, W.-B., and STERNBACH, H. (1969) Mikro-Disk-elektrophoretische Analyse reiner DNA abhängiger RNA-Polymerase aus *Escherichia coli*. IV. Isolierung und Charakterisierung elektrophoretisch getrennter Enzymfraktionen. *Hoppe-Seyler's Z. Physiol. Chem.* **350**, 767–774.

NEUHOFF, V., SCHILL, W.-B. and STERNBACH, H. (1970) Microanalysis of pure deoxyribonucleic acid-dependent ribonucleic acid polymerase from *Escherichia coli*. *Biochem. J.* **117**, 623–631.

NIELANDS, J. B. (1952) Lactic dehydrogenase of heart. I. Purity, kinetics, and equilibrium. *J. Biol. Chem.* **199**, 373–381.

NIR, I., BRIEL, G., DAMES, W. and NEUHOFF, V. (1973) Micro-disc electrophoretic pattern of the proteins and the free amino acids in rat pineals during ontogenesis. *Arch. Internat. de Physiol. Biochemie.*

ORNSTEIN, L. (1964) Disc-electrophoresis. I. Background and theory. *Ann. N.Y. Acad. Sci.* **121**, 321–349.

OSBORNE, N. N., ANSORG, R. and NEUHOFF, V. (1971) Micro-disc electrophoretic separation of soluble proteins from nervous and other tissues of *Helix* (Pulmonate Mollusca). *Intern. J. Neuroscience*, **1**, 259–264.

PEAKCOCK, A. C. and DINGMAN, C. W. (1967) Resolution of multiple ribonucleic acid species by polyacrylamide gel electrophoresis. *Biochemistry*, **6**, 1818–1827.

POULIK, M. D. (1957) Starch gel electrophoresis in a discontinuous system of buffers. *Nature*, **180**, 1477–1479.

PUN, J. Y. and LOMBROZO, K. (1964) Microelectrophoresis of brain and pineal protein in polyacrylamide gel. *Analyt. Biochem.* **9**, 9–20.

QUENTIN, C.-D. and NEUHOFF, V. (1972) Micro-isoelectric focusing for the detection of LDH isoenzymes in different brain regions of rabbit. *Int. J. Neuroscience*, **4**, 17–24.

RAYMOND, S. and WEINTRAUB, L. (1959) Acrylamide gel as a supporting medium for zone electrophoresis. *Science*, **130**, 711.

RILEY, R. F. and COLEMAN, M. K. (1968) Isoelectric fractionation of proteins on a micro-scale in polyacrylamide and agarose matrices. *J. Lab. Clin. Med.* **72**, 714–720.

RINGBORG, U., EGYHAZY, E., DANEHOLT, B. and LAMBERT, B. (1968) Agarose acrylamide composition gels for microfractionation of RNA. *Nature*, **220**, 1037–1039.

ROBERTS, S., ZOMZELY, C. E. and BONDY, S. C. (1970) Protein synthesis in nervous tissue. In: *Protein Metabolism of the Nervous System* (Ed. A. Lajtha), pp. 3–37. Plenum Press, London.

RODBARD, D. and CHRAMBACH, A. (1971) Estimation of molecular radius, free mobility, and valence using polyacrylamide gel electrophoresis. *Analyt. Biochem.* **40**, 95–134.

ROSE, I. A. (1961) The use of kinetic isotope effects in the study of metabolic control. I. Degradation of glucose-1-D by the hexose monophosphate pathway. *J. Biol. Chem.* **236**, 603–609.

ROSE, S. P. R. (1971) Neurochemical correlates of learning and environmental change. In: *Short-term Changes in Neural Activity and Behaviour* (Eds. G. Horn and R. A. Hinde), pp. 517–551. Cambridge University Press, Cambridge.

RÜCHEL, R. (1973) Personal communication.

RÜCHEL, R., MESECKE, S., WOLFRUM, D. I. and NEUHOFF, V. (1973a) Mikroelektrophorese an kontinuierlichen Polyacrylamid Gradienten Gelen. I. Herstellung und Eigenschaften von Gelgradienten in Kapillaren; ihre Anwendung zur Proteinfraktionierung und Molgewichtsbestimmung. *Hoppe-Seyler's Z. Physiol. Chem.* **354**, 1351–1368.

RÜCHEL, R., MESECKE, S., WOLFRUM, D. I. and NEUHOFF, V. (1973b) Mikroelektrophorese an kontinuierlichen Polyacrylamid Gradienten Gelen. II. Mikroelektrophorese und elektrophoretische Zerlegung von SDS-protein-Komplexen in Polyacrylamidgel-Gradienten *Hoppe-Seyler's Z. Physiol. Chem.* (in press).

SHOOTER, E. M. and ROBOZ-EINSTEIN, E. (1971) Proteins of the nervous system. *Ann. Rev. Biochem.* **40**, 635–652.

SMEDS, S. and BJÖRKMAN, U. (1972) Micro-scale protein separation by electrophoresis in continuous polyacrylamide concentration gradients. *J. Chromatog.* **71**, 499–505.

SVENSSON, H. (1961) Isoelectric fractionation, analysis, and characterisation of ampholytes in natural pH gradients. I. The differential equation of state of solute concentrations at a steady state and its solution for simple cases. *Acta Chem. Scand.* **15**, 325–341.

UNGAR, G. (1970) Molecular mechanisms in information processing. In: *International Review in Neurobiology* (Eds. C. C. Pfeiffer and J. R. Smythies), Vol. 13, pp. 223–254. Academic Press, New York.

VAN DER HELM, H. J. (1961) A simple method for demonstrating lactic acid dehydrogenase isoenzymes. *Lancet*, ii, 108–109.

VESTERBERG, O. and SVENSSON, H. (1966) Isoelectric fractionation, analysis, and characterisation of ampholytes in natural pH gradients. IV. Further studies on the resolving power in connection with separation of myoglobins. *Acta Chem. Scand.* **20**, 820–834.

WAEHNELDT, T. V. and NEUHOFF, V. (1974) (in preparation).

WIEGERS, U. and HILZ, M. (1971) Rapid isolation of undegraded polysomal RNA without phenol *FEBS Letters*, **23**, 77–82.

WIEGERS, U. and HILZ, M. (1972) A new method using proteinase K to prevent m-RNA degradation during isolation from HeLa cells. *Biochem. Biophys. Res. Commun.* **44**, 513–519.

WIELAND, T. and PFLEIDERER, G. (1957) Nachweis der Heterogenität von Michlsäuredehydrogenase verschiedenen Ursprungs durch Trägerelektrophoresis. *Biochem. Z.* **329**, 112–116.

WILSON, D. L. (1971) Molecular weight distribution of proteins synthesised in single, identified neurons of *Aplysia. J. Gen. Physiol.* **57**, 26–40.

WOLFRUM, D. I., RÜCHEL, R., MESECKE, S. and NEUHOFF, V. (1974) (in preparation).

WRIGLEY, C. W. (1968) Analytical fractionation of plant and animal proteins by gel electrofocusing. *J. Chromatog.* **36**, 362–365.

YOSHIDA, A. (1966) Glucose-6-phosphate dehydrogenase of human erythrocytes. I. Purification and characterisation of normal (B) enzyme. *J. Biol. Chem.* **241**, 4966–4976.

CHAPTER 10

General Comments

THE development of suitable microprocedures has allowed the study of a very important field in microbiology, namely, the analysis of single neurons. As demonstrated in this monograph, a number of crucial data have already been established through the use of microtechniques; for example, it now seems likely that the enzyme capable of hydroxylating tryptophan (tryptophan-hydroxylase) is localised only in serotonin-containing cells (see page 82). The issue which arises is not so much therefore whether in fact biochemical studies on single neurons are justified, but rather the necessity of recognising the limitations of such studies and not forgetting other neuronal preparations. One limitation is the degree of accuracy required. Working with smaller quantities of tissue creates the problem of error caused by contamination, dirty glassware, efficiency of procedure, efficiency of experimenter, etc., as a microerror in the microscale is in effect similar to a macroerror in the normal scale. Another difficulty to be considered is whether the cells analysed are still in their 'normal' state, after the often lengthy isolation procedure, and whether the normal, biological variations among the animals are reflected at the cellular level.

It is partly for the reasons mentioned that it has often been argued that a microprocedure is only useful when it is quantitative, since the method would theoretically be 100% efficient. While quantitative micromethods should always be aimed at, often a very difficult achievement simply because of the scale of the system, too much effort spent on technical modification in an attempt to make the procedure quantitative can only really be justified provided other sources of error are acknowledged. Indeed it is possible to dismiss the microdansyl procedure for analysing amines and amino acids as being theoretically an unsuitable method compared with, for example, the microphospholipids and micro-electrophoresis techniques on the grounds that it is not quantitative. This, as pointed out (see page 59), is due to the reaction conditions which depend on a number of parameters difficult to control, viz. pH, reaction time and concentration of dansyl-chloride. Because of this, attempts are being made in this laboratory (Urban *et al.*, 1973) to synthesise an intensely fluorescing compound, which stems from pseudosaccharine chloride (Hettler, 1966, 1968), with better reaction properties than dansyl-chloride. If successful, this will allow amino groups to be quantitatively analysed. However, from the results so far obtained by applying the microdansyl procedure (see pages 79–113), it is evident that although like all methods it has its limitations, a great number of important findings have come to light already, e.g. the ability to follow the metabolism of radioactive substances in single neurons or to note the occurrence of certain substances in only one neuron-type.

As a way towards understanding the co-ordination of brain and behaviour, micro-analysis of single cells has its disadvantages, even assuming the problem of error could be completely overcome. For a start, studies are normally carried out on single-cell bodies or somata rather than on individual neurons. This distinction is often overlooked and is exceedingly important, since the complete neuron is a very complex structure and the cell body is only part of it and has very different properties from the dendrites, axons and synapses. It is therefore likely that minute biochemical changes occurring in the whole neuron, which would occur at any of the neuron's loci (receptive loci, parameter loci, mixing and integrating loci and transmitter loci, see page 1), and which may be electrophysiologically detectable, can be missed when analysing only the biochemistry of cell bodies. Moreover, the question arises whether it is justifiable to assume that the morphologically distinct cell bodies of different neurons (even from a single nervous system) are distinct functional units too. Caution is definitely required when comparing the biochemical properties of invertebrate and vertebrate cell bodies, as a number of crucial differences do exist between their nervous systems (Cohon, 1970; Horridge, 1968).

Criticism can also be applied in other areas. In the dissection of neurons (cell bodies with parts of axons) or glia, there occurs in almost every instance some contamination of one by the other. The cell is often damaged besides, through the dissection procedure, with the result that the integrity and the biochemistry of the neuron are destroyed, and although the dissection of cells may appear from electrophysiological and electron microscopy assessment to be viable, it remains a question whether this is really so.

On the whole, however, biochemical studies on single neurons or rather cell bodies have produced a number of interesting and important data, as illustrated by the examples given in this monograph. It may well be that the only way of detecting possible chemical alterations in nervous tissue during the process of learning would also be through similar microstudies. Griffiths (1966, 1967, 1970) has considered the extent of neuronal alterations which may be involved in memory. If in a lifetime T seconds, N is the total alteration involved in the memory storage, then the alterations per second will be N/T. For mammals T is in the range of 3×10^7 to 3×10^9, so that the fraction N/T will be a very small fraction of N. From this he concludes that the amount of changes occurring in the brain in one training procedure will be very minute. This is an important point and should be taken into account regardless of the way in which the process of learning is being investigated. Now if it is assumed that learning is restricted to only a few neuronal elements, then the expression N/T would be a valid description of the expected change per second, and if the correct time and right neuronal element were chosen, the change might be large enough to be detected biochemically. Clearly it is in this field that the analysis of minute brain areas or cell parts will be particularly useful. As already mentioned, attempts have now been made to relate this biochemistry with the physiology, morphology and functional role of individual cells, though for technical reasons these studies have been carried out on the largest cells available. Since presumably relatively few neurons actually 'learn', the majority having integrating and coding functions on the input and output side of the system, it is possible that the ideal 'learning' cells are in fact some very small neurons which, because of their size, have not yet been discovered.

Perhaps another question to be asked by neurochemists analysing the brain with micro- or normal techniques, is whether the information acquired by studying particular parameters actually manifests the uniqueness of the brain. For example, since the synthesis of ATP from glycolysis and oxidase phosphorylation in the brain is not fundamentally different from that in other tissues, studies pursued along these lines do not really reveal the brain as a specialised organ. More interesting would be to study enzyme systems and substrates which may contribute to the very special nature of nervous tissue, i.e. the ability to transmit and store information. The objective is the type of compounds and enzyme system which can illuminate the biochemical uniqueness of the brain and these may perhaps be found by searching for

(a) Compounds which can be shown to have a demonstrated electrophysiological activity.
(b) Compounds in which a change in concentration or turn-over in the brain can be shown to correlate with a change in the behaviour of the animal.
(c) Compounds in which a change in concentration or turn-over in certain cells can be shown to correlate with a change in the electrophysiological activity of the neuron.
(d) Enzyme systems and compounds which occur only in certain types of neurons.
(e) Enzyme systems and compounds which either occur only in nervous tissue or the concentration of which in nervous tissue is extraordinarily high compared with other tissues.

Finally, attention will be drawn to types of preparations of neuronal tissues suitable for studying the biochemistry of nervous systems, since there is no universal preparation, e.g. single cells, which alone will clarify the biochemical uniqueness of nervous tissue. The following preparations, progressing from whole animal preparations to part of brain tissue are not intended to be a complete list, but rather some suggestions.

(1) *Sampling of blood.* This method, pioneered by Schmidt (see Schmidt *et al.*, 1945), makes use of sampling arterial blood (generally the carotid artery) and venous blood (generally the jugular vein) and measuring the O_2 and CO_2 contents. Other compounds, like glucose and its metabolites, can also be measured easily. This technique can be particularly useful for studying the effect of drugs.

(2) *Brain lesions.* Using stereotactic localisation, relatively discrete areas of brain tissue can be destroyed. It was this type of approach that allowed Heller *et al.* (1962) to study the monoamine pathway of the brain.

(3) *Perfusion of vertebrate brain.* This procedure, exploited mainly by Geiger (1958), is not widely used because of the difficulty in the operation techniques and because of abnormal EEG and values of glucose utilisation. These findings cast some doubt upon the usefulness of this preparation.

(4) *Perfusion of invertebrate brain (e.g. snail).* This technique, employed by Osborne (1972) is particularly useful in that, unlike the vertebrate brain, invertebrate preparations retain their functional activity and survive for several hours or even days. Moreover, the preparation is relatively easy to prepare. Also, because of the large size of the neurons, it is possible to record the functional activity of particular neurons while the brain is being perfused. With this technique, the uptake, storage and turn-over rates of transmitter substances can be measured in a similar way to McIntosh (1963), who measured the turn-over rate of acetylcholine in the perfused cervical ganglion of the cat.

(5) *Perfusion of spinal cord.* In this procedure (see, for example, Angelucci, 1956), the vertebrate canal of intact bull frogs is cannulated so that experiments can be carried out to stimulate the sciatic nerve, thereby stimulating the nerve cord and measuring the increased efflux of neuroactive substances.

(6) *Perfusion of vertebrate peripheral nervous tissue.* Numerous perfused isolated pheripheral nerves such as the sciatic or vagus are particularly suitable and have been used in the study of the effect of drugs on conduction. The bulk of information on turn-over of acetylcholine in nervous tissue comes from studies performed on the perfused cervical ganglion (McIntosh, 1963). By using one ganglion to assay the resting level of acetylcholine, and by perfusing the contralateral superior ganglion, the amount of transmitter synthesised and released could be analysed under a variety of experimental conditions, e.g. electrical stimulation, addition of acetylcholinesterase to the perfusion fluid, and a perfusion medium of varying ion composition.

(7) *Cortical cups.* In this procedure plastic rings are mounted on the exposed surface of the cerebral hemispheres or cerebellum to collect, for example, putative neurotransmitters. This technique has been used by Mitchell (1963) to analyse the spontaneous and evoked release of acetylcholine from the cerebral cortex.

(8) *Regional cannulation.* This technique, developed by Feldberg and Lotti (1967), and which can be applied to a variety of animals, involves cannulation with two cylinders; one for injecting material and one for withdrawing it. Both intraventricular and intracerebral cannulae are available.

(9) *Brain–nerve–muscle preparation.* This preparation, developed by Kerkut *et al.* (1965), is useful because it consists of a brain connected via peripheral nerves to a muscle which does not require any circulating system to survive. It is possible to place a lanolin barrier between the brain and the muscle with the nerve trunk running through the lanolin. The brain or the muscle can then be isolated in a Ringer solution containing a radioactive material and the transport of the radioactive material studied by, for example, electrical stimulation of one of the organs.

(10) *Culturing of nervous tissue.* In this procedure a variety of neuronal elements, which include superior cervical ganglia and spinal neurons, can be cultured (Murray, 1965), and their biochemistry is studied in relation to the effect of drugs, for example.

(11) *Nerve tissue slices.* Tissue slices are a particularly popular preparation for study because they are comparatively easy to prepare (McIlwain, 1966). The thin tissue slices are easily penetrated by added substrates, though this can also be a disadvantage in that leakage of material from the slice will occur.

(12) *Hand-dissected neurons.* This very important and comparatively difficult technique has been discussed on page 8.

(13) *Centrifugation 'dissection'.* In this technique, developed by Rose (1967), brain fractions enriched by neuronal and glial elements are separated using Ficol density gradient centrifugation. A similar procedure has also been developed by Norton and Podluso (1970). These techniques have the disadvantage that the particles isolated are heterogeneous in nature.

(14) *Nerve endings (synaptosomes) and synaptic vesicles.* These preparations developed independently by Whittaker *et al.* (1964) and De Robertis *et al.* (1962) make use of sucrose density gradient centrifugation to separate certain fractions of the brain. Following centrifugation to separate certain fractions of the brain several distinct bands of material can be identified by the electron-microscope, e.g. myelin, fragments of membrane and synaptosomes, etc. The synaptosomes may then be isolated and ruptured by osmotic shock so as to release the synaptic vesicles, the particles containing the putative transmitters. Like the centrifugation 'dissection' procedure, the major disadvantage of this technique is that the particles isolated are heterogeneous so that it is not known just what percentage contain serotonin, for example. Moreover the changes occurring during the isolation procedure are by no means clear. Nevertheless this approach has already contributed much to our understanding of nervous mechanisms.

References

ANGELUCCI, L. (1956) Experiments with perfused frog's spinal cord. *Brit. J. Pharmacol.* **11**, 161–170.

COHON, M. J. (1970) A comparison of invertebrate and vertebrate central neurons. In: *The Neurosciences: 2nd Study Program* (Ed. F. O. Schmitt), pp. 798–812. The Rockefeller University Press, New York.

De ROBERTIS, E., ARNAIZ, G. R. L. and De IRALDI, A. D. (1962) Isolation of synaptic vesicles from nerve endings of the rat brain. *Nature*, **194**, 794–795.

FELDBERG, W. and LOTTI, V. J. (1967) Temperature responses to monoamines and an inhibition of MAO-injected into the cerebral ventricles of rats. *Brit. J. Pharmacol.* **31**, 152–161.

GEIGER, A. (1958) Correlation of brain metabolism and function by the use of a brain perfusion method *in situ. Physiol. Rev.* **38**, 1–20.

GRIFFITHS, J. S. (1966) A theory of the nature of memory. *Nature*, **211**, 1160–1163.

GRIFFITHS, J. S. (1967) *A View of the Brain.* Oxford University Press, Oxford.

GRIFFITHS, J. S. (1970) The transition from short to long term memory. In: *Short-term Changes in Neural Activity and Behaviour* (Eds. G. Horn and R. A. Hinde), pp. 499–516. Cambridge University Press, Cambridge.

HELLER, A., HARVEY, J. A. and MOORE, R. Y. (1962) A demonstration of a fall in brain serotonin following CNS lesions in rat. *Biochem. Pharmacol.* **11**, 859–880.

HETTLER, H. (1966) Charakterisierung von primären und sekundären Aminen mit Pseudosaccharinchlorid. Fresenius *Z. anal. Chem.* **220**, 9–15.

HETTLER, H. (1968) Chapman-Mumm rearrangement of pseudosaccharinesters. *Tetrahedron Letters*, **15**, 1793–1796.

HORRIDGE, G. A. (1968) *Interneurons.* Freeman & Co., London.

KERKUT, G. A., LEAKE, L. D., SHAPIRA, A., COWAN, S. and WALKER, R. J. (1965) The presence of glutamate in nerve muscle perfusates of *Helix*, *Carcinus* and *Periplaneta. Comp. Biochem. Physiol.* **15**, 485–502.

McINTOSH, F. C. (1963) Synthesis and storage of acetylcholine in nervous tissue. *Can. J. Biochem. Physiol.* **41**, 2555–2570.

McILWAIN, H. (1966) *Biochemistry of the Central Nervous System.* Churchill, London.

MITCHELL, J. F. (1963) The spontaneous and evoked release of acetylcholine from the cerebral cortex. *J. Physiol.* **165**, 98–116.

MURRAY, M. R. (1965) *Cells and Tissues in Culture* (Ed. E. W. Willmer), Vol. 2. Academic Press, New York.

NORTON, W. T. and PODLUSO, S. E. (1970) Neuronal soma and whole neuroglia of rat brain; a new isolation technique. *Science*, **167**, 1144–1146.

OSBORNE, N. N. (1972) Effect of electrical stimulation on the *in vivo* metabolism of glucose and glutamic acid in an identified neuron. *Brain Research*, 237–241.

ROSE, S. P. R. (1967) Preparation of enriched fractions from cerebral cortex containing isolated metabolically active neuronal and glial cells. *Biochem. J.* **102**, 33–43.

SCHMIDT, C. F., KETY, S. S. and PENNES, H. H. (1945) The gaseous metabolism of the brain of the monkey. *Am. J. Physiol.* **143**, 33–41.

URBAN, I. S., JASTORFF, B., HETTLER, H. and NEUHOFF, V. (1973) Synthesis of a new reagent for microanalysis of amino acids and related compounds (in preparation).

WHITTAKER, V. P., MICHAELSON, J. A. and KIRKLAND, R. J. (1964) The separation of synaptic vesicles from nerve ending particles (synaptosomes). *Biochem. J.* **90**, 293–303.

Index